4th International Conference on Mathematical Applications in Engineering 2017 (ICMAE'17)

Journal of Physics: Conference Series Volume 949

Kuala Lumpur, Malaysia
8 - 9 August 2017

Editors:

Abdumalik Rakhimov
Bekbaev Ural
Jamal I. Daoud
Khikmat Saburov
Md. Sazzad H. Chowdhury

ISBN: 978-1-5108-5521-2
ISSN: 1742-6588

Printed from e-media with permission by:

Curran Associates, Inc.
57 Morehouse Lane
Red Hook, NY 12571

Some format issues inherent in the e-media version may also appear in this print version.

Printed by Curran Associates, Inc. (2018)

For permission requests, please contact the Institute of Physics
at the address below.

Institute of Physics
Dirac House, Temple Back
Bristol BS1 6BE UK

Phone: 44 1 17 929 7481
Fax: 44 1 17 920 0979

techtracking@iop.org

Additional copies of this publication are available from:

Curran Associates, Inc.
57 Morehouse Lane
Red Hook, NY 12571 USA
Phone: 845-758-0400
Fax: 845-758-2633
Email: curran@proceedings.com
Web: www.proceedings.com

Table of contents

Volume 949

4th International Conference on Mathematical Applications in Engineering 2017 (ICMAE'17)

8–9 August 2017, International Islamic University Malaysia, Kuala Lumpur, Malaysia

Accepted papers received: 20 December 2017
Published online: 2 January 2018

Preface

Papers

IOP Conf. Series: Journal of Physics: Conf. Series **949** (2017) 011001 doi:10.1088/1742-6596/949/1/011001

Preface

Mastery and perfection in mathematics is a mandatory pre-requisite in advancing the frontiers of science. Mathematics provides the methodology and framework of rigorous thoughts in understanding and unveiling the beauty of the physical universe. It is often astonished how simple mathematical expressions are sufficed in describing complex physical phenomena.

Today, the rapid development in science and technology poses new challenges to mathematicians and scientists that requires innovative approach and methodology or eventually new mathematical theories. As such, the Kulliyyah of Engineering, International Islamic University Malaysia (IIUM) in collaboration with MIMOS Bhd. are therefore proud to present the 4th International Conference on Mathematical Modelling in Engineering 2017 (ICMAE4). We aspire ICMAE to serve as an effective platform for academicians and researchers internationally to share ideas, state-of-the-art innovations and future developments of mathematical methods in diverse field of studies. The specific conference topics include:

- Mathematical physics and differential equations
- Algebra & analysis
- Mathematical modelling & simulation
- Probability & mathematical statistics
- Analytical & numerical methods

This publication contains papers presented during ICMAE4 which was held on the 8-9 August 2017 in the Faculty of Engineering, International Islamic University Malaysia. All 25 papers selected have been subjected to peer reviewing process. As such, we would like to thank all reviewers, type-setting editors and last but not least all committee members for the successful publication of this special issue of ICMAE4.

Guest Editors,

Abdumalik Rakhimov
Bekbaev Ural
Jamal I. Daoud
Khikmat Saburov
Md. Sazzad H. Chowdhury

IOP Conf. Series: Journal of Physics: Conf. Series **949** (2017) 011002 doi:10.1088/1742-6596/949/1/011002

Peer review statement

All papers published in this volume of *Journal of Physics: Conference Series* have been peer reviewed through processes administered by the proceedings Editors. Reviews were conducted by expert referees to the professional and scientific standards expected of a proceedings journal published by IOP Publishing.

Equiconvergence in Summation Associated with Elliptic Polynomial

A Fargana[1], A A Rakhimov[2], A A Khan[3] and T B H Hassan[4]
[1,3]Department of Manufacturing and Material Engineering, International Islamic University Malaysia (IIUM), Kuala Lumpur, Malaysia.
[2,4]Department of Science in Engineering, International Islamic University Malaysia (IIUM), Kuala Lumpur, Malaysia.

E-mail: [1]nipafarjana94@gmail.com, [2]abdumalik@iium.edu.my, [3]aakhan@iium.edu.my, [4]torla@iium.edu.my

Abstract. In this paper we prove a precise equiconvergence relation between index of the Bochner-Riesz means of the expansions and power of the singularity of the distributions with compact support in summation associated with the elliptic operator.

1. Introduction

Localization principle for the Fourier series at certain point means dependence of the convergence or divergence only from the small neighbourhood of that point [1]. Equiconvergence of the Fourier series and the Fourier integral means convergence of both at the same and at the term. Same for equisummmability when summation is going by regular method. Here in this paper we consider the Reisz method of summation.

Note that, in N dimensional case, when N > 1, equiconvergence and localization principles for the Fourier series and the integral is not valid by the Pringsheim convergence [2]. Equiconvergence of a spectral expansion corresponding to a schrodinger operator with summable potential, with Fourier integral is studied in [3]. A comparison theorem on equiconvergence of the Fourier Jacobi series with certain trigonometric Fourier series is proved [4]. In [5] it is showed that uniform equiconvergence of the expansions of the integrable functions in eigenfunctions of the Sturm-Liouville operator. And the more general expanded expansions of distributions is studied in [6]-[12].

In this paper we discussed about equiconvergence in summation of the Fourier series and integral of the linear continuous functional which associated with an elliptic polynomial.

2. Preliminaries

Let the space of infinitely differentiable function φ: $T^N \rightarrow C$ is defined by $\varepsilon(T^N)$. The locally convex topology formed from the system of semi norms

$$P_{k,\gamma}(\phi) = \sup_{x \in k} \left| D^\gamma \phi(x) \right|,$$

where, K is a compact subset of $T^N = [-\pi, \pi]^N$, $\gamma = (\gamma_1, \gamma_2, ...\gamma_N)$ is N dimensional vector with the non-negative integer components γ_j (j = 1, 2 ...., N). By $|\gamma| = (\gamma_1 + \gamma_2 + ... + \gamma_N)$ we denote a length of the multi-index γ. For instance, For instance, $D^{\gamma} = D_1^{\gamma_1} D_2^{\gamma_2} .... D_N^{\gamma_N}$, where $D_j = \frac{1}{i}\frac{\partial}{\partial x_j}, j = 1,2,.....N$.

The conjugate space $\varepsilon'(T^N)$ to the local convex topological space $\varepsilon(T^N)$ is the set of all distributions with the compact support in T^N. Any functional $f \in \varepsilon'(T^N)$ can be written as

$$f = (2\pi)^{-\frac{N}{2}} \sum_{n \in Z^N} f_n \exp(inx), \quad (1)$$

where Z^N is the set of all vectors with integer components, f_n is the Fourier coefficient which is defined as the value of f on the test function on $f = (2\pi)^{-N/2} \exp(-inx)$ and $x \in T^N$. Consider the following elliptic polynomial:

$$A(n) = \left(\sum_{j=1}^{r+1} n_j^2\right)^{m+1} + \left(\sum_{j=r+2}^{N} n_j^2\right)^{m} \left(\sum_{j=1}^{N} n_j^2\right),$$

where $n = (n_1, n_2, ... n_N) \in Z^N$. *m* is a positive integer number, and *r= 0, 1, 2, ……, N-1.*
The Riesz means of order *s* (*s* is non-negative real number) of the Fourier series (1) is define as

$$\sigma_\lambda^s f(x) = (2\pi)^{-\frac{N}{2}} \sum_{A(n)<\lambda} \left(1 - \frac{A(n)}{\lambda}\right)^s f_n \exp(inx). \quad (2)$$

Now, we extend a distribution f from N - dimensional torus T^N to the whole space R^N by zero. For the extended distribution use again symbol f. Then the Bochner-Riesz means of order s of the Fourier integral of f is,

$$R_\lambda^s f(x) = (2\pi)^{-\frac{N}{2}} \int_{A(\xi)<\lambda} \left(1 - \frac{A(\xi)}{\lambda}\right)^s \hat{f}(y) \exp(iA(\xi)x) d\xi, \quad (3)$$

where, $\hat{f}(y) = < f, (2\pi)^{-N/2} \exp(-iA(\xi)x) >$ is the Fourier transformation of the extended functional f and it acts on, $(2\pi)^{-N/2} \exp(-iA(\xi)x)$ via x.

Alimov [13] has showed the sufficient condition for localization in the Liouville space $L_2^l(T^N)$ for arbitrary self-adjoint elliptic operators. He showed that for Reisz means of order s of multiple Fourier series and integrals the localization is valid under the conditions,

$$l + s \geq \max\left\{\frac{N-1}{2}, \frac{N-1}{p}\right\}, \qquad 1 \leq p \leq \infty .$$

Il'in [14] is proved that this condition cannot be improved even for the Laplace operator. Actually, the sufficient localization condition for any elliptic operator is $l + s \geq \frac{N-1}{p}$ [see in 15]. Ashurov [16] is proved that the conditions $l+s \geq \frac{N-1}{p} - r\left(\frac{1}{p}-\frac{1}{2}\right)$, $1 \leq p \leq 2$ are sufficient of the expansion for localization of multiple Fourier integrals.

3. Main Theorem

Let l be any real number and $L_2^l(T^N)$ denote the Liouville space of distributions

$$L_2^l(T^N) = \{f \in \varepsilon' : \sum_{n \in Z^N} (1+|n|^2)^l f_n^2 < \infty\}.$$

3.1. Theorem 1

Let $l>0$.Then for any $f \in L_2^{-l}(T^N)$

$$\sigma_\lambda^s f(x) = R_\lambda^s f(x) + O(1)\|f\|_{-l},$$

where $\|.\|_{-l}$ is a norm in $L_2^{-l}(T^N)$:

$$\|f\|_{-l} = (2\pi)^{-\frac{N}{2}} \sqrt{\sum_{n \in Z^N} (1+|n|^2)^{-l} f_n^2}\,.$$

4. Estimation of the Regularized Direchlet Kernel

The regularised Direchlet kernel $D_\lambda^s(x)$ is the Riesz means of the partial sums of the Fourier series of the Dirac delta function can be written as:

$$D_\lambda^s(x) = (2\pi)^{-N} \sum_{A(n)<\lambda} \left(1 - \frac{A(n)}{\lambda}\right)^s f_n \exp(inx)\ . \tag{4}$$

From (4) it follows that for any linear continuous functional f with the compact support formula (2) can be stated as

$$\sigma_\lambda^s f(x) = < f, D_\lambda^s(x-y) >, \tag{5}$$

where f acts to the regularized kernel $D_\lambda^s(x-y)$ with the respect to the variable y. In the same way, for the Fourier integral (3) can be written as

$$R_\lambda^s f(x) = < f, \Theta_\lambda^s(x-y) > , \tag{6}$$

where $\Theta_\lambda^s(x)$ is the Bochner-Riesz means of the Fourier integral of the Dirac delta function can be expressed as:

$$\Theta_\lambda^s f(x) = (2\pi)^{-\frac{N}{2}} \int_{A(\xi)<\lambda} \left(1 - \frac{A(\xi)}{\lambda}\right)^s e^{iA(\xi)x} d\xi = \frac{c\mu^N \cos\left(\mu|x'| + \left(\frac{r}{2} - s\right)\left(\frac{\pi}{2}\right)\right)}{\left(\mu|x'|\right)^{\left(\frac{r+2}{2}\right)+s+(N-1-r)/2m}} \times\left(1 + O\left(\frac{1}{\mu|x'|} + O(|x''|)\mu^{1-1/2m}\right)\right), \tag{7}$$

IOP Conf. Series: Journal of Physics: Conf. Series **949** (2017) 012001 doi:10.1088/1742-6596/949/1/012001

where x' and x'' are the vectors of the spaces R^{r+1} and R^{N-r-1} and here $\mu = \lambda^{1/2(m+1)}$

Let $\hat{\Theta}_{\lambda}^{s}(\xi)$ be the Fourier transformation of the Bochner- Riesz kernel (7). Then

$$\left|\hat{\Theta}_{\lambda}^{s}(\xi)\right| \leq cons\left((1+|\xi|\right)^{-N-l}, \tag{8}$$

$$\left|\Theta_{\lambda}^{s}(x)\right| \leq cons\left((1+|x|\right)^{-N-l}, \tag{9}$$

From the definition of the kernel $\Theta_{\lambda}^{s}(x)$, it is apparent that

$$\hat{\Theta}_{\lambda}^{s}(\xi)=\begin{cases}\left(1-\dfrac{A(\xi)}{\lambda}\right)^{s}, & if\, A(\xi)<\lambda, \\ 0, & otherwise\ .\end{cases} \tag{10}$$

The Poisson summation formula is valid if the function $g(x)$ and its Fourier transformation $\hat{g}(\xi)$satisfy the inequality (8) and (9).Then

$$\sum_{n\in Z^N} g(x+2\pi n)=(2\pi)^{-\frac{N}{2}}\sum_{n\in Z^N}\hat{g}(n)\exp(inx). \tag{11}$$

Applying (11) for the function $g(x)=\Theta_{\lambda}^{s}(x)$ we obtain

$$\sum_{n\in Z^N}\Theta_{\lambda}^{s}(x+2\pi n)=(2\pi)^{-\frac{N}{2}}\sum_{n\in Z^N}\hat{\Theta}_{\lambda}^{s}(n)\exp(inx). \tag{12}$$

Further from (7) and (12) we have

$$D_{\lambda}^{s}(x)=\sum_{n\in Z^N}\Theta_{\lambda}^{s}(x+2\pi n). \tag{13}$$

Separating the right hand side of (13) by $n = 0$, we obtain

$$D_{\lambda}^{s}(x)=\Theta_{\lambda}^{s}(x)+\Theta_{*,\lambda}^{s}(x), \tag{14}$$

where the function $\Theta_{*,\lambda}^{s}(x)$ defined as

$$\Theta_{*,\lambda}^{s}(x)=\sum_{n\in Z^N, n\neq 0}\Theta_{\lambda}^{s}(x+2\pi n). \tag{15}$$

5. Proof of the Theorem 1

Now, from formula (15) we can write

$$\sigma_{\lambda}^{s}f(x)-R_{\lambda}^{s}f(x)=<f,\Theta_{*,\lambda}^{s}(x-y)>. \tag{16}$$

Equality (16) and the bellowing lemma ensue the assertion of the Theorem 1.

Lemma 1. Let $l > 0$, $f \in L_2^{-l}(T^N) \cap \varepsilon'(T^N)$ and let $\sup f \subset \Omega \subset\subset T^N$.
Then

$$< f, \Theta_{*,\lambda}^s(x-y) > = O(1)\|f\|_{-l},$$

uniformly in any compact set $K \subset T^N \setminus \overline{\Omega}$.

Proof. Let Ω_0 is a proper sub domain of the domain Ω. Then

$$\left|< f, \Theta_{*,\lambda}^s(x-y) >\right| \leq \|f\|_{-l} \left\|\Theta_{*,\lambda}^s(x-y)\right\|_{l,0}, \tag{17}$$

where $\|.\|_{l,0}$ is a norm of in the space $L_2^l(\Omega_0)$ via the variable $y \in \Omega_0$.

When $|x-y| > c$, we have [17]

$$\left\|\Theta_{*,\lambda}^s(x-y)\right\|_0 = O\left(\lambda^{1/4(m+1)}\right)^{-(N-s-1-(r/2)-(N-1-r)/2m)}, \tag{18}$$

where $\|.\|_0$ is a norm in $L_2(\Omega_0)$.

Then Lemma 1 follows from (17) and

$$\left\|\Theta_{*,\lambda}^s(x-y)\right\|_{l,0} = O\left(\lambda^{1/4(m+1)}\right)^{(N-s-1-(r/2)-(N-1-r)/2m)} \left\|\Theta_{*,\lambda}^s(x-y)\right\|_0 \tag{19}$$

6. Summary

In this paper the uniform convergence of eigenfunction expansions of the distributions from the generalized Sobolev spaces. Note, that in [18] it is studied localization of the Riesz mean of the Fourier trigonometric series in summations associated with the elliptic polynomials. The results of the current paper generalized the results of the papers [19]-[21] where equisummmability studied for the spherical summations. In [22] the uniform convergence of the Fourier series studied on a closed domain.

7. Acknowledgment

Ongoing research on the topics of the paper supported by the IIUM Fundamental Research Grant Scheme, FRGS14-142-0383.

8. References

[1] Khavin V P and Nikol'skiı N K 1992 Commutative Harmonic Analysis. IV, Encyclopaedia of Mathematical; Sciences, vol. **42**.

[2] Alimov Sh A, Il'in V A and Nikishin E M 1977 Problems of Convergence of Multiple Trigonometric Series and Spectral Decompositions. Russian Mathematical Surveys **32**, 115-139.

[3] Denisov S A 1998 Equiconvergence of a Spectral Expansion Corresponding to a Schrodinger Operator with Summable Potential, with Fourier Integral. Differential Equations. **34**,1043-1048.

[4] Marcokova M 1995 Equiconvergence of Two Fourier Series. Journal of Approximation Theory. **80** 151-163. http://dx.doi.org/10.1006/jath.1995.1012

[5] Sadovnichaya V 2010 Equiconvergence of Eigenfunction Expansions for Sturm-Liouville Operators with a Distributional Potential. Sbornik: Mathematics. **201**,61-76.

[6] Alimov Sh A 1993 On the Spectral Decompositions of Distributions. Doklady Mathematics. **331** 661-662.

[7] Alimov Sh A and Rakhimov A A 1996 Localization of Spectral Expansions of Distributions. Difference Equations. **32**, 798-802.

[8] Alimov Sh A and Rakhimov A A 1997 Localization of Spectral Expansions of Distributions in a Closed Domain. Difference Equations. **33**, 80-82.

[9] Rakhimov A A 2000 On the Localization of Multiple Trigonometric Series of Distributions. Dokl. Math. 62,163-165. (translation from Dokl. Akad. Nauk, Ross. Akad. Nauk, **374,** 20-22).

[10] Rakhimov A A 1996 Localization Conditions for Spectral Decompositions Related to Elliptic Operators from Class Ar. Mathematical Notes. **59**, 298-302. (translation from Mat. Zametki, **59**, 421-427).

[11] Rakhimov A, Ahmedov A and Zainuddin H 2012 On the Spectral Expansions of Distributions Connected with Schrodinger Operator. Applied Mathematics Letters. **25,**921-924.

[12] Rakhimov A A 1996 Spectral Decompositions of Distributions from Negative Sobolev Classes. Difference Equations. **32**, 1011-1013. https://zbmath.org/?q=an:0890.47031.

[13] Ashurov R R 1983 On conditions for localization of spectral expansions corresponding to elliptic operators with constant coefficients. Mat. Zametki. 33(6). 847-856. English translation (1983). Math. Notes **33**, 434-439, Zbl. 588.35065.

[14] Il'in V A 1968 Problems of localization and convergence for Fourier series in fundamental systems of functions of the Laplace operator. Russian Mathematical Surveys. **23(2)**, 59.

[15] Alimov Sh A 1974 On the localization of spectral decompositions. Differ. Uravn. **10(4)**, 744-746. English translation (1975). Differ. Equations **10**, 576-578, Zbl. 284.35059.

[16] Ashurov R R 1983 On conditions for localization of spectral expansions corresponding to elliptic operators with constant coefficients. Mat. Zametki. **33(6)**, 847-856. English translation (1983). Math. Notes **33,** 434-439, Zbl. 588.35065.

[17] Bastis A Y 1983 On the Asymptotic of the Riesz-Bochner Kernel. Analysis Mathematica, **9** 247-258. http://dx.doi.org/10.1007/BF01910305

[18] Rakhimov A.A. 2011 On the localization of the riesz means of multiple fourier series of distributions. Abstract and Applied Analysis, DOI: 10.1155/2011/731249

[19] Rakhimov A.A., 2017 Torla Bin Hj Hassan1, Ahmad Fadly Nurullah bin Rasedee, On Equiconvergence of Fourier Series and Fourier Integral, IOP Conf. Series: Journal of Physics: Conf. Series 819(2017) 012025 doi:10.1088/1742-6596/819/1/012025

[20] Rakhimov A.A. On the equiconvergence of the Fourier series and the Fourier integral of distributions, AIP Conference Proceedings 1739, 020060 (2016); doi: http://dx.doi.org/10.1063/1.4952540

[21] Rakhimov A.A. 2015 On the equiconvergence of the Fourier series and integral of distributions, Journal of Applied Mathematics and Physics, 3, 1361-1366.

[22] Rakhimov A.A. 2015 On the uniform convergence of the Fourier series on a closed domain, Eurasian Mathematical Journal, **8 (3),** 60-69.

IOP Conf. Series: Journal of Physics: Conf. Series **949** (2017) 012002 doi:10.1088/1742-6596/949/1/012002

Parametric Model Based On Imputations Techniques for Partly Interval Censored Data

Abdallah Zyoud[1], F. A. M. Elfaki[2,*], and Meftah Hrairi[1]
[1]Department of Mechanical Engineering, Faculty of Engineering, IIUM, P.O. Box 10, 50728 Kuala Lumpur, Malaysia
[2]Department of Mathematics, Statistics and Physics, College of Arts and Science, Qatar University, P.O. Box 2713, Doha, Qatar

[*]E-mail: felfaki@qu.edu.qa

Abstract: The term 'survival analysis' has been used in a broad sense to describe collection of statistical procedures for data analysis. In this case, outcome variable of interest is time until an event occurs where the time to failure of a specific experimental unit might be censored which can be right, left, interval, and Partly Interval Censored data (PIC). In this paper, analysis of this model was conducted based on parametric Cox model via PIC data. Moreover, several imputation techniques were used, which are: midpoint, left & right point, random, mean, and median. Maximum likelihood estimate was considered to obtain the estimated survival function. These estimations were then compared with the existing model, such as: Turnbull and Cox model based on clinical trial data (breast cancer data), for which it showed the validity of the proposed model. Result of data set indicated that the parametric of Cox model proved to be more superior in terms of estimation of survival functions, likelihood ratio tests, and their P-values. Moreover, based on imputation techniques; the midpoint, random, mean, and median showed better results with respect to the estimation of survival function.

1. Introduction

One of the main objectives in the survival studies is the comparison of survival functions, which is due to the fact that it may be needed to compare two different treatments for one disease. In this paper, the comparison problem for several imputation techniques were discussed based on the existence of PIC failure time data. PIC data is common in numerous applications in medicine, engineering, biology and economy. One of its most common occurrences is in medical or health studies that have periodic follow-up hospital visits, where the event, usually a patient displaying a certain disease symptoms or recovery signs, may occur between visits, but it also may occur in the hospital, hence partly interval censoring. For some subjects, PIC data indicates the exact failure times which are observed. For the remaining subjects, survival time of interest is observed only for a certain time interval instead of being exact [1] [2] [3]. [4] provided an example of this kind of data that is; Framingham Heart Disease Study which follows a group of women diagnosed with breast cancer and separates them into two different groups based on the type of treatment that they receive. Another example about Fatigue Failure (Crack size data) which measures the number of cycles it takes before a crack appears in the subjects may be considered as PIC data. Using imputation greatly simplifies the calculation process especially when dealing with larger sets of data. Applying several imputation techniques to our data was relatively easy to implement. Using R software, we were able to develop a programming code to apply different imputation methods to the data sets and proceed with the parametric analysis. In this study, several imputation techniques were used to estimate survival function and compared with the one that was obtained by Turnbull based on interval censored and PIC failure time data. In the next two sections, parametric model and imputation techniques will be discussed.

2. Parametric Model

While nonparametric methods have been the standard approach for analyzing simple homogeneous survival data without covariate information for making fewer assumptions, parametric survival models are sometimes used for inference. Let's take $(X_1, X_2, ..., X_n)$ to be the true survival times, and also are assumed to be independent and identically distributed with survival function $S(t;\beta)$ and hazard function $h(t;\beta)$, but only a right censored sample $(\widetilde{X}_i, D_i)$, $i = 1,...,n$ is observed. If we assume independent censoring, then the likelihood function for the parameter is presented by Equation (1).

$$L(\beta) = \prod_{i=1}^{n} (h(\widetilde{X}_i, \beta))^{D_i} S(\widetilde{X}_i, \beta) \tag{1}$$

where β is a parameter, that needs to be estimated.

The values of β can be estimated by maximizing the likelihood as shown in Equation (2). (The estimation of β 's is not address it here in the paper, reader refer to [6])

$$L(\beta) = \prod_{i=1}^{n} \left[\frac{\exp(\beta' z_i)}{\sum_{j \in R_i} \exp(\beta' z_j)} \right]^{D_i} \tag{2}$$

where $R_i = \{j : \widetilde{X}_j \geq \widetilde{X}_i\}$ is the risk set at time $\widetilde{X}_i$, which is a set of subjects which are still alive and uncensored at the time. We can use the Breslow estimator to estimate the cumulative baseline hazard $\Lambda_0(t)$ as presented in Equation (3).

$$\hat{\Lambda}_0(t) = \sum_{\widetilde{X}_i \leq t}^{n} \left[\frac{D_i}{\sum_{j \in R_i} \exp(\hat{\beta}' z_j} \right] \tag{3}$$

Nelson Aalen estimator could be used if β is known and is also equal to the maximum likelihood estimate $\hat{\beta}$. The estimator based on Equations (2) and (3) for which have a parametric maximum likelihood interpretation. Equation (2) can be used to estimate β in a large sample, in which $\hat{\beta}$ is normally distributed with the proper mean and with a covariance that can be estimated by information matrix.

The function of Equation (1) may be analyzed using standard large sample theory. Thus, standard tests, that are: Wad test, Score test, and likelihood test, are used as inferential tools. Mostly distributions used in estimation parametric survival models that are Weibull distribution and exponential distribution. Both distributions contain very simplest model.

IOP Conf. Series: Journal of Physics: Conf. Series **949** (2017) 012002 doi:10.1088/1742-6596/949/1/012002

3. Imputation Methods

Several imputation techniques were used in this study and compared with Turnbull's model. The performances of the techniques were also compared with each other. The imputation techniques are based on Kaplan-Meir estimator.

The imputation techniques can be classified into two different categories, which are:

1. ***Simple imputation techniques***

   There are three main types of simple imputation techniques:

   1. Right imputation where the event time is imputed by the right limit of the censoring interval R_i .

   2. Left imputation where the event time is imputed by the left limit of the censoring interval L_i

   3. Mid imputation which refers to imputing the event time by the midpoint of the censoring interval $[L_i, R_i]$ as $(L_i + R_i)/2$.

2. ***Probability-based imputation techniques***

   It is usually require estimating the distribution of the partly interval censored data based on the observed intervals and using our knowledge of the distribution to impute the missing data. The most common probability-based imputation methods are, conditional mean imputation, conditional median imputation, and random Imputation.

4. An Example

The proposed method was applied to the modified breast cancer data which was found in some studies [6] [7]. The data set here consist of 46 patients who were treated by the means of Radiation therapy (R) only and 48 other patients who were treated by the means of adjuvant Chemotherapy next to Radiation (R+C). This study aimed to compare cosmetic effects of the treatment through Radiation therapy alone (R) against the combination of Radiation therapy with Chemotherapy (R+C) on women who suffer from breast cancer in its early stages. The event of interest in this study is the time when breast retraction is first observed. The patients were checked in the clinic once every 4 to 6 months, where the actual dates of the event were recorded exactly if available. If the dates are not available, interval of events was noted. To set up data as partly interval censored data, for instance: for radiation, 25 observations were set up as right censored, 21 as interval censored, and 20 as exact. Likewise, for R+C, the set up was for 13 observations as right censored, 35 as interval censored, and 20 as exact, reader refer to [6]. Results of this data set were analyzed as PIC as presented next.

Based on parametric analysis for cancer PIC data, Figures 1, 2, 3, 4, 5, and 6 show the results obtained by random imputation, median imputation, midpoint, left point, right point, and mean imputation techniques respectively. The figures look like Turnbull method especially the one which was obtained by mean and median imputation. Likelihood ratio test shows better result in the case of midpoint, right point, mean, and random imputations as shown in Table 1.

Table 1: Likelihood Ratio Test and their P-value based on parametric model from cancer PIC data

Imputation Techniques	Likelihood Ratio Test	P-value
Midpoint	4.6	0.032
Left-point	3.03	0.082
Right-point	4.74	0.029
Mean	4.49	0.034
Median	4.22	0.04
Random	5.13	0.023

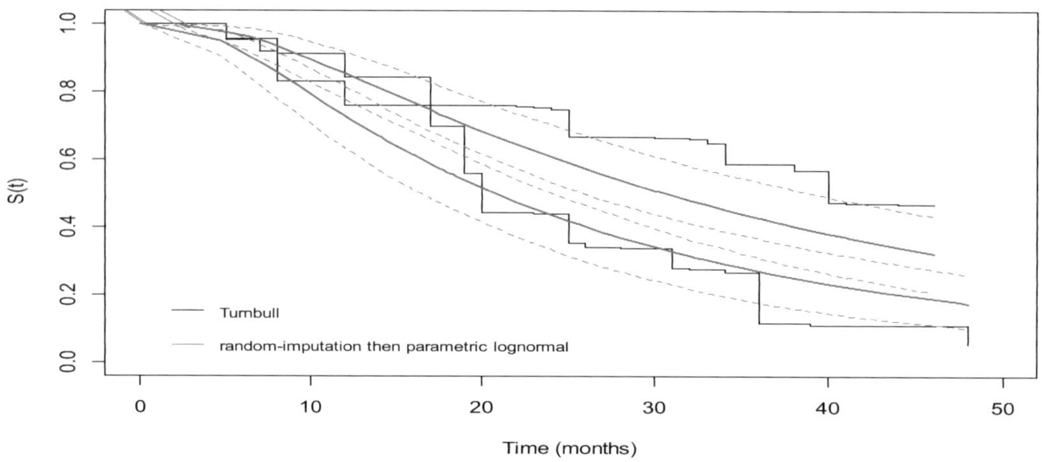

Figure 1: Estimated survival function obtained by random imputation vs Turnbull based on parametric model from cancer PIC data.

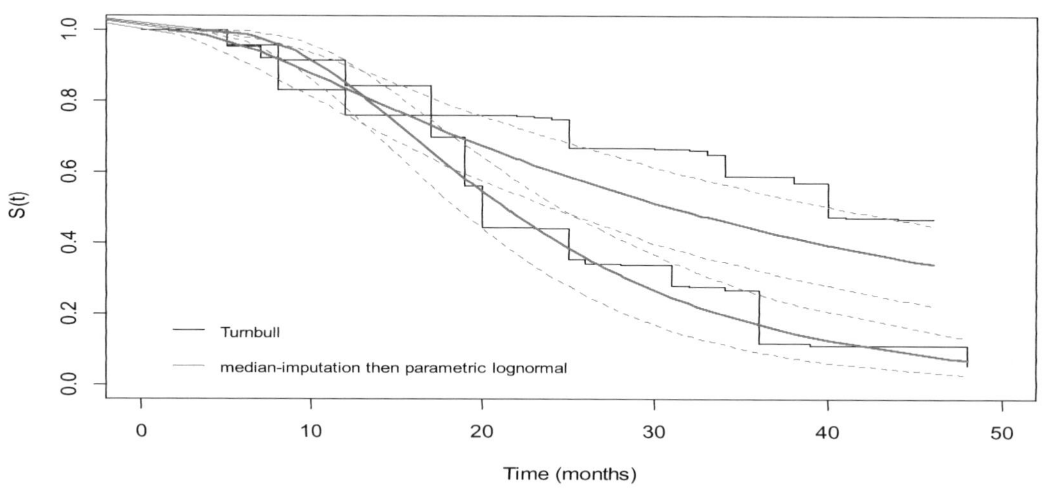

Figure 2: Estimated survival function obtained by median imputation vs Turnbull based on parametric model from cancer PIC data.

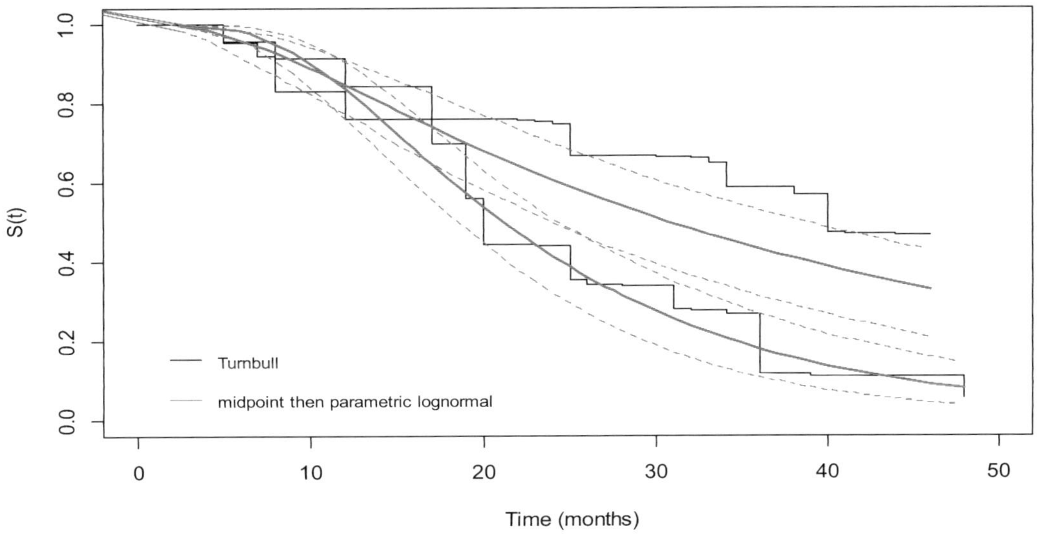

Figure 3: Estimated survival function obtained by midpoint imputation vs Turnbull based on parametric model from cancer PIC data.

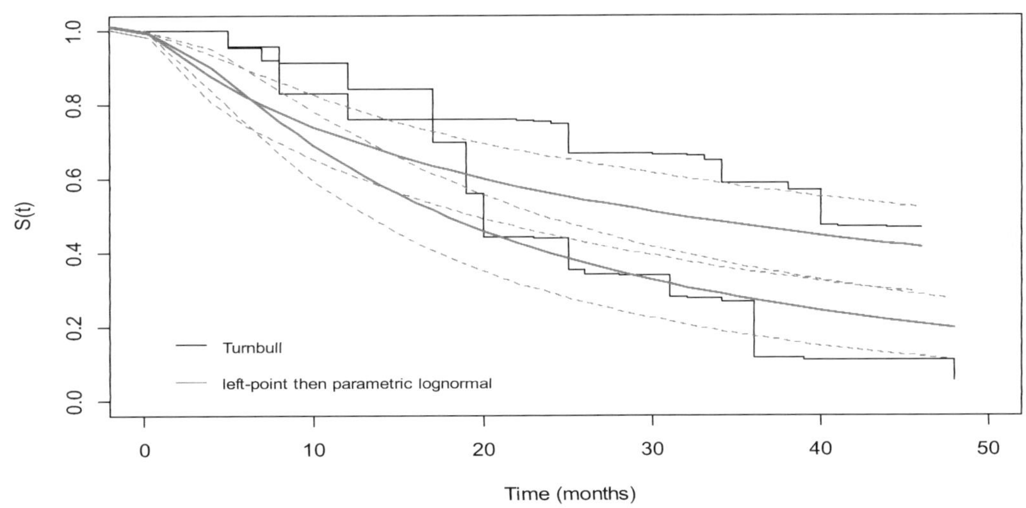

Figure 4: Estimated survival function obtained by left imputation vs Turnbull based on parametric model from cancer PIC data.

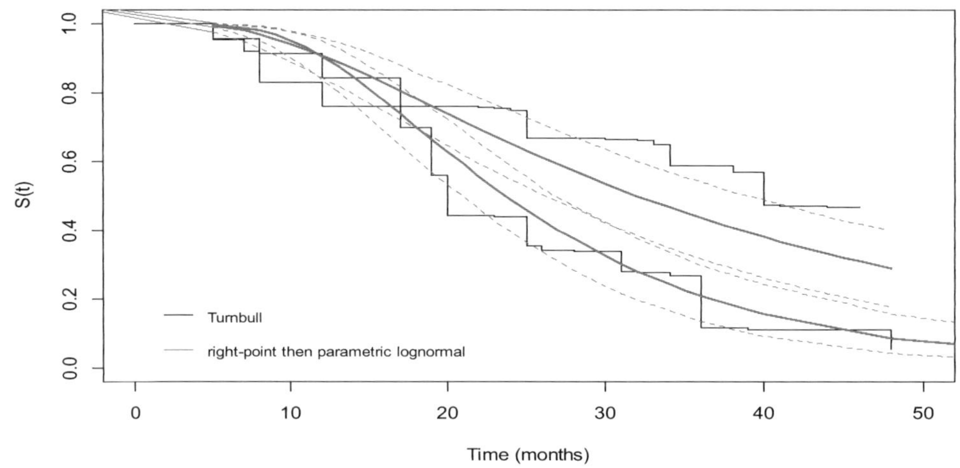

Figure 5: Estimated survival function obtained by right imputation vs Turnbull based on parametric model from cancer PIC data.

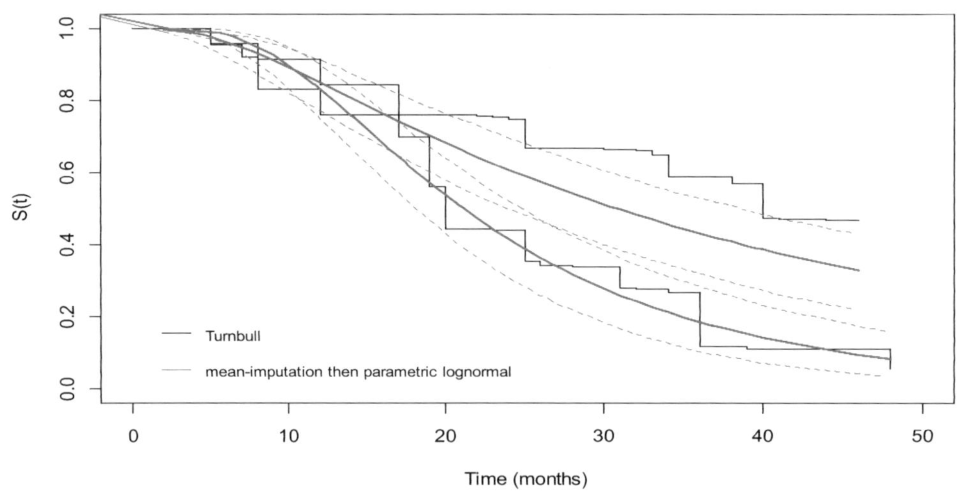

Figure 6: Estimated survival function obtained by mean imputation vs Turnbull based on parametric model from cancer PIC data.

5. Conclusion

In this paper, a simple modification of estimating survival function for PIC data using parametric model based on imputation techniques was proposed. Modification of breast cancer data is used and R software is also used to generate the results. Obtained results are same as the result which is obtained by Turnbull. However, based on partly interval censored data, random imputation and mean & median imputation show better results compared with others imputation techniques as well as Turnbull with respect to the smallest P-value (Table 1).

The parametric Cox model based on imputation techniques gives essentially better estimations of survival function as a Turnbull method, and is also much easier in term of predictions.

References

[1] Huang, J. Asymptotic properties of nonparametric estimation based on partly interval-censored data. *Statistica Sinia*, (1999): 9; 501 – 519.
[2] Kim, J. S. . Maximum likelihood estimation for the proportional hazards model with partly interval-censored data. Journal of the Royal Statistical Society, (2003): 65; 489- 502.
[3] Zhao, X. Zhao, Q.J. Sun. Generalized log-rank test for partly interval-censored failure time data. Biometrical Journal. (2008):3;375-385.
[4] Odell, P. M., Anderson, K. M. and D'Agostino, R. B. Maximum likelihood estimation for interval-censored data using a Weibull-based accelerated failure time model. Biometrics. (1992):48;951-959.
[5] Kumar, P. Modeling and analysis of degradation data. Indian Institute of TechnologyBombay,Available:http://www.academia.edu/266427/Modelling_and_Analysis_of_Degradation_Data [Accessed: 5/1/2013]. (2010).
[6] Abdallah M T Zyoud. Imputation Techniques for Reliability Analysis Based On Partly Interval Censored Data. M.Sc., thesis. International Islamic University Malaysia. (March 31, 2017).
[7] Finkelstein, D. M., and Wolfe, R. A. A semi parametric model for regression analysis of interval-censored failure time data. Biometrics. (1985): 41;933-945.

IOP Conf. Series: Journal of Physics: Conf. Series **949** (2017) 012003 doi:10.1088/1742-6596/949/1/012003

Some effects of perturbed flight schedules to the the performance of aircraft routings

Khusnul Novianingsih

Department of Mathematics education, Universitas Pendidikan Indonesia, Jl. Dr. Setiabudhi No. 229 Bandung 40154, Indonesia

E-mail: k_novianingsih@upi.edu

Abstract. In this paper, we observe some effects of schedule perturbations to the performance of aircraft routings. We show that moving departure time of flights to adding their block-time do not guarantee that the departure delay of the flights will be reduced. Contrarily, moving arrival time of flights always improve the performance of aircraft routings, especially for reducing arrival delay. The computational results also show that by adding more slacks, we obtain better performance of aircraft routings.

1. Introduction

Airline schedules are operated under uncertain conditions. Some factors such as bad weather and airport congestion can cause some flight schedule can be delayed or canceled. These affect to airline on-time performance which it is related to passenger satisfaction. To reduce the number of flight delays, airlines need to construct robust aircraft routings. Robust aircraft routings are aircraft route schedules that are arranged to minimize a specific robustness measure, as the expectation of departure delay, arrival delay, or propagated delay.

Recently, some researchers have shown that improving the robustness of aircraft routings can be done by perturbed flights. That is by moving the departure time or arrival time of flights earlier or later. Chiraphadanakul et al. [4] derive an optimization model to re-time departure time or arrival time of flight for minimizing the expectation of total arrival delay. AhmadBeygi et al. [2] construct an optimization model to re-time the departure time of flights for minimizing the expectation of total propagated delay. The similar model to minimize the expectation of total departure delay is proposed by Novianingsih et al [7]. An optimization model to built new aircraft routings that allow re-timing flights is built by Aloulou et al. [3].

The research mention above improve the robustness of aircraft routing through optimization models to determine the optimal perturbation. This research characterize the influence of perturbed flights analytically. As we know, only Schaefer et al. [8] who investigate the topic analytically. Unlike Schaefer et al. [8] who observe the perturbation of the original flight schedule to improve the operational performance of a given crew schedule, this paper inspect the effects of perturbed flight schedules to the aircraft routing operations. We obtain that moving departure time of flights does not always improve the performance of aircraft routings. But the arrival delay of flights will be reduced when the block-times are added. By simulation, we also show that more slack means better performance.

This paper is organized as follows. We discuss flight delay on an aircraft schedule in Section 2. Section 3 discuss the effect of perturbed flights to the performance of aircraft routings. We present computational results in Section 4. We give a conclusion in the last section.

2. Flight delay on an aircraft schedule

An aircraft schedule consists a set of aircraft routings that partitions the flights to be flow by a single aircraft. The aircraft scheduling problem is usually modeled as:

$$\min\{\mathbf{cx} : \mathbf{Ax} = \mathbf{1}, \mathbf{x} \in \{0,1\}\}, \tag{2.1}$$

where its columns matrix $\mathbf{A} = (a_{fr})$ represent all possible aircraft routings in flight schedules, in which $a_{fr} = 1$ if flight f in routing r, and $a_{fr} = 0$ otherwise. One routing consists a sequence of flights and it complies a number of maintenance requirements. The element x_r of $\mathbf{x}$ is a decision variable that has value 1 if routing r is selected as a solution, and $x_r = 0$ otherwise. The element c_r of $\mathbf{c}$ is cost associated to routing r.

Let F and R be a set of flights and a set of aircraft routings, respectively. For $f \in F$, let dep_f and arr_f be the planned departure time of flight f and the planned arrival time of flight f, respectively. Given an aircraft route r which consists a sequence of flights $f_1, ..., f_n$.

Definition 2.1 *Slack between two consecutive flights f_i and f_{i+1} in r is defined as*

$$s_{i,i+1} = c_{i,i+1} - m_{i,i+1},$$

where $ct_{i,i+1} = dep_{i+1} - arr_i$ is the planned connecting time between flight f_i and f_{i+1}, and $m_{i,j}$ is the minimum required connecting time to serve flight f_{i+1} after flight f_i.

Assume that in flight schedule operations we use a push-back recovery. Using this assumption, if any flight in r is disrupted, the next flights in r will be delayed until the planes that serve the flights are available [1]. Assume that the planes are always available to serve the flights in schedules. Suppose that we divide sources of a delay for flight f_i into two categories:ground delay and block-time delay. A ground delay is a delay before an aircraft take-off, and it does not include propagated delay caused by the previous flight in the same routing. A block-time delay is delay in the air, and we can view as the difference between the planned block-time of a flight with its actual block-time.

For aircraft routing r, let $\alpha_1, ..., \alpha_n$ and $\beta_1, ..., \beta_n$ be nonnegative random variables where α_i indicates departure delay of flight f_i and β_i indicates arrival delay of flight f_i, respectively. Then, the random departure time and the random arrival time of flight f_i are given by $\gamma_i = dep_i + \alpha_i$ and $\delta_i = arr_i + \beta_i$, respectively. Let $\lambda_1, ..., \lambda_n$ be nonnegative random variables, where λ_i represents ground delay of flight f_i in r. Let $\mu_1, ..., \mu_n$ be nonnegative random variables, where μ_i indicates the block-time delay of flight f_i relative to its planned block-time. Assume that the distributions of ground-time delay and block-time delay are independent of the time of day.

According Lan et al. [5], a delay can be decomposed into a propagated delay and a non-propagated delay. A propagated delay is a flight delay caused by waiting for incoming aircraft, and a non-propagated delay is a delay caused by other reasons. While a propagated delay is influenced by previous flights in an aircraft routing, a non-propagated delay more describes airport conditions where the flight depart from, such as weather condition and traffic intensities. Consider a departure delay of flight f as decomposition of a propagated delay and a non-propagated. Let $pd_{i,i+1}$ be the propagated delay to flight f_{i+1} caused by flight f_i. Since we do not have propagated delay for the first flight in r, then the departure delay of the first flight is calculated as

$$\alpha_1 = \lambda_1.$$

For $i = 2, ..., n$, the departure delay of $f_i \in r$ is determine as

$$\alpha_i = \lambda_i + pd_{i-1,i}.$$

A propagated delay to flight f_i occurs only if the arrival delay of flight f_{i-1} make the new connection $ct_{i-1,i}$ is less than the minimum required connecting time. So, we calculate the propagated delay $pd_{i-1,i}$ as

$$pd_{i-1,i} = \max\{\beta_{i-1} - s_{i-1,i}, 0\}. \tag{2.2}$$

On the other hand, the departure delay of flight f_{i-1} and its block-time delay will contribute to its arrival delay. So, we have

$$\beta_{i-1} = \alpha_{i-1} + \mu_{i-1}. \tag{2.3}$$

By substituting (2.3) to (2.2), we get

$$pd_{i-1,i} = \max\{\alpha_{i-1} + \mu_{i-1} - s_{i-1,i}, 0\}. \tag{2.4}$$

Finally, we have the following proposition as a formula to calculate the departure delay of flights along a path of an aircraft routing.

Proposition 2.2 *The departure delay of flight f_i in routing r is given by*

$$\alpha_1 = \lambda_1,$$

and

$$\alpha_i = \lambda_i + \max\{\alpha_{i-1} + \mu_{i-1} - s_{i-1,i}, 0\},$$

for $i = 2, ..., n$.

Given $\lambda_1, ..., \lambda_n$ of arbitrary ground delays and $\mu_1, ..., \mu_n$ of arbitrary block-time delays for flights $f_1, ..., f_n$ in r. The following proposition gives the bound of departure delay of flights in an aircraft routing.

Proposition 2.3 *For each f_i in r,*

$$\lambda_i \leq \alpha_i \leq \sum_{k=1}^{i} \lambda_k + \sum_{k=2}^{i} |\mu_{k-1} - s_{k-1,k}|.$$

PROOF. Recall from Proposition 2.2,

$$\alpha_i = \lambda_i + \max\{\alpha_{i-1} + \mu_{i-1} - s_{i-1,i}, 0\}.$$

Since

$$\max\{\alpha_{i-1} + \mu_{i-1} - s_{i-1,i}, 0\} \leq |\alpha_{i-1} + \mu_{i-1} - s_{i-1,i}|,$$

and

$$|\alpha_{i-1} + \mu_{i-1} - s_{i-1,i}| \leq \alpha_{i-1} + |\mu_{i-1} - s_{i-1,i}|,$$

then

$$\alpha_i \leq \lambda_i + \alpha_{i-1} + |\mu_{i-1} - s_{i-1,i}|. \tag{2.5}$$

Substituting for α_{i-1} in Equation(2.5),

$$\begin{aligned}
\alpha_i &\leq \lambda_i + (\lambda_{i-1} + \alpha_{i-2} + |\mu_{i-2} - s_{i-2,i-1}|) + |\mu_{i-1} - s_{i-1,i}| \\
&\leq \lambda_i + \lambda_{i-1} + \lambda_{i-2} + \alpha_{i-3} + |\mu_{i-3} - s_{i-3,i-2}| + |\mu_{i-2} - s_{i-2,i-1}| + |\mu_{i-1} - s_{i-1,i}| \\
&\vdots \\
&\leq \sum_{k=1}^{i} \lambda_k + \sum_{k=2}^{i} |\mu_{k-1} - s_{k-1,k}| \qquad (2.6)
\end{aligned}$$

If $pd_{i-1,i} = 0$, then $\alpha_i = \lambda_i$. □

Since $\beta_i = \alpha_i + \mu_i$, we established the following corollary.

Corollary 2.4 *For each f_i in r,*

$$\lambda_i \le \beta_i \le \sum_{k=1}^{i} \lambda_k + \mu_i + \sum_{k=2}^{i} |\mu_{k-1} - s_{k-1,k}| \,. \tag{2.7}$$

Especially, if $pd_{i-1,i} > 0$, for all $i = 2, ..., n$, then we have

$$\alpha_i = \sum_{k=1}^{i} \lambda_k + \sum_{k=2}^{i} (\mu_{k-1} - s_{k-1,k}) \,, \tag{2.8}$$

and

$$\beta_i = \sum_{k=1}^{i} \lambda_k + \mu_i + \sum_{k=2}^{i} (\mu_{k-1} - s_{k-1,k}) \,. \tag{2.9}$$

3. The Perturbed Schedule

Let $\mathbf{x}$ and $\mathbf{y}$ be a nonnegative vector in $\mathbb{R}^{|F|}$.

Definition 3.1 *The perturbed schedules $F + (\mathbf{x}, \mathbf{y})$ is defined as the new flight schedules which obtain from F where the departure time and the arrival time of each flight f_i in F are changed by $(dept_i - x_i)$ and $(arrt_i + y_i)$, respectively. Perturbed schedules $F + (\mathbf{x}, \mathbf{y})$ is called as feasible perturbed schedules if the planned aircraft routings and the planned crew pairing in F remain feasible under perturbed schedules $F + (\mathbf{x}, \mathbf{y})$.*

For flight f_i in $F + (\mathbf{x}, \mathbf{y})$, the new block time of flight f_i is increased to $bt_i + x_i + y_i$. Hence, the perturbed random block-time delay of flight f_{i-1} becomes $\mu'_i = \mu_i - x_i - y_i$. Similarly, the new slack for connecting flight f_{i-1} and f_i is reduced to $s'_{i-1,i} = s_{i-1,i} - x_i - y_{i-1}$.

Let r' be the feasible perturbed routing of r, that is the aircraft routing which all flight in r are elements of $F + (\mathbf{x}, \mathbf{y})$. Let α'_i and β'_i be the random departure time and the random arrival delay of flight f_i in r, respectively. Given a sequence of random ground delay $\lambda_1, ..., \lambda_n$ and a sequence of random block-time delay $\mu_1, ..., \mu_n$ for $f_1, ..., f_n$. According Proposition 2.2, the departure delay of flight $f_i \in r'$ is

$$\alpha'_i = \lambda_i + \max\{\alpha'_{i-1} + \mu'_{i-1} - s'_{i-1,i}, 0\}.$$

Since

$$s'_{i-1,i} = s_{i-1,i} - x_i - y_{i-1},$$

and

$$\mu'_{i-1} = \mu_{i-1} - x_{i-1} - y_{i-1},$$

then

$$\alpha'_i = \lambda_i + \max\{\alpha'_{i-1} + \mu_{i-1} - s_{i-1,i} + x_i - x_{i-1}, 0\} \tag{3.10}$$

Equation (3.10) shows that re-timing arrival time of flights will not affect to their departure delays. Since the arrival delay is the sum of the departure delay and the block-time delay, then the shorter time of departure delay is the shorter time of arrival delay.

Now assume that the propagated delays of each flight in r' is always positif. According Equation (2.8), the departure delay of flight f_i in the perturbed routing is given by the following relation.

$$\begin{aligned}\alpha'_i &= \sum_{k=1}^{i}\lambda_k + \sum_{k=1}^{i-1}(\mu_k - x_k - y_k) - \sum_{k=2}^{i}(s_{k-1,k} - y_{k-1} - x_k) \\ &= \sum_{k=1}^{i}\lambda_k + \sum_{k=1}^{i-1}\mu_k - \sum_{k=2}^{i}s_{k-1,k} - \sum_{k=1}^{i-1}(x_k + y_k) + \sum_{k=2}^{i}(y_{k-1} + x_k) \\ &= \sum_{k=1}^{i}\lambda_k + \sum_{k=1}^{i-1}\mu_k - \sum_{k=2}^{i}s_{k-1,k} + \sum_{k=1}^{i-1}(-x_k - y_k + y_k + x_{k+1}) \\ &= \sum_{k=1}^{i}\lambda_k + \sum_{k=1}^{i-1}\mu_k - \sum_{k=2}^{i}s_{k-1,k} + (x_i - x_1) \end{aligned} \tag{3.11}$$

Using Equation (2.9), we can also derive the formulation of the arrival delay of flight f_i in a perturbed routing as follow.

$$\begin{aligned}\beta'_i &= \sum_{k=1}^{i}(\lambda_i + (\mu_i - x_k - y_k)) - \sum_{k=2}^{i}(s_{i-1,i} - x_k - y_{k-1}) \\ &= \sum_{k=1}^{i}(\lambda_i + \mu_i) - \sum_{k=2}^{i}s_{i-1,i} - \sum_{k=1}^{i}(x_k + y_k) + \sum_{k=2}^{i}(x_k + y_{k-1}) \\ &= \sum_{k=1}^{i}(\lambda_i + \mu_i) - \sum_{k=2}^{i}s_{i-1,i} - \sum_{k=1}^{i}(x_k + y_k) + \sum_{k=1}^{i-1}(x_{k+1} + y_k) \\ &= \sum_{k=1}^{i}(\lambda_i + \mu_i) - \sum_{k=2}^{i}s_{i-1,i} - (x_1 + y_1 + ... + x_i + y_i) + (x_2 + y_1 + ... + x_i + y_{i-1}) \\ &= \sum_{k=1}^{i}(\lambda_i + \mu_i) - \sum_{k=2}^{i}s_{i-1,i} - (x_1 + y_i) \end{aligned} \tag{3.12}$$

Since x_1 and y_i are the non negative number, we can conclude that the moving flights in F will reduce the arrival delay of the flights in the routing. This condition is not true for moving departure time of flights. Equation (3.11) tell us that as long as the propagated delay is positive, the departure delay of flight f_i is influenced by the moving of both its flight and the first flight in the perturbed routing. It means that re-timing departure time of flights does not always improve the performance of aircraft routing. For a special case when of the departure time and arrival time of each flight in a routing move in a constant number, we obtain that the departure delay before and after the re-timing are the equal. This condition is explain in the following proposition.

Proposition 3.2 *For $\lambda_1, ..., \lambda_n$ of arbitrary ground delays and $\mu_1, ..., \mu_n$ of arbitrary block-time delays for flights $f_1, ..., f_n$, if $x_i = c$, and $y_i = c$, for $i = 1, ...n$ and c is any positif real number, then $\alpha'_i = \alpha_i$ for $i = 1, ...n$.*

PROOF. Using Proposition 2.3, we have $\alpha_1 = \lambda_1$ and $\alpha'_1 = \lambda_1$. Now, we get $\alpha'_1 = \alpha_1$. Suppose that $\alpha'_{i-1} = \alpha_{i-1}$. Since flight the departure time and the arrival time change of flight f_{i-1}

IOP Conf. Series: Journal of Physics: Conf. Series **949** (2017) 012003 doi:10.1088/1742-6596/949/1/012003

is x_{i-1} and y_{i-1}, respectively and we have $x_{i-1} = y_{i-1} = c$, the arrival delay of flight f_i is determined by

$$\begin{aligned}\beta'_{i-1} &= \max\{\alpha'_{i-1} + \mu_{i-1} - x_{i-1} - y_{i-1}, 0\}\\ &= \max\{\alpha'_{i-1} + \mu_{i-1} - 2c, 0\}.\end{aligned}$$

Since the propagated delay of flight f_i caused by flight f_{i-1} in r' is

$$pd'_{i-1,i} = \max\{\beta'_{i-1} - s'_{i-1,i}, 0\},$$

and $s'_{i-1,i} = s_{i-1,i} - y_{i-1} - x_i = s_{i-1,i} - 2c$, then

$$\alpha'_i = \lambda_i + \max\{\beta'_{i-1} - s_{i-1,i} + 2c, 0\}.$$

Next, we will consider two cases for β'. If we have $\beta'_{i-1} = 0$, then $\alpha'_i = \lambda_i$ for this case. Since $\alpha'_{i-1} + \mu_{i-1} \le 2c$ and $\alpha'_{i-1} = \alpha_{i-1}$ according our assumption, then we get $\alpha_{i-1} + \mu_{i-1} \le 2c$. Let consider two cases when $\alpha_{i-1} + \mu_{i-1} \le 2c$. If $\alpha_{i-1} + \mu_{i-1} \le 0$, then $pd_{i-1,i} = 0$. As a result, we have $\alpha_i = \lambda_i = \alpha'_i$. Otherwise, let we have $\alpha_{i-1} + \mu_{i-1} \le> 0$. Since $s_{i-1,i} = s'_{i-1,i} + 2c$ and $\beta_{i-1} = \alpha_{i-1} + \mu_{i-1} \le< 2c$, then we have $pd_{i-1,i} = 0$. Now, we obtain $\alpha_i = \lambda_i = \alpha'_i$.

Let $\beta'_{i-1} > 0$. Then

$$\begin{aligned}pd'_{i-1,i} &= \max\{\alpha'_{i-1} + \mu'_{i-1} - s'_{i-1,i}, 0\}\\ &= \max\{\alpha'_{i-1} + \mu_{i-1} - 2c - (s_{i-1,i} - 2c), 0\}\\ &= \max\{\alpha'_{i-1} + \mu_{i-1} - s_{i-1,i}, 0\}.\end{aligned}$$

Since $\alpha'_{i-1} = \alpha_{i-1}$, we obtain $pd'_{i-1,i} = \max\{\alpha_{i-1} + \mu_{i-1} - s_{i-1,i}, 0\} = pd_{i-1,i}$. As the consequence, $\alpha'_i = \alpha_i$. Now, we complete the proof. □

4. Computational results

We consider one-day flight schedules of an airline in Indonesia for computational study. The schedules consist of 287 flights which are covered by 92 aircraft routings. We construct a flight delay simulation by the following steps.

(1) Generate a number of flight delays, each with its ground delay and block-time delay.

(2) For each flight delay, apply push-back recovery strategy. If the new connecting time between the delayed flight and the connecting flight is less then the minimum required connecting time, the connecting flight will be delayed until the minimum connecting time is fulfilled.

(3) Calculate total departure delays and total arrival delays

(4) Repeat step (1)-(3) for a number of iterations.

(5) Calculate the average of total departure delays and total arrival delays.

The number of flight delays in Step (1) is generated randomly. We model the probability distribution of the ground delay and the block-time delay according one-year historical delay data of the airline. We refer the probability distribution of ground delays in Novianingsih et al. [7]. The distribution of ground delays is modeled as log-normal distribution:

$$f(x; \mu, \sigma) = \frac{1}{x\sigma\sqrt{2\pi}} \exp(-\frac{(\ln x - \mu)^2}{2\sigma^2}), x > 0, \tag{4.13}$$

where μ and σ are mean and standard deviation, respectively. In our case study, the logarithm of parameters in the log-normal distributions are $\mu \in [2.8 - 3.4]$ and $\sigma \in [0.4 - 0.6]$. The distribution of block-time delays is modeled as uniformly distributed random numbers on the interval $[0 - 30]$.

To perturb schedules, we allow departure times and arrival times of planned flight schedules to be moved earlier or later no more than 15 minutes in order to preserve passenger projection in flight schedule design. To observe the effect of additional slack to the performance of aircraft routing, we modify the departure times and the arrival times of the original flight schedules such that the new flight schedules will produce the aircraft routings with more slack then the original aircraft routings, but the original block-times do not change. The same technique is also applied to obtain the new flight schedules that will produce aircraft routings with more length of block-times than the original block-times and the same length of slacks with the original slacks. For both cases, we record total departure delay, total arrival delay, and 15-minute on time performance (15-OTP). According 15-OTP, a flight is delay if depart on a gate in more than 15 minutes after its schedule departure time.

The effect of additional slack and additional block-time are summarized on Table 1 and 2, respectively. From Table 1, we can see that the larger slack will result to the better aircraft routing performance. The same condition also occurs for the additional of block-time. The increasing block-time in aircraft routings will affect to the increasing of the aircraft routing performance. More block-times are the better performance.

Table 1. The effect of additional slacks to the aircraft routings performance.

Robustness measure	Additional slack (mins)			
	0	5	10	15
Total departure delay (mins)	4320	3995	3422	2931
Total arrival delay (mins)	4320	3995	3422	2931
15-OTP (percent)	71	78	85	91

Table 2. The effect of additional block-time to the aircraft routings performance.

Robustness measure	Additional block-time (mins)			
	0	5	10	15
Total departure delay (mins)	4320	3880	3600	3212
Total arrival delay (mins)	4450	4450	4450	4450
15-OTP (percent)	71	79	87	92

We also consider four other cases of perturbations for each aircraft routing. In the first case, we set $x_1 = 10$ minutes and $x_i = x_{i-1}$ for $i = 2, 3, \ldots$. In Case 2, we set $x_1 = 5, x_i = \frac{3}{2}x_{i-1}$ for $i = 2, 3, \ldots$. In Case 3, we set $x_1 = 5, x_i = x_{i-1}$ for $i = 2, 3, \ldots$. In Case 4, we set $x_1 = 15, x_i = \frac{2}{3}x_{i-1}$ for $i = 2, 3, \ldots$. For all cases, we set $x_i = y_i$ for $i = 1, 2, \ldots$. We perform the same flight delay simulation for all cases.

Table 3. The effect of perturbed flights to the aircraft routings performance.

Robustness measure	Original	Case 1	2	3	4
Total departure delay (mins)	4320	4320	4502	4320	3701
Total arrival delay (mins)	4405	4380	4798	4391	4302
15-OTP (percent)	71	71	68	71	84

Using the results on the Table 3, we can see that the performance of aircraft routings are not improve when we move departure time and arrival time of flights in a constant number, and it ratifies the Proposition 3.2. The performance will continue to increase, if the additional of block-time of each flight in the aircraft routings is large than the reducing slack in its connection.

5. Conclusion

In this paper, we study some effects of flight schedule perturbations to the performance of aircraft routing. This research suggest that we can add block-times by moving departure time or arrival time of flights so that slacks in aircraft connections is reduced. We show that adding block-time of flights will not increase the arrival delay. However, the departure delay does not always decrease when the departure time of flights is changed. In reality, adding more block-time can affect to the increasing of crew costs. To obtain the perturbed schedules in aircraft routings that will not increase the crew costs, we should include the crew scheduling problem in the perturbed scheduled. This is our further research object.

Acknowledgment

This research was supported by Penelitian Produk Terapan from Ministry of Technology, and Higher Education Indonesia with contract number: 008/SP2H/LT/DRPM/IV/2017. The authors thank to Garuda Indonesia Airline for supporting the data.

References

[1] Abdelghany A, Ekollu G, Narasimhan R and Abdelghany K 2004 A proactive crew recovery decision support tool for commercial airline during irreguler operations *Ann. Oper. Res.* **127** 309

[2] AhmadBeygi S, Cohn A and Lapp M 2008 *Decreasing airline delay propagation by re-allocating schedule slack*(Mechigan: University of Mechigan)

[3] M.A. Aloulou M A, Haouari M and Mansour F Z 2013 A model for enhanching robustness of aircraft and passenger *Transportation Research Part C* **32** 48

[4] Chiraphadanakul V and Bernhard C 2013 Robust flight schedules through slack re-allocation *Euro J. Transport. Log.* **2** 277

[5] Lan S, Carke J P and Bernhart C 2006 Planning for robust airline operations: Optimizing aircraft routings and flight departure times to minimize passenger disruptions *Transport. Sci.* **40** 15

[6] Lettovsky L, Johnson E L and Nemhauser G L 2000 Planning for robust airline operations: Airline crew recovery *Transport. Sci.* **34** 337

[7] Novianingsih K and Hadianti R 2016 Flight re-timing models to improve the robustness of airrline schedules *Thai J. Math.* **14** 49

[8] Schaefer A J and Nemhauser G L 2006 Improving airline operational performance through schedule perturbation *Ann. Oper. Res.* **114** 3

[9] Sohoni M, Lee Y and Klabjan D 2011 Robust airline scheduling under block time uncertainty *Transport. Sci.* **45 451**

IOP Conf. Series: Journal of Physics: Conf. Series **949** (2017) 012004 doi:10.1088/1742-6596/949/1/012004

A note on a separating system of rational invariants for finite dimensional generic algebras

U. Bekbaev

Deparment of Science in Engineering, KOE, IIUM, Malaysia

E-mail: bekbaev@iium.edu.my

Abstract. The paper deals with a construction of a separating system of rational invariants for finite dimensional generic algebras. In the process of dealing an approach to a rough classification of finite dimensional algebras is offered by attaching them some quadratic forms.

1. Introduction

In [1] we have offered an approach to classification problem of finite dimensional algebras with respect to basis changes. It has been shown that if there is a special map with certain properties then it is possible to classify (to list canonical representatives) of all algebras who's set of structural constants do not nullify a certain polynomial. In this case one can provide a separating system of rational invariants for the algebras considered. The approach was successfully applied in [2] to get a complete classification of all 2-dimensional algebras over algebraically closed fields.

Unfortunately, so far we have no example of such a map in 3-dimensional case. Therefore in the current paper we deal with a weaker problem, namely with a construction of separating system of rational invariants for finite dimensional generic algebras. The existence such of system of invariants is known [3]. By generic algebras we mean the set of all algebras who's system of structural constants does not nullify a fixed nonzero polynomial in structural variables, over the basic field F. In process of dealing with the problem we show a way for a rough classification of finite dimensional algebras by attaching them some quadratic forms.

2. Main results

Further, whenever $A = (a_{ij}) \in Mat(p \times q, F)$, $B \in Mat(p' \times q', F)$ we use $A \otimes B$ for the matrix

$$\begin{pmatrix} a_{11}B & a_{12}B & ... & a_{1q}B \\ a_{21}B & a_{22}B & ... & a_{2q}B \\ . & . & \vdots & . \\ a_{p1}B & a_{p2}B & ... & a_{pq}B \end{pmatrix},$$ where F is a field of characteristic not 2.

Let us consider an m-dimensional algebra $\mathbf{A}$ with multiplication $\cdot$ given by a bilinear map $(\mathbf{u}, \mathbf{v}) \mapsto \mathbf{u} \cdot \mathbf{v}$. If $e = (e_1, e_2, ..., e_m)$ is a basis for $\mathbf{A}$ then one can represent the bilinear map by a matrix

$$A_e = (A^i_{ejk})_{i,j,k=1,2,...,m} \in Mat(m \times m^2; F),$$

where $e_j \cdot e_k = e_1 A^1_{ejk} + e_2 A^2_{ejk} + ... + e_m A^m_{ejk}$, $j, k = 1, 2, ..., m$, such that $\mathbf{u} \cdot \mathbf{v} = eA_e(u \otimes v)$ for any $\mathbf{u} = eu, \mathbf{v} = ev$, where $u = (u_1, u_2, ..., u_m), v = (v_1, v_2, ..., v_m)$ are column vectors. So

IOP Conf. Series: Journal of Physics: Conf. Series **949** (2017) 012004 doi:10.1088/1742-6596/949/1/012004

the algebra **A** is presented by the matrix $A_e \in Mat(m \times m^2; F)$ called the matrix of structure constants (MSC) of **A** with respect to the basis e.

If $e' = (e'_1, e'_2, ..., e'_m)$ is also a basis for **A**, $g \in GL(m, F)$, $e'g = e$ then it is well known that

$$A_{e'} = gA_e(g^{-1})^{\otimes 2}.$$

Further a basis e is fixed and therefore instead of A_e we use A, we do not make difference between **A** and its matrix A. Let $X = (X^i_{jk})_{i,j,k=1,2,...,m}$ stand for a variable matrix and $Tr_1(X)$, $Tr_2(X)$ stand for the row vectors

$$\left(\sum_{i=1}^m X^i_{i1}, \sum_{i=1}^m X^i_{i2}, ..., \sum_{i=1}^m X^i_{im}\right), \quad \left(\sum_{i=1}^m X^i_{1i}, \sum_{i=1}^m X^i_{2i}, ..., \sum_{i=1}^m X^i_{mi}\right),$$

respectively.

We use τ for the representation of $GL(m, F)$ on the m^3- dimensional vector space $V = Mat(m \times m^2; F)$ defined by $\tau : (g, A) \mapsto B = gA(g^{-1} \otimes g^{-1})$.

For simplicity instead of *τ-equivalent* , *τ-invariant* we use the terms *equivalent* and *invariant*, respectively.

We represent each MSC A as a row vector with entries from $Mat(m, F)$ by parting it consequently into elements of $Mat(m, F)$:

$$A = (A_1, A_2, ..., A_m), \; A_1, A_2, ..., A_m \in Mat(m, F).$$

If C is a block matrix with blocks from $Mat(m, F)$ we use notation $C^{\overline{*}}$, where $*$ is the tensor product or transpose operation, to mean that the operation $*$ with C is done "$Mat(m, F)$"(not over F), for example for the above presented "row" matrix A the notation $A^{\overline{t}}$ means "column" vector over $Mat(m, F)$ with the entries $A_1, A_2, ..., A_m$,

$$A^{\overline{\otimes}2} = (A_1^2, A_1A_2, ..., A_1A_m, A_2A_1, A_2^2, ..., A_2A_m, ..., A_mA_1, A_mA_2, ..., A_m^2).$$

One can see that the equality $B = gA(g^{-1} \otimes g^{-1})$ can be represented as

$$B = (B_1, B_2, ..., B_m) = gA(g^{-1})^{\otimes 2} = (gA_1g^{-1}, gA_2g^{-1}, ..., gA_mg^{-1})(g^{-1} \otimes I),$$

where I is $m \times m$ identity matrix. Moreover, for any matrices C and D the equality

$$(C \otimes I)\overline{\otimes}(D \otimes I) = (C \otimes D) \otimes I$$

holds true. Therefore, the following equalities hold true

$$(B_1, B_2, ..., B_m)^{\overline{\otimes}k} = (gA_1g^{-1}, gA_2g^{-1}, ..., gA_mg^{-1})^{\overline{\otimes}k}((g^{-1})^{\otimes k} \otimes I),$$

$$\begin{pmatrix} B_1 \\ B_2 \\ \vdots \\ B_m \end{pmatrix}^{\overline{\otimes}k} (B_1, B_2, ..., B_m)^{\overline{\otimes}k} = \begin{pmatrix} B_1^2 & B_1B_2 & \cdots & B_1B_m \\ B_2B_1 & B_2^2 & \cdots & B_2B_m \\ \vdots & \vdots & \cdots & \vdots \\ B_mB_1 & B_mB_2 & \cdots & B_m^2 \end{pmatrix}^{\overline{\otimes}k},$$

$$\begin{pmatrix} B_1^2 & B_1B_2 & \cdots & B_1B_m \\ B_2B_1 & B_2^2 & \cdots & B_2B_m \\ \vdots & \vdots & \cdots & \vdots \\ B_mB_1 & B_mB_2 & \cdots & B_m^2 \end{pmatrix}^{\overline{\otimes}k} =$$

$$(((g^t)^{-1})^{\otimes k} \otimes I) \begin{pmatrix} gA_1^2g^{-1} & gA_1A_2g^{-1} & \cdots & gA_1A_mg^{-1} \\ gA_2A_1g^{-1} & gA_2^2g^{-1} & \cdots & gA_2A_mg^{-1} \\ \vdots & \vdots & \cdots & \vdots \\ gA_mA_1g^{-1} & gA_mA_2g^{-1} & \cdots & gA_m^2g^{-1} \end{pmatrix}^{\overline{\otimes} k} ((g^{-1})^{\otimes k} \otimes I).$$

Component-wise application of trace to this equality, which is denoted by $\tilde{Tr}$ results in

$$\tilde{Tr}\left(\begin{pmatrix} B_1^2 & B_1B_2 & \cdots & B_1B_m \\ B_2B_1 & B_2^2 & \cdots & B_2B_m \\ \vdots & \vdots & \cdots & \vdots \\ B_mB_1 & B_mB_2 & \cdots & B_m^2 \end{pmatrix}^{\overline{\otimes} k}\right) =$$

$$((g^t)^{-1})^{\otimes k}\tilde{Tr}\left(\begin{pmatrix} gA_1^2g^{-1} & gA_1A_2g^{-1} & \cdots & gA_1A_mg^{-1} \\ gA_2A_1g^{-1} & gA_2^2g^{-1} & \cdots & gA_2A_mg^{-1} \\ \vdots & \vdots & \cdots & \vdots \\ gA_mA_1g^{-1} & gA_mA_2g^{-1} & \cdots & gA_m^2g^{-1} \end{pmatrix}^{\overline{\otimes} k}\right)(g^{-1})^{\otimes k} =$$

$$((g^{-1})^{\otimes k})^t\tilde{Tr}\left(\begin{pmatrix} A_1^2 & A_1A_2 & \cdots & A_1A_m \\ A_2A_1 & A_2^2 & \cdots & A_2A_m \\ \vdots & \vdots & \cdots & \vdots \\ A_mA_1 & A_mA_2 & \cdots & A_m^2 \end{pmatrix}^{\overline{\otimes} k}\right)(g^{-1})^{\otimes k},$$

as far as for any matrices C, D and E, where D is a block matrix with blocks from $Mat(m, F)$ and $(C \otimes I)D(E \otimes I)$ has a meaning, the equality

$$\tilde{Tr}((C \otimes I)D(E \otimes I)) = C\tilde{Tr}(D)E$$

is valid. One can represent the above obtained matrix equality in a compact form as follows

$$\tilde{Tr}((B^{\bar{t}}B)^{\overline{\otimes} k}) = ((g^{-1})^{\otimes k})^t\tilde{Tr}((A^{\bar{t}}A)^{\overline{\otimes} k})(g^{-1})^{\otimes k}.$$

Note that $\tilde{Tr}((A^{\bar{t}}A)^{\overline{\otimes} k})$ is a symmetric matrix. The obtained equality yields the following theorem.

Theorem 1. *Invariants of the quadratic form given by the matrix* $\tilde{Tr}((X^{\bar{t}}X)^{\overline{\otimes} k})$ *are invariants of the* m*-dimensional algebra defined by MSC* X*.*

This result can be used for a rough classification of finite dimensional algebras: Two m-dimensional algebras A, B are roughly equivalent if the quadratic forms given by matrices

$$\tilde{Tr}(A^{\bar{t}}A) = \begin{pmatrix} Tr(A_1^2) & Tr(A_1A_2) & \cdots & Tr(A_1A_m) \\ Tr(A_2A_1) & Tr(A_2^2) & \cdots & Tr(A_2A_m) \\ \vdots & \vdots & \cdots & \vdots \\ Tr(A_mA_1) & Tr(A_mA_2) & \cdots & Tr(A_m^2) \end{pmatrix},$$

$$\tilde{Tr}(B^{\bar{t}}B) = \begin{pmatrix} Tr(B_1^2) & Tr(B_1B_2) & \cdots & Tr(B_1B_m) \\ Tr(B_2B_1) & Tr(B_2^2) & \cdots & Tr(B_2B_m) \\ \vdots & \vdots & \cdots & \vdots \\ Tr(B_mB_1) & Tr(B_mB_2) & \cdots & Tr(B_m^2) \end{pmatrix}$$

are equivalent.

It is clear that entries of $\tilde{Tr}(X^{\bar{t}}X)$ are polynomials in components of X and there exists nonsingular matrix $Q(X^{\bar{t}}X)$ with rational entries in X such that the matrix

$$\tilde{Tr}(\overline{X}^{\bar{t}}\overline{X}) = (Q(X^{\bar{t}}X)^{-1})^t\tilde{Tr}(X^{\bar{t}}X)Q(X^{\bar{t}}X)^{-1} = D(X)$$

is a diagonal matrix and $Q(g) = I$, whenever g is a nonsingular diagonal matrix and $\overline{X} = \tau(Q(X^{\bar{t}}X), X)$.

Therefore in algebraically closed field F case one can define a nonempty invariant open subset $V_0 \subset V$ such that $\tilde{Tr}(\overline{A}^{\bar{t}}\overline{A}) = D(A)$ and $D(A)$ is nonsingular whenever $A \in V_0$.

Theorem 2. *Two algebras $A, B \in V_0$ are isomorphic if and only if*

$$\overline{B} = \tau(g_0, \overline{A}) \text{ for some } g_0 \in GL(m, F) \text{ such that } g_0^t D(B)g_0 = D(A).$$

Proof. If $B = \tau(g, A)$ then $\overline{B} = \tau(Q(B^{\bar{t}}B), B) = \tau(Q(B^{\bar{t}}B), \tau(g, A)) =$

$$\tau(Q(B^{\bar{t}}B)g, A) = \tau(Q(B^{\bar{t}}B)gQ(A^{\bar{t}}A)^{-1}, \tau(Q(A^{\bar{t}}A), A) = \tau(Q(B^{\bar{t}}B)gQ(A^{\bar{t}}A)^{-1}, \overline{A}),$$

and for $g_0 = Q(B^{\bar{t}}B)gQ(A^{\bar{t}}A)^{-1}$ one has

$$g_0^t D(B)g_0 = (Q(B^{\bar{t}}B)gQ(A^{\bar{t}}A)^{-1})^t D(B)Q(B^{\bar{t}}B)gQ(A^{\bar{t}}A)^{-1} =$$

$$(Q(A^{\bar{t}}A)^{-1})^t(g^t(Q(B^{\bar{t}}B)^t D(B)Q(B^{\bar{t}}B))g)Q(A^{\bar{t}}A)^{-1} =$$

$$(Q(A^{\bar{t}}A)^{-1})^t(g^t(\tilde{Tr}(B^{\bar{t}}B))g)Q(A^{\bar{t}}A)^{-1} = (Q(A^{\bar{t}}A)^{-1})^t\tilde{Tr}(A^{\bar{t}}A)Q(A^{\bar{t}}A)^{-1} = D(A).$$

Visa versa if $\overline{B} = \tau(g_0, \overline{A})$ for some g_0 such that $g_0^t D(B)g_0 = D(A)$ then for $g = Q(B^{\bar{t}}B)^{-1}g_0Q(A^{\bar{t}}A)$ one has

$$\tau(g, A) = \tau(Q(B^{\bar{t}}B)^{-1}g_0Q(A^{\bar{t}}A), A) = \tau(Q(B^{\bar{t}}B)^{-1}g_0, \tau(Q(A^{\bar{t}}A), A) =$$

$$\tau(Q(B^{\bar{t}}B)^{-1}g_0, \overline{A}) = \tau(Q(B^{\bar{t}}B)^{-1}, \tau(g_0, \overline{A})) = \tau(Q(B^{\bar{t}}B)^{-1}, \overline{B}) = B.$$

Assume that there exists a matrix $P(X)$, with entries consisting of rational functions of components X, such that $P(\overline{A})$ is nonsingular for any $A \in V_0$ and the equality

$$P(\tau(g, \overline{A})) = P(\overline{A})g^{-1} \text{ holds true whenever } g^t D(\tau(g, A))g = D(A). \tag{1}$$

Theorem 3. *For A,B $\in V_0$ there exists $g_0 \in GL(m, F)$ such that $g_0^t D(B)g_0 = D(A)$ and $\overline{B} = \tau(g_0, \overline{A})$ if and only if*

$$\tau(P(\overline{B}), \overline{B}) = \tau(P(\overline{A}), \overline{A}),\ (P(\overline{B})^{-1})^t D(B)P(\overline{B})^{-1} = (P(\overline{A})^{-1})^t D(A)P(\overline{A})^{-1}.$$

Proof. If $\overline{B} = \tau(g_0, \overline{A})$ and $g_0^t D(B)g_0 = D(A)$ then

$$\tau(P(\overline{B}), \overline{B}) = \tau(P(\tau(g_0, \overline{A})), \tau(g_0, \overline{A})) = \tau(P(\overline{A})g_0^{-1}, \tau(g_0, \overline{A})) = \tau(P(\overline{A}), \overline{A})$$

and $(P(\overline{B})^{-1})^t D(B)P(\overline{B})^{-1} = ((P(\overline{A})g_0^{-1})^{-1})^t D(B)(P(\overline{A})g_0^{-1})^{-1} =$

$$((P(\overline{A})^{-1})^t g_0^t D(B)g_0P(\overline{A})^{-1} = (P(\overline{A})^{-1})^t D(A)P(\overline{A})^{-1}.$$

Visa versa, if equalities

$$\tau(P(\overline{B}),\overline{B})=\tau(P(\overline{A}),\overline{A}),\ (P(\overline{B})^{-1})^tD(B)P(\overline{B})^{-1}=(P(\overline{A})^{-1})^tD(A)P(\overline{A})^{-1}$$

are valid then for $g_0=P(\overline{B})^{-1}P(\overline{A})$ one has $g_0^tD(B)g_0=D(A)$ and

$$\tau(g_0,\overline{A})=\tau(P(\overline{B})^{-1}P(\overline{A}),\overline{A})=\tau(P(\overline{B})^{-1},\tau(P(\overline{A}),\overline{A}))=\tau(P(\overline{B})^{-1},\tau(P(\overline{B}),\overline{B}))=\overline{B}.$$

So Theorems 2 and 3 imply that the system of entries of matrices

$$\tau(P(\overline{X}),\overline{X}),\ \ (P(\overline{X})^{-1})^t\tilde{Tr}(\overline{X}^{\bar{t}}\overline{X})P(\overline{X})^{-1}$$

is a separating system of rational invariants for algebras from V_0.

The above presented results show the importance of construction of matrix $P(X)$ with property (1). Further we discuss a construction of such matrix by the use of rows $r(\overline{A})$ for which the equality $r(\tau(g,\overline{A}))=r(\overline{A})g^{-1}$ is valid, whenever $g^tD(\tau(g,A))g=D(A)$. To construct such rows one can use the following approach.

Assume that the equalities $\overline{B}=g\overline{A}(g^{-1})^{\otimes 2}$, $\tilde{C}=gCg^t$ are true, where $C^t=C$ and C is a nonsingular matrix. In this case

$$\tilde{C}^{\otimes 2}=g^{\otimes 2}C^{\otimes 2}(g^{\otimes 2})^t,\ \overline{B}\tilde{C}^{\otimes 2}=g\overline{A}C^{\otimes 2}(g^{\otimes 2})^t,\ \tilde{C}^{\otimes 2}\overline{B}^t=g^{\otimes 2}C^{\otimes 2}\overline{A}^tg^t$$

and $\overline{B}\tilde{C}^{\otimes 2}\overline{B}^t=g\overline{A}C^{\otimes 2}\overline{A}^tg^t$.

On induction it is easy to see that for any natural k the equality

$$\overline{B}^{\otimes 2^0}\overline{B}^{\otimes 2^1}...\overline{B}^{\otimes 2^{k-1}}\tilde{C}^{\otimes 2^k}(\overline{B}^{\otimes 2^0}\overline{B}^{\otimes 2^1}...\overline{B}^{\otimes 2^{k-1}})^t=g\overline{A}^{\otimes 2^0}\overline{A}^{\otimes 2^1}...\overline{A}^{\otimes 2^{k-1}}C^{\otimes 2^k}(\overline{A}^{\otimes 2^0}\overline{A}^{\otimes 2^1}...\overline{A}^{\otimes 2^{k-1}})^tg^t$$

holds true. Therefore due to the equalities

$$\overline{B}^{\otimes 2^0}\overline{B}^{\otimes 2^1}...\overline{B}^{\otimes 2^{k-1}}\tilde{C}^{\otimes 2^k}(\overline{B}^{\otimes 2^0}\overline{B}^{\otimes 2^1}...\overline{B}^{\otimes 2^{k-1}})^t\tilde{C}^{-1}=$$
$$g\overline{A}^{\otimes 2^0}\overline{A}^{\otimes 2^1}...\overline{A}^{\otimes 2^{k-1}}C^{\otimes 2^k}(\overline{A}^{\otimes 2^0}\overline{A}^{\otimes 2^1}...\overline{A}^{\otimes 2^{k-1}})^tC^{-1}g^{-1},$$
$$\overline{B}^{\otimes 2^0}\overline{B}^{\otimes 2^1}...\overline{B}^{\otimes 2^{k-1}}\tilde{C}^{\otimes 2^k}(\overline{B}^{\otimes 2^0}\overline{B}^{\otimes 2^1}...\overline{B}^{\otimes 2^{k-1}})^t\tilde{C}^{-1}\overline{B}=$$
$$g\overline{A}^{\otimes 2^0}\overline{A}^{\otimes 2^1}...\overline{A}^{\otimes 2^{k-1}}C^{\otimes 2^k}(\overline{A}^{\otimes 2^0}\overline{A}^{\otimes 2^1}...\overline{A}^{\otimes 2^{k-1}})^tC^{-1}\overline{A}(g^{-1})^{\otimes 2}$$

one has

$$Tr_i(\overline{B}^{\otimes 2^0}\overline{B}^{\otimes 2^1}...\overline{B}^{\otimes 2^{k-1}}\tilde{C}^{\otimes 2^k}(\overline{B}^{\otimes 2^0}\overline{B}^{\otimes 2^1}...\overline{B}^{\otimes 2^{k-1}})^t\tilde{C}^{-1}\overline{B})=$$
$$Tr_i(\overline{A}^{\otimes 2^0}\overline{A}^{\otimes 2^1}...\overline{A}^{\otimes 2^{k-1}}C^{\otimes 2^k}(\overline{A}^{\otimes 2^0}\overline{A}^{\otimes 2^1}...\overline{A}^{\otimes 2^{k-1}})^tC^{-1}\overline{A})g^{-1},\ i=1,2.$$

The last equality shows that in our algebra case on can try to construct the needed matrix $P(X)$ by the use of rows

$$Tr_i(X^{\otimes 2^0}X^{\otimes 2}...X^{\otimes 2(k-1)}((\tilde{Tr}(X^{\bar{t}}X))^{-1})^{\otimes 2^k}(X^{\otimes 2^0}X^{\otimes 2}...X^{\otimes 2(k-1)})^t\tilde{Tr}(X^{\bar{t}}X)X),$$

where $i=1,2,\ k=0,1,2,....$ What is left here unjustified is that one should justify the existence, in general, of a linear independent system consisting of m such rows.

Acknowledgments
The author acknowledges MOHE for a support by grant FRGS14-153-0394.

References

[1] Bekbaev U 2017, *IOP Conf. Series: Journal of Physics: Conf. Series*, **819**), pp. 2–9.
[2] H. Ahmed, U. Bekbaev, I. Rakhimov 2017, *AIP Conference Proceedings*, **1830**, 070016; doi: 10.1063/1.4980965).
[3] V. Popov 2014, *arXiv: 1411.6570v2[math.AG]*, pp. 1–20.

IOP Conf. Series: Journal of Physics: Conf. Series **949** (2017) 012005 doi:10.1088/1742-6596/949/1/012005

The Multistage Homotopy Perturbation method for solving Hyperchaotic Chen system

M. S. H. Chowdhury [1,*], **Nur Isnida Razali** [1], **Waqar Asrar** [2] **and Zanariah Abdul Majid** [3]

[1] Department of Science in Engineering, Faculty of Engineering, International Islamic University Malaysia, Jalan Gombak, 53100 Kuala Lumpur, Malaysia of Mathematics,
[2]Department of Mechanical Engineering, Faculty of Engineering, International Islamic University Malaysia, Jalan Gombak, 53100 Kuala Lumpur, Malaysia
[3]Department of Mathematics, Faculty of Science, Universiti Putra Malaysia, 43400 Selangor, Malaysia

Email: [1,*]sazzadbd@iium.edu.my

Abstract. Finding accurate and efficient methods for solving nonlinear hyper chaotic problems has long been an active research undertaking. Like the well-known Adomian decomposition method (ADM) and based on observations, it is hopeless to find HPM solutions of hyper chaotic systems valid globally in time. To overcome this shortcoming, we employ the multistage homotopy-perturbation method (MHPM) to the nonlinear hyperchaotic Chen system. Based on the cases investigated, MHPM is more stable for a longer time span than the standard HPM. Comparisons with the standard HPM and the well-known purely numerical fourth-order Runge-Kutta method (RK4) suggest that the MHPM is a powerful alternative for nonlinear hyper chaotic system. The new algorithm and the new technique for choosing the initial approximations were shown to yield rapidly convergent series solutions.

1. Introduction

Nonlinear problems modelled by differential equations play a prominent role in various fields such as physics, chemistry, biology, mathematics, engineering and other disciplines. There are few phenomena in different fields of science occurring linearly. Most problems are essentially nonlinear and are described by nonlinear differential equations. If the considered problem is linear, the corresponding set of differential equations is also linear and can be solved without mathematical difficulties; if the researched problem is nonlinear, the obtained set of differential equations is generally nonlinear. Up to now, there are few effective numerical methods to solve this kind of nonlinear differential equations with a large number of unknowns and strong nonlinearity. Generally, iterative techniques are used to solve these nonlinear problems; however, the convergence of iterations seems sometimes just luck, especially when the nonlinearity of the considered problems is strong. Chaos is very interesting nonlinear phenomenon and has been intensively studied in the last three decades. The dynamical systems that exhibit chaotic behavior are sensitive to initial conditions. Chaotic behavior can be found in a variety of systems such as electrical circuits, lasers, fluid dynamics, mechanical devices, time evolution of the magnetic field of celestial bodies, population growth in ecology, the dynamics of molecular vibrations, and not forgetting the weather [1,2]. In this paper, we consider the hyperchaotic Chen system [3] defined as

$$\begin{aligned} \dot{x} &= a(y-x)+w, \\ \dot{y} &= dx - xz + cy, \\ \dot{z} &= xy - bz, \\ \dot{w} &= yz + rw. \end{aligned} \tag{1}$$

The above system is hyperchaotic for the parameters of $a = 35$, $b = 3$, $c = 12$, $d = 7$, and $r = 0.5$. Hyperchaotic system was first discovered by O.E. Rossler in 1979 [4]. Details of other hyperchaotic systems can be found in many references [5-9, 36-37].

The Homotopy technique has been applied in many nonlinear problems in engineering field for the reason that homotopy method deforms continuously from a difficult into a simpler problem which can be easily solved. Ji-huan He, founder of Homotopy Perturbation method in 1998 [10, 11], in his paper showed that the Homotopy Perturbation Method is more powerful than Homotopy Analysis Method by Liao which is can be categorized into a generalized Taylor expansion method [12]. HPM was further developed and improved by He [13- 15]. Changbum and Rathinasamy [16] solved the linear and nonlinear two- point boundary value problems using HPM and compared the result with extended Adomian Decomposition Method (EADM) and the shooting method. As a result, HPM yields more accurate results with rapid convergence than other methods. Recently, Chowdhury et al. [17], Chowdhury and Hashim [18], Hashim and Chowdhury [19] and Hashim et al. [20] were the first to successfully apply the multistage homotopy-perturbation method (MHPM) to the chaotic Lorenz system, chaotic Chen system and a class of ODEs. Other successfully solved problems by HPM can be found in references [21- 24].

Inspired from the previous works done by others, in this present paper, we are interested to solve hyperchaotic Chen systems using multistage homotopy perturbation method (MHPM). Based on the past researches, chaotic Lorenz system has been solved using Adomian decomposition method (ADM) [25-27] and differential quadrature method (DQM) [28]. Fractional Lorenz system also has been solved using ADM [29] and homotopy analysis method (HAM) [30]. Meanwhile for chaotic Chen system, a lot of work has been done to find its solution using numerical method such as homotopy perturbation method (HPM) [31], variational iteration method (VIM) [32], solutions by HAM [33] and also by ADM [27]. Differential transformation method (DTM) is used by Alomari and Mossa et al. for the fractional Chen system [34, 35].

2. The Solution Procedure

Consider a general system of first-order ODEs:

$$\begin{aligned} \frac{du_1}{dt} + g_1(t, u_1, u_2, \dots, u_m) &= f_1(t), \\ \frac{du_2}{dt} + g_2(t, u_1, u_2, \dots, u_m) &= f_2(t), \\ &\vdots \\ \frac{du_m}{dt} + g_m(t, u_1, u_2, \dots, u_m) &= f_m(t), \end{aligned} \tag{2}$$

subject to the initial conditions

$$u_1(t_0) = c_1, u_2(t_0) = c_2, \dots, u_m(t_0) = c_m. \tag{3}$$

First we write the system (3) in operator form

$$\begin{aligned} &L(u_1) + N_1(u_1, u_2, \dots, u_m) - f_1 = 0, \\ &L(u_2) + N_2(u_1, u_2, \dots, u_m) - f_2 = 0, \\ &\vdots \\ &L(u_m) + N_m(u_1, u_2, \dots, u_m) - f_m = 0. \end{aligned} \tag{4}$$

Subject to the initial conditions (4), where $L = d/dt$ is a linear operator and $N_1, N_2, \dots, N_m$ are the nonlinear operators. We will next present the solution approaches for (4) based on standard HPM and MHPM separately.

2.1 Solution by HPM

According to HPM, we construct the homotopy for (4) which satisfies the following relations:

$$\begin{aligned} &L(u_1) - L(v_1) + pL(v_1) + p[N_1(u_1, u_2, \dots, u_m) - f_1] = 0, \\ &L(u_2) - L(v_2) + pL(v_2) + p[N_2(u_1, u_2, \dots, u_m) - f_2] = 0, \\ &\vdots \\ &L(u_m) - L(v_m) + pL(v_m) + p[N_m(u_1, u_2, \dots, u_m) - f_m] = 0. \end{aligned} \tag{5}$$

Where $p \in [0,1]$ is an embedding parameter and $v_1, v_2, \dots, v_m$ are initial approximations satisfying the given conditions. It is obvious that when the perturbation parameter $p = 0$, Eqs. (5) become a linear system and when $p = 1$ we get the original nonlinear system. Let take the approximations as follows:

$$\begin{aligned} &u_1(t) = u_{1,0}(t) + pu_{1,1}(t) + p^2 u_{1,2}(t) + p^3 u_{1,3}(t) + \dots, \\ &u_2(t) = u_{2,0}(t) + pu_{2,1}(t) + p^2 u_{2,2}(t) + p^3 u_{2,3}(t) + \dots, \\ &\vdots \\ &u_m(t) = u_{m,0}(t) + pu_{m,1}(t) + p^2 u_{m,2}(t) + p^3 u_{m,3}(t) + \dots, \end{aligned} \tag{6}$$

and

$$\begin{aligned} &u_{1,0}(t) = v_1(t) = u_1(t_0) = c_1, \\ &u_{2,0}(t) = v_2(t) = u_2(t_0) = c_2, \\ &\vdots \\ &u_{m,0}(t) = v_m(t) = u_m(t_0) = c_m, \end{aligned} \tag{7}$$

where $u_{i,j}$, $(i = 1, 2, \dots, m; j = 1, 2, \dots)$ are functions yet to be determined. Substituting (6)-(7) into (5) and arranging the coefficient of same powers of p, we get

$$\begin{aligned}
&L(u_{1,1})+L(v_1)+N_1(u_{1,0},u_{2,0},\dots u_{m,0})-f_1=0,\\
&\quad u_{1,1}(t_0)=0,\\
&L(u_{2,1})+L(v_2)+N_2(u_{1,0},u_{2,0},\dots u_{m,0})-f_2=0,\\
&\quad u_{2,1}(t_0)=0,\\
&\quad\vdots\\
&L(u_{m,1})+L(v_m)+N_m(u_{1,0},u_{2,0},\dots u_{m,0})-f_m=0,\\
&\quad u_{m,1}(t_0)=0,\\
&L(u_{1,2})+N_1(u_{1,1},u_{2,1},\dots u_{m,1})=0,\\
&\quad u_{1,2}(t_0)=0,\\
&L(u_{2,2})+N_2(u_{1,1},u_{2,1},\dots u_{m,1})=0,\\
&\quad u_{2,2}(t_0)=0,\\
&\quad\vdots\\
&L(u_{m,2})+N_m(u_{1,1},u_{2,1},\dots u_{m,1})=0,\\
&\quad u_{m,2}(t_0)=0,
\end{aligned} \tag{8}$$

etc. solve the above systems of equations for the unknown $u_{i,j}(i=1,2,\dots,m;\, j=1,2,\dots)$ by applying the inverse operator

$$L^{-1}(.)=\int_{t_0}^{t}(.)\,dt. \tag{9}$$

Therefore, according to HPM the n-term approximations for the solutions of (4) can be expressed as

$$\begin{aligned}
\phi_{1,n}(t)&=u_1(t)=\lim_{p\to 1}u_1(t)=\sum_{k=o}^{n-1}u_{1,k}(t),\\
\phi_{2,n}(t)&=u_2(t)=\lim_{p\to 1}u_2(t)=\sum_{k=o}^{n-1}u_{2,k}(t),\\
&\vdots\\
\phi_{m,n}(t)&=u_m(t)=\lim_{p\to 1}u_m(t)=\sum_{k=o}^{n-1}u_{m,k}(t).
\end{aligned} \tag{10}$$

2.2 *Solution by MHPM*

The solution by HPM is not valid for large t. A simple way of ensuring validity of the approximations for large t is to treat the algorithm of HPM in a sequence of intervals choosing the initial approximations as

$$\begin{aligned}
u_{1,0}(t)&=v_1(t)=u_1(t^*)=c_1^*,\\
u_{2,0}(t)&=v_2(t)=u_2(t^*)=c_2^*,\\
&\vdots\\
u_{m,0}(t)&=v_m(t)=u_m(t^*)=c_m^*,
\end{aligned} \tag{11}$$

where t^* is the left-end point of each subinterval. Then solve the (8) for the unknowns $u_{i,j}(i=1,2,\dots,m;\, j=1,2,\dots)$ by applying the inverse linear operator

$$L^{-1}(.)=\int_{t^*}^{t}(.)\,dt. \tag{12}$$

In order to carry the iterations in every subinterval of equal length $\Delta t,[0,t_1),[t_1,t_2)\ldots[t_{j-1},t)$, we need to know the values of

$$u^*_{1,0}(t)=u_1(t^*),\qquad u^*_{2,0}(t)=u_2(t^*),\ldots,\qquad u^*_{m,0}(t)=u_m(t^*). \tag{13}$$

But in general, we do not have this information on our clearance except at the initial point $t^*=t_0$. A simple way for obtaining necessary values could be my means of the previous n-terms of approximations $\phi_{1,n},\phi_{2,n},\ldots,\phi_{m,n}$ of the preceding subinterval given by Eqs. (11), i.e.

$$u^*_{1,0}\approx\phi_{1,n}(t^*),\qquad u^*_{2,0}\approx\phi_{2,n}(t^*),\ldots,\qquad u^*_{m,0}\approx\phi_{m,n}(t^*). \tag{14}$$

3. Application

In this section, we will study the hyperchaotic Chen system (1) subject to the initial conditions:

$$x(t_0)=c_1,\qquad y(t_0)=c_2,\qquad z(t_0)=c_3,\qquad w(t_0)=c_4. \tag{15}$$

According to HPM, we can construct a homotopy which satisfies the following relations

$$\begin{aligned}
&\dot v_1-\dot x_0+p(\dot x_0-av_2+av_1-v_4)=0,\\
&\dot v_2-\dot y_0+p(\dot y_0+v_1v_3-cv_2)=0,\\
&\dot v_3-\dot z_0+p(\dot z_0-v_1v_2+bv_3)=0,\\
&\dot v_4-\dot w_0+p(\dot w_0-v_1v_3-rv_4)=0.
\end{aligned} \tag{16}$$

We take the initial approximations as

$$\begin{aligned}
&v_{1,0}(t)=x_0(t)=x(t^*)=c_1,\\
&v_{2,0}(t)=y_0(t)=y(t^*)=c_2,\\
&v_{3,0}(t)=z_0(t)=z(t^*)=c_3,\\
&v_{4,0}(t)=w_0(t)=w(t^*)=c_4.
\end{aligned} \tag{17}$$

and

$$\begin{aligned}
&v_1(t)=v_{1,0}(t)+pv_{1,1}(t)+p^2v_{1,2}(t)+p^3v_{1,3}(t)+\cdots,\\
&v_2(t)=v_{2,0}(t)+pv_{2,1}(t)+p^2v_{2,2}(t)+p^3v_{2,3}(t)+\cdots,\\
&v_3(t)=v_{3,0}(t)+pv_{3,1}(t)+p^2v_{3,2}(t)+p^3v_{3,3}(t)+\cdots,\\
&v_4(t)=v_{4,0}(t)+pv_{4,1}(t)+p^2v_{4,2}(t)+p^3v_{4,3}(t)+\cdots.
\end{aligned} \tag{18}$$

where $u_{i,j}(t), i=j=1,2,3,\ldots$ are functions yet to be determined. Substitute (17)-(18) into (16) and arrange the terms with the same powers of p, we have;

$$\begin{aligned}
&\dot v_{1,1}-av_{2,0}+av_{1,0}-v_{4,0}=0,\\
&\dot v_{1,2}-av_{2,1}+av_{1,1}-v_{4,1}=0,\\
&\dot v_{1,3}-av_{2,2}+av_{1,2}-v_{4,2}=0,
\end{aligned}$$

$$\begin{aligned}
&\dot{v}_{2,1} - dv_{1,0} + v_{1,0}v_{3,0} - cv_{2,0} = 0,\\
&\dot{v}_{2,2} - dv_{1,1} + v_{1,0}v_{3,1} + v_{1,1}v_{3,0} - cv_{2,1} = 0,\\
&\dot{v}_{2,3} - dv_{1,2} + v_{1,0}v_{3,2} + v_{1,1}v_{3,1} + v_{1,2}v_{3,0} - cv_{2,2} = 0,\\
&\dot{v}_{3,1} - v_{1,0}v_{2,0} + bv_{3,0} = 0,\\
&\dot{v}_{3,2} - v_{1,0}v_{2,1} - v_{1,1}v_{2,0} + bv_{3,1} = 0,\\
&\dot{v}_{3,3} - v_{1,0}v_{2,2} - v_{1,1}v_{2,1} - v_{1,2}v_{2,0} + bv_{3,2} = 0,\\
&\dot{v}_{4,1} - v_{2,0}v_{3,0} + rv_{4,0} = 0,\\
&\dot{v}_{4,2} - v_{2,0}v_{3,1} - v_{2,1}v_{3,0} + rv_{4,1} = 0,\\
&\dot{v}_{4,3} - v_{2,0}v_{3,2} - v_{2,1}v_{3,1} - v_{2,2}v_{3,0} + rv_{4,2} = 0.
\end{aligned} \tag{19}$$

Solve the unknowns of the above Eqs. (19) by taking the initial conditions as $u_{i,j} = 0, i, j = 1,2,3$

$$\begin{aligned}
&v_{1,1}(t) = [ac_2 - ac_1 + c_4](t - t^*),\\
&v_{2,1}(t) = [dc_1 - c_1c_3 + cc_2](t - t^*),\\
&v_{3,1}(t) = [c_1c_2 - bc_3](t - t^*),\\
&v_{4,1}(t) = [c_2c_3 + rc_4](t - t^*),\\
&v_{1,2}(t) = \frac{1}{2}[a(dc_1 - c_1c_3 + cc_2) - a(ac_2 - ac_1 + c_4) + c_2c_3 + rc_4](t - t^*)^2,\\
&v_{2,2}(t) = \frac{1}{2}[d(ac_2 - ac_1 + c_4) - (c_1c_2 - bc_3) - c_3(ac_2 - ac_1 + c_4)\\
&\quad + c(dc_1 - c_1c_3 + cc_2)](t - t^*)^2,\\
&v_{3,2}(t) = \frac{1}{2}[c_1(dc_1 - c_1c_3 + cc_2) + c_2(ac_2 - ac_1 + c_4) - b(c_1c_2 - bc_3)](t - t^*)^2,\\
&v_{4,2}(t) = \frac{1}{2}[c_2(c_1c_2 - bc_3) + c_3(dc_1 - c_1c_3 + cc_2) + r(c_2c_3 + rc_4)](t - t^*)^2,\\
&v_{1,3}(t) = \frac{1}{6}[a(d(ac_2 - ac_1 + c_4) - c_1(c_1c_2 - bc_3) - c_3(ac_2 - ac_1 + c_4)\\
&\quad + c(dc_1 - c_1c_3 + cc_2))\\
&\quad - a(a(dc_1 - c_1c_3 + cc_2) - a(ac_2 - ac_1 + c_4) + c_2c_3 + rc_4)\\
&\quad + c_2(c_1c_2 - bc_3) + c_3(dc_1 - c_1c_3 + cc_2) + r(c_2c_3 + rc_4)](t - t^*)^3,\\
&v_{2,3}(t) = \frac{1}{6}[d(a(dc_1 - c_1c_3 + cc_2) - a(ac_2 - ac_1 + c_4) + c_2c_3 + rc_4)\\
&\quad - c_1(c_1(dc_1 - c_1c_3 + cc_2) + c_2(ac_2 - ac_1 + c_4) - b(c_1c_2 - bc_3))\\
&\quad - 2(ac_2 - ac_1 + c_4)(c_1c_2 - bc_3)\\
&\quad - (a(dc_1 - c_1c_3 + cc_2) - a(ac_2 - ac_1 + c_4) + c_2c_3 + rc_4)c_3\\
&\quad + c(d(ac_2 - ac_1 + c_4) - c_1(c_1c_2 - bc_3) - c_3(ac_2 - ac_1 + c_4)\\
&\quad + c(dc_1 - c_1c_3 + cc_2))](t - t^*)^3,\\
&v_{3,3}(t) = \frac{1}{6}[c_1(d(ac_2 - ac_1 + c_4) - c_1(c_1c_2 - bc_3) - c_3(ac_2 - ac_1 + c_4)\\
&\quad + c(dc_1 - c_1c_3 + cc_2)) + 2(ac_2 - ac_1 + c_4)(dc_1 - c_1c_3 + cc_2)\\
&\quad + (a(dc_1 - c_1c_3 + cc_2) - a(ac_2 - ac_1 + c_4) + c_2c_3 + rc_4)c_2\\
&\quad - b(c_1(dc_1 - c_1c_3 + cc_2) + c_2(ac_2 - ac_1 + c_4) - b(c_1c_2 - bc_3))](t - t^*)^3,
\end{aligned} \tag{20}$$

$$v_{4,3}(t) = \frac{1}{6}[c_2(c_1(dc_1 - c_1c_3 + cc_2) + c_2(ac_2 - ac_1 + c_4) - b(c_1c_2 - bc_3))$$
$$+2(dc_1 - c_1c_3 + cc_2)(c_1c_2 - bc_3) + (d(ac_2 - ac_1 + c_4) - c_1(c_1c_2 - bc_3)$$
$$-c_3(ac_2 - ac_1 + c_4) + c(dc_1 - c_1c_3 + cc_2))c_3$$
$$+r(c_2(c_1c_2 - bc_3) + c_3(dc_1 - c_1c_3 + cc_2) + r(c_2c_3 + rc_4))](t-t^*)^3.$$

Thus, the solution of system (1) is

$$\begin{aligned} x(t) &= \sum_{k=0}^{\infty} v_{1,k}(t), \\ y(t) &= \sum_{k=0}^{\infty} v_{2,k}(t), \\ z(t) &= \sum_{k=0}^{\infty} v_{3,k}(t), \\ w(t) &= \sum_{k=0}^{\infty} v_{4,k}(t). \end{aligned} \quad (21)$$

To carry out the iterations on every subinterval of equal length Δt, we need to know the values of the following initial conditions:

$$c_1 = x(t^*), \quad c_2 = y(t^*), \quad c_3 = z(t^*), \quad c_4 = w(t^*). \quad (22)$$

In general, we do not have these information at our clearance except at the initial point $t^* = t_0 = 0$, but we can obtain these values following the MHPM as discussed earlier. We note that the 15-term approximation of *x, y, z* and *w* are denoted as

$$x(t) \approx \phi_{15}(t) = \sum_{i=0}^{14} v_{1,i}, \quad y(t) \approx \phi_{15}(t) = \sum_{i=0}^{14} v_{2,i}, \quad z(t) \approx \phi_{15}(t) = \sum_{i=0}^{14} v_{3,i},$$
$$w(t) \approx \phi_{15}(t) = \sum_{i=0}^{14} v_{4,i}.$$

4. Results and Discussions

The system (1) is hyperchaotic for the parameters of $a = 35$, $b = 3$, $c = 12$, $d = 7$, and $r = 0.5$ with the initial conditions as $x(0) = -20, y(0) = 0, z(0) = 0$ and $w(0) = 15$. The series solutions for 15-term HPM for hyperchaotic Chen system (1) are obtained as :

$x(t) =$ $-20.0 + 715.0t - 14958.75000\, t^2 + 193915.2083\, t^3 - 1862311.641\, t^4 + 11648014.04\, t^5 - 782382.0271\, t^6 - 1397601846\, t^7 + 27028163520\, t^8 - 334919921200\, t^9 + 2586282504000\, t^{10} + + 3406869902000\, t^{11} - 581768028700000\, t^{12} + 13372041570000000\, t^{13} - 212029685400000000\, t^{14}$

$y(t) =$ $-140.0\, t + 1662.500000\, t^2 - 18920.41667\, t^3 - 196909.6354\, t^4 + 11464051.41\, t^5 - 279184644\, t^6 + 4764117419\, t^7 - 58999397170\, t^8 + 406793603700\, t^9 +$

$3539842879000\, t^{10} - 193395949000000\, t^{11} + 4341484559000000 t^{12} - 70942516200000000\, t^{13} + 898449650500000000\, t^{14}$

$$z(t) = 1400.0\, t^2 - 45850.00000\, t^3 + 949717.7083\, t^4 - 12891221.88\, t^5 + 89122615.75\, t^6 + 1151838885\, t^7 - 56664558560\, t^8 + 1277368685000\, t^9 - 20929183260000\, t^{10} + 260102072300000\, t^{11} - 2099638152000000\, t^{12} - 2959658978000000\, t^{13} + 533351445800000000\, t^{14}$$

$$w(t) = 15.0 + 7.5\, t + 1.875000000\, t^2 + 0.3125000000\, t^3 - 48999.96092\, t^4 + 1744400.004\, t^5 - 39133747.92\, t^6 + 565133927.6\, t^7 - 3314666453\, t^8 - 97148329530\, t^9 + 4100955790000\, t^{10} - 93117682730000\, t^{11} + 1522699888000000\, t^{12} - 17406073610000000\, t^{13} + 73528243810000000\, t^{14}$$

There are 3 methods used in this paper to solve the hyperchaotic Chen which are HPM, MHPM and the well-established RK4 as the reference. The algorithm discussed is coded in the computer algebra package Maple together with the Maple's built-in fourth-order Runge-Kutta. The Maple environment variable Digits controls the number of significant digits to 16 in all calculations done in this paper. Also, we fixed the term used to 15 and step size to $\Delta t = 0.001$ by comparing the solution of RK4 at different step sizes i.e. $\Delta t=0.01$, $\Delta t=0.001$ *and* $\Delta t=0.0001$. From the results presented in table 1 for hyperchaotic Chen, the absolute errors between $\Delta t=0.01$ with $\Delta t=0.001$ are still large comparing to $\Delta t=0.001$ with $\Delta t=0.0001$ which the minimum difference recorded is to the value of 10^{-5}. Therefore, the step size of $\Delta t=0.001$ is chosen for this study because it gives small error and computationally costly as the time taken is reasonable. The time range in this work is taken to be from $t=0$ *to* $t=10$.

Table 1: Hyperchaotic Chen system, differences between RK4 solutions on $\Delta t=0.01$, $\Delta t=0.001$and $\Delta t=0.0001$

	$\Delta = \lvert RK4_{0.01} - RK4_{0.001} \rvert$					$\Delta = \lvert RK4_{0.001} - RK4_{0.0001} \rvert$			
t	Δx	Δy	Δz	Δw	t	Δx	Δy	Δz	Δw
1	0.0004482	5.998E-06	0.00132	0.0003517	1	6.956E-09	1.711E-08	6.184E-08	2.865E-08
2	0.003455	0.002203	0.006475	0.01659	2	2.405E-07	1.553E-07	4.548E-07	1.061E-06
3	0.002024	0.003053	0.0005701	0.01326	3	1.228E-07	1.982E-07	7.074E-09	9.202E-07
4	7.054E-05	0.0007172	0.003162	0.02676	4	1.373E-08	3.838E-08	2.089E-07	1.836E-06
5	0.005558	0.007073	0.001747	0.007351	5	3.710E-07	4.749E-07	1.235E-07	5.514E-07
6	0.04084	0.08134	0.0463	0.2095	6	2.770E-06	5.468E-06	3.047E-06	1.400E-05
7	0.044	0.05512	0.004727	0.05878	7	2.962E-06	3.717E-06	3.362E-07	3.859E-06
8	0.01109	0.02319	0.05064	0.02173	8	7.865E-07	1.589E-06	3.354E-06	1.205E-06
9	0.0128	0.003577	0.1543	0.3116	9	8.382E-07	2.363E-07	1.016E-05	2.041E-05
10	0.1429	0.1313	0.1916	0.7805	10	8.919E-06	8.120E-06	1.256E-05	5.031E-05

Table 2: Differences between 15-term HPM and 15-term MHPM with RK4 solutions on Δt=0.001.

$\Delta = \lvert HPM - RK4_{0.001} \rvert$					$\Delta = \lvert MHPM_{0.001} - RK4_{0.001} \rvert$				
t	Δx	Δy	Δz	Δw	t	Δx	Δy	Δz	Δw
1	1.992E+17	8.317E+17	5.285E+17	5.756E+16	1	6.956E-09	1.711E-08	6.184E-08	2.866E-08
2	3.367E+21	1.416E+22	8.706E+21	1.068E+21	2	2.405E-07	1.553E-07	4.549E-07	1.061E-06
3	9.931E+23	4.186E+24	2.545E+24	3.247E+23	3	1.228E-07	1.982E-07	7.073E-09	9.203E-07
4	5.603E+25	2.365E+26	1.429E+26	1.859E+25	4	1.374E-08	3.838E-08	2.090E-07	1.836E-06
5	1.278E+27	5.398E+27	3.251E+27	4.279E+26	5	3.710E-07	4.750E-07	1.235E-07	5.515E-07
6	1.644E+28	6.949E+28	4.175E+28	5.538E+27	6	2.770E-06	5.468E-06	3.048E-06	1.400E-05
7	1.425E+29	6.025E+29	3.614E+29	4.820E+28	7	2.962E-06	3.717E-06	3.363E-07	3.860E-06
8	9.252E+29	3.913E+30	2.344E+30	3.139E+29	8	7.866E-07	1.590E-06	3.354E-06	1.205E-06
9	4.817E+30	2.037E+31	1.219E+31	1.638E+30	9	8.382E-07	2.363E-07	1.016E-05	2.042E-05
10	2.107E+31	8.914E+31	5.330E+31	7.180E+30	10	8.920E-06	8.121E-06	1.256E-05	5.032E-05

According to the data in table 2 for hyperchaotic Chen system, we see that the 15-term MHPM solutions at Δt=0.001 highly agree with the solutions of RK4 to at least 5 decimal places while HPM solutions are not valid even at t=1. This demonstrates that the hyperchaotic Chen is solved accurately by MHPM.

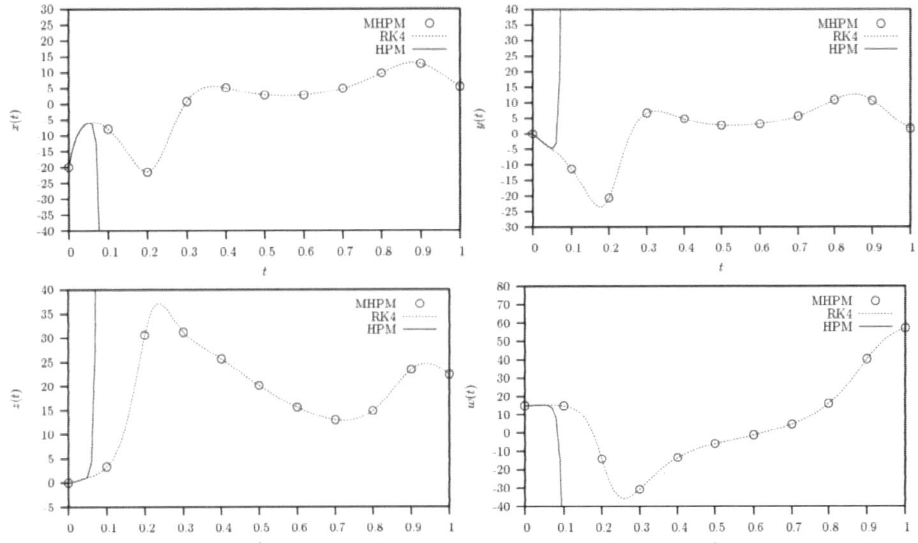

Figure 1: The comparison between 15-term HPM, MHPM and RK4 on Δt=0.001. Solid line represents HPM, dash line is RK4 while circle is for MHPM.

In Figure 1, when plotting solutions of HPM, MHPM and RK4 together, we find that the HPM solution deviate rapidly from RK4 and MHPM solutions at t<0.1s. It is obvious that HPM is not valid for the solution of hyperchaotic systems as time goes by. But, solution of MHPM and RK4 agreed to each other as time increases. Next, the *x-y-z, x-y-w, x-z-w* and *y-z-w* phase portraits of the hyperchaotic Chen system using 15-term MHPM solution on Δt=0.001 are illustrated in figure 2.

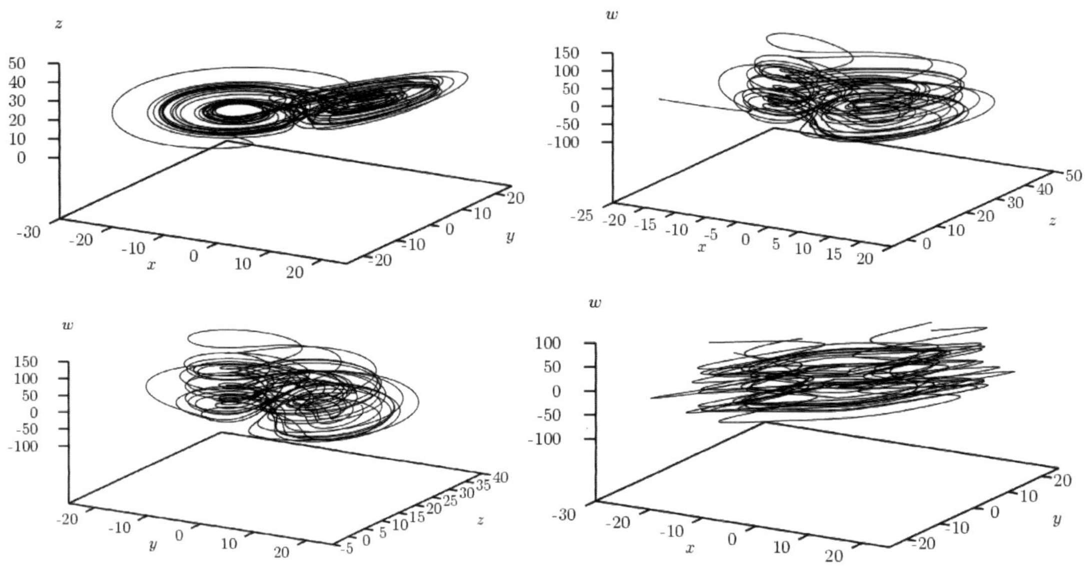

Figure 2: Phase portraits using 15-term MHPM on Δt=0.001 for hyperchaotic Chen system.

5. Conclusion

In this study, HPM and MHPM have been successfully employed to obtain the approximate analytical solutions for hyper chaotic system. . The MHPM has the advantage of giving an analytical form of the solution within each time interval which is not possible in purely numerical techniques like RK4. The present technique offers an explicit time-marching algorithm that works accurately over such a bigger time span and fewer terms than the standard HPM. The results presented in this paper suggest that MHPM is also readily applicable to more complex nonlinear hyper chaotic systems.

Acknowledgments

The authors would like to acknowledge the financial supports received from the International Islamic University Malaysia, Ministry of Higher Education Malaysia through the research grant FRGS-14-143-0384.

6. References

[1] Chen G, Dong X. From chaos to order: methodologies, perspectives and applications. Singapore: World Scientific; 1998.
[2] Perez G, Cerdeira H. Extracting messages masked by chaos. Phys Rev Lett 1995; 74 (11):1970.
[3] Yuxia L, Wallace K, Chen G. Generating hyperchaos via state feedback control. Int J Bifurcat Chaos 2005; 15:3367-75.
[4] Rossler O. An equation for hyperchaos. Phys Lett A 1979; 71(2-3):155.
[5] Li C, Chen G, Chaos and hyperchaos in the fractional-order Rossler equations. Physica A 2004; 341-55.
[6] Wang G, Zhang X, Zheng Y, Li Y. A new modified hyperchaotic Lu system. Physica A 2006; 371: 260.
[7] Li Y, Tang KS, Chen G. Hyperchaotic Chen system and its generation. Dyn Contin Discret Impul Syst, Ser B: Appl Algorithms 2007; 14:97.
[8] Thamilmaran K, Lakshmanan M, Venkatesan A. A hyperchaos in modified canonical Chua's circuit. Int J Bifurc Chaos 2004; 14 (1):221.
[9] Jia Q. Hyperchaos generated from the Lorenz chaotic system and its control. Phys Lett A 2007; 366:217.

[10] J.-H He. An approximate solution technique depending upon an artificial parameter, Commun. Nonlinear Sci. Numer. Simulat. 3 (2) (1998) 92-97.
[11] J.-H He. Newton-like iteration method for solving algebraic equations, Commun. Nonlinear Sci. Numer. Simulat. 3 (2) (1998) 106-109.
[12] J.-H He. Comparison of homotopy perturbation method and homotopy analysis method. Appl. Math. And Comput 156 (2004) 527-539.
[13] J.-H He. A coupling method of homotopy technique and perturbation technique for nonlinear problems, Int. J. Nonlinear Mech. 35 (1) (2000) 37-43.
[14] J.-H He. Homotopy perturbation technique. Comput. Methods Appl. Mech. Engrg. 178(3/4) (1999) 257-262.
[15] J.H He. Homotopy perturbation method for solving boundary value problems. Physics Letter A 350 (1-2) (2006) 87-88.
[16] Changbun Chun, Rathinasamy Sakthivel. Homotopy perturbation technique for solving two-point boundary value problems- comparison with other methods. Computer Physics Communications 181 (2010) 1021-1024.
[17] M. S. H Chowdhury, I. Hashim. Application of multistage homotopy perturbation method for the solutions of the Chen system. Nonlinear Anal: Real World Appl 2009; 10 (1) :381-391.
[18] M. S. H Chowdhury, I. Hashim, S. Momani. The multistage homotopy perturbation method: a powerful scheme for handling the Lorenz system. Chaos, Solitons and Fractals 2009; 40 (4):1929-37.
[19] I. Hashim, M. S. H Chowdhury. Adaption of homotopy perturbation method for numeric-analytic solution of system of ODEs. Physics Letters A, vol. 372, no.4, pp. 470-481 (2008).
[20] I. Hashim, M. S. H Chowdhury, S. Mawa. On multistage homotopy perturbation method applied to nonlinear biochemical reactor model. Chaos, Solitons and Fractals, vol. 36, no. 4, pp 823-827 (2008).
[21] Yongguang Yu, Han-Xiong Li. Application of the multistage homotopy perturbation method to solve a class of hyperchaotic systems. Chaos, Solitons and Fractals 42 (2009) 2330-2337.
[22] M. Rafei, D.D.Ganji, H. Daniali. Solution of the epidemic model by homotopy perturbation method. Appl Math and Comput 187 (2007) 1056-1062.
[23] Jafar Saberi-Nadjafi, Asghar Ghorbani. He's homotopy perturbation method: An effective tool for solving nonlinear integral and integro-differential equations. Compututers and Mathematics with Aplications 58 (2009) 2379-2390.
[24] Edyta Hetmaniok, Damian Slota, Roman Witula. Convergence and error estimation of homotopy-perturbation method for Fredholm and Volterra integral equations. Appl Math and Comput 218 (2012) 10717-10725.
[25] Peter Vadasz, Shmuel Olek. Convergence and accuracy of Adomian's Decomposition method for the solution of Lorenz equations. International Journal of Heat and Mass Transfer 43 (10) 2000 1715-1734.
[26] I. Hashim, M. S. M. Noorani, R. Ahmad, S. A. Bakar, E. S. Ismail, A. M. Zakaria. Accuracy of the Adomian decomposition method applied to the Lorenz system. Chaos, Solitons and Fractals 28 (5) 2006 1149-1158.
[27] O. Abdulaziz, N. F. M. Noor, I. Hashim, M. S. M. Noorani. Further accuracy on Adomian decomposition method for chaotic systems. Chaos, Solitons and Fractals 36 (2008) 1405-1411.
[28] S. A. Eftekhari, A. A. Jafari. Numerical solution of chaotic dynamical systems by the method of differential quadrature. Scientia iranica B (2012) 19 (5), 1299-1315.
[29] Changpin Li, Yihong Wang. Numerical algorithm based on Adomian decomposition method for fractional differential equations. Computers and Mathematics with Applications 57 (2009) 1672-1681.
[30] A. K. Alomari, M. S. M. Noorani, R. Nazar, C. P. Li. Homotopy Analysis Method for solving fractional Lorenz system. Communications in Nonlinear Science and Numerical Simulation 15 (7) 2010 1864-1872.

[31] M. S. H. Chowdhury, I. Hashim. Application of multistage homotopy perturbation method for the solutions of the Chen system. Nonlinear Analysis, Real World Applications 10 (1) 2009 381-391.
[32] S. M. Goh, M. S. M. Noorani, I. Hashim. On solving the chaotic Chen system: A new time marching design for the variational iteration method using Adomian's polynomial. Chaos, Solitons and Fractals 54(2) 245-260.
[33] A. K. Alomari, M. S. M. Noorani, R. Nazar. Adaption of homotopy analysis method for the numeric-analytic solution of Chen system. Communications in Nonlinear Science and Numerical Simulation 14 (5) 2009 2336-2346.
[34] M. Mossa Al- Sawalha, M. S. M. Noorani. A numeric-analytic method for approximating the chaotic Chen system. Chaos, Solitons and Fractals 42 (3) 2009 1784-1791.
[35] A. K. Alomari. A new analytic solution for fractional chaotic dynamical systems using the differential transformation method. Computers and Mathematics with Applications 61 (2011) 2528-2534.
[36] Chen A, Lu J, Yu S. Generating hyperchaotic Lu attractor via state feedback control. Physica A 2006;364:103-10.
[37] Qiang Jia. Projective synchronization of a new hyperchaotic Lorenz system. Physics Letters A 370 (2007) 40–45.

IOP Conf. Series: Journal of Physics: Conf. Series **949** (2017) 012006 doi:10.1088/1742-6596/949/1/012006

Analytical approximations for the oscillators with anti-symmetric quadratic nonlinearity

Md. Alal Hosen[1, 3], M S H Chowdhury [2,*], Mohammad Yeakub Ali [3] and Ahmad Faris Ismail [4]

[1] Department of Mathematics, Rajshahi University of Engineering and Technology (RUET), Rajshahi-6204, Bangladesh

[2] Department of Science in Engineering, Faculty of Engineering, International Islamic University Malaysia, Jalan Gombak, 53100 Kuala Lumpur, Malaysia

[3] Department of Manufacturing and Material Engineering, Faculty of Engineering, International Islamic University Malaysia, Jalan Gombak, 53100 Kuala Lumpur, Malaysia

[4] Department of Mechanical Engineering, Faculty of Engineering, International Islamic University Malaysia, Jalan Gombak, 53100 Kuala Lumpur, Malaysia

Email: sazzadbd@iium.edu.my (Corresponding author: M. S. H. Chowdhury)

Abstract. A second-order ordinary differential equation involving anti-symmetric quadratic nonlinearity changes sign. The behaviour of the oscillators with an anti-symmetric quadratic nonlinearity is assumed to oscillate different in the positive and negative directions. In this reason, Harmonic Balance Method (HBM) cannot be directly applied. The main purpose of the present paper is to propose an analytical approximation technique based on the HBM for obtaining approximate angular frequencies and the corresponding periodic solutions of the oscillators with anti-symmetric quadratic nonlinearity. After applying HBM, a set of complicated nonlinear algebraic equations is found. Analytical approach is not always fruitful for solving such kinds of nonlinear algebraic equations. In this article, two small parameters are found, for which the power series solution produces desired results. Moreover, the amplitude-frequency relationship has also been determined in a novel analytical way. The presented technique gives excellent results as compared with the corresponding numerical results and is better than the existing ones.

1. Introduction

In recent past, the mathematical interpretation of the nonlinear oscillations has received extensive attention in the field of physics, applied mathematics engineering and other discipline. The mathematical model of oscillation of the human eardrum is the quadratic nonlinear oscillator [1]. Researchers have been investigated many analytical approaches to solve nonlinear differential equations. Perturbation technique [2-5] is the versatile technique, whereby the solution is expanded in powers of a small parameter. However, for the strongly nonlinear regime the perturbation technique is almost fail. Several authors employed many other powerful analytical methods (non-perturb) to drive approximate periodic solutions especially for the strongly nonlinear oscillators, include as, the Max-

Min Approach [6], Parameter Expansion Method [7], Variational Iteration Method [8], Amplitude Frequency Formulation [9], Energy Balance Method [10-12], He's Energy Balance Method [13], Global Residue Harmonic Balance Method [14] and so on.

Cveticanin [15] has used the Jacobi elliptic functions to derive the exact solution of the anti-symmetric quadratic equation. However, the author only assumed that its exact solution is given by an equation which includes the Jacobian elliptic function and five unknown parameters that need to be determined. The author did not solve the nonlinear differential equation. Mickens and Mixon [16] obtained accurate analytical approximate solutions to an anti-symmetric quadratic non-linear oscillator using the generalized harmonic balance method and Chen et. al. [17] using the elliptic perturbation method. Beléndez et. al. approximately solved this nonlinear oscillator using a modified He's homotopy perturbation method [18] as well as a novel rational harmonic balance approach [19]. Hu [20] has used Harmonic Balance Method to determine an approximate solution of a quadratic nonlinear oscillator; but the method is not a simple one. Hu [20] has obtained two separate harmonic balance solutions respectively for two regions $x > 0$ and $x < 0$. The solution is continuous, but the derivative does not exist when it cuts the axis. Momani et. al. [21] introduced the modified homotopy perturbation method for solving strongly nonlinear oscillators with anti-symmetric quadratic nonlinearity. Recently, Cveticanin et. al. [22] obtained approximate solutions for the oscillators with symmetric and asymmetric quadratic nonlinearity in the form of Jacobi elliptic functions. Mahtab et. al. [23] has also approximated the quadratic nonlinear oscillators using Harmonic balance method. An iteration procedure has been applied to determine approximate frequencies for the Quadratic Nonlinear Oscillator by Haque et. al. [24].

Solving strongly nonlinear systems, the method of harmonic balance [25-32] is another efficient method. As a rule, a set of complex nonlinear algebraic equations are appeared when Harmonic Balance Method is imposed. In this paper, such nonlinear algebraic equations are approximated using a power series solutions by a new small parameter. Moreover, an analytical technique has also been applied to obtain the amplitude-frequency relationship. The approximated results have compared with the existing results and the corresponding numerical solutions (Runge-Kutta fourth order method).

2. The Solution Procedure

A second-order nonlinear differential equation can be considered as

$$y'' + \omega_0^2 y = -\varepsilon f(y, y') \text{ and the initial conditions } [y(0) = A_0,\ y'(0) = 0], \quad (1)$$

where $f(y, y')$ is a nonlinear function such that $f(-y, -y') = f(y, y')$, $\omega_0 \geq 0$ and ε is a constant.

A n-th order periodic solution of Eq. (1) can be supposed as

$$y = A_0(\rho\cos(\omega t) + u\cos(3\omega t) + v\cos(5\omega t) + w\cos(7\omega t) + z\cos(9\omega t) + \cdots), \quad (2)$$

where A_0, ρ and ω are constants. If $\rho = 1 - u - v - \cdots$, the solution Eq. (2) readily satisfies the initial condition given in Eq. (1).

Substituting Eq. (2) into Eq. (1) and expanding $f(y, y')$ in a Fourier series, it can be converted to an algebraic identity as follows

$$\begin{aligned} A_0[\rho(\omega_0^2 - \omega^2)\cos(\omega t) + u(\omega_0^2 - 9\omega^2)\cos(3\omega t) + \cdots] = -\varepsilon[F_1(A_0, u, \cdots)\cos(\omega t) \\ + F_3(A_0, u, \cdots)\cos(3\omega t) + \cdots] \end{aligned} \quad (3)$$

By comparing the coefficients of equal harmonics of Eq. (3), the following nonlinear algebraic equations can be obtained as

$$\rho(\omega_0^2 - \omega^2) = -\varepsilon F_1, \quad u(\omega_0^2 - 9\omega^2) = -\varepsilon F_3, \quad v(\omega_0^2 - 25\omega^2) = -\varepsilon F_5, \cdots \quad (4)$$

With the help of the first equation, ω^2 is eliminated from all the rest of Eq. (4). Thus Eq. (4) takes the following form

$$\rho\omega^2 = \rho\omega_0^2 + \varepsilon F_1, \quad 8\omega_0^2 u\rho = \varepsilon(\rho F_3 - 9uF_1), \quad 24\omega_0^2 v\rho = \varepsilon(\rho F_5 - 25vF_1), \cdots \quad (5)$$

Substituting $\rho = 1 - u - v - \cdots$, and simplification, second-, third- equations of Eq. (5) can be transformed into

$$u = G_1(\omega_0, \varepsilon, A_0, u, v, \cdots, \lambda_0), \quad v = G_2(\omega_0, \varepsilon, A_0, u, v, \cdots, \lambda_0), \cdots, \tag{6}$$

where $G_1, G_2, \cdots$ exclude respectively the linear terms of $u, v, \cdots$.

Whatever the values of ε, ω_0 and A_0 there exists a parameter $\lambda_0(\varepsilon, \omega_0, A_0) << 1$, such that $u, v, \cdots$ are expandable in following power series in terms of λ_0 as

$$u = U_1\lambda_0 + U_2\lambda_0^2 + \cdots, \quad v = V_1\lambda_0 + V_2\lambda_0^2 + \cdots, \quad \cdots \tag{7}$$

where $U_1, U_2, \cdots, V_1, V_2, \cdots$ are constants.

Finally, substituting the values of $u, v, \cdots$ from Eq. (7) into the first equation of Eq. (5), the angular frequency ω is determined. This completes the determination of all related functions for the proposed n-th order periodic solution as given in Eq. (2).

3. Example

3.1 Differential equation with anti-symmetric quadratic nonlinearity

Let us consider the following equation which has investigated in Momani et. al. [21] is

$$y'' + 2y + y^2 = 0 \text{ and the initial conditions } [y(0) = a_0,\ y'(0) = 0]. \tag{8}$$

As we know, an asymmetric behavior of the nonlinear oscillator is different in positive and negative directions. An asymmetric limit zone $[-b, a]$, for positive a and negative b has been considered where the system is assumed to oscillate. Both $y = a$ and $y = -b$ represent the turning points in which $\dot{y} = 0$, a and b are an unknown amplitude to be determined.

The method is applicable to determine approximate solutions of various differential equations whose nonlinear function satisfies the relation, $f(-y, -y') = -f(y, y')$. However, the method is also useful when nonlinear function satisfies the relation, $f(-y, -y') = f(y, y')$ (quadratic nonlinearities as well as to some nonlinear problems with mixed parity).

We use the solution of the form of Eq. (2) in both regions where $y > 0$ and $y < 0$. In this regard, we usually match the solution satisfying the conditions, $y_+(\varphi) = y_-(\varphi)$, $y'_+(\varphi) = y'_-(\varphi)$. In article [20], such type of approximate solution (only first approximation) was obtained satisfying the first condition. Since the second condition was ignored, the solution was not clearly matched at the point considered. But the higher approximation gradually coincides with the numerical solution.

For the region $y > 0$, a second approximate solution of Eq. (8), can be considered as

$$y_+(t) = a_0(\rho\cos(\omega_+ t) + u_+\cos(3\omega_+ t)). \tag{9}$$

Let us consider y^2, can be expanded in a Fourier series as

$$b_1\cos(\omega_+ t) + b_3\cos(3\omega_+ t) + \cdots. \tag{10}$$

Herein b_1, b_3, $\cdots$ are evaluated as

$$\begin{aligned} b_1 &= \frac{8a_0^2}{105\pi}\left(35 - 56u_+ + 48u_+^2\right), \\ b_3 &= \frac{8a_0^2}{315\pi}\left(21 + 120u_+ - 176u_+^2\right), \\ &\cdots\cdots\cdots \end{aligned} \tag{11}$$

Now substituting $\rho = 1 - u_+$, using Eq. (9)-(11) into Eq. (8) and then equating the coefficients of $\cos(\omega_+ t)$ and $\cos(3\omega_+ t)$, the following equations can be obtained as

$$-(1 - u_+)\omega_+^2 + 2 - 2u_+ + \frac{8a_0}{105\pi}\left(35 - 56u_+ + 48u_+^2\right) = 0, \tag{12}$$

and

$$-9\,u_{+}\,\omega_{+}^{2}+2u_{+}+\frac{8\,a_{0}}{315\pi}\left(21+120u_{+}-176u_{+}^{2}\right)=0\,. \tag{13}$$

After simplification, Eq. (12) can be expressed into another form as

$$\omega_{+}^{2}=2+\frac{8\,a_{0}}{3\pi}-\frac{8a_{0}u_{+}}{5\pi}+\frac{72\,a_{0}u_{+}^{2}}{35\pi}+\frac{72\,a_{0}u_{+}^{3}}{35\pi}+\frac{72\,a_{0}u_{+}^{4}}{35\pi}+\cdots. \tag{14}$$

By elimination of ω_{+} from Eq. (13), with the help of Eq. (14), the equation of u_{+} can be expressed as

$$u_{+}=\lambda_{+}\left(1+\frac{30\,\pi\,u_{+}^{2}}{a_{0}}+\frac{1216\,u_{+}^{2}}{21}-\frac{160\,u_{+}^{3}}{3}\right),\quad \text{where } \lambda_{+}=\frac{7a_{0}}{210\pi+282a_{0}}. \tag{15}$$

Therefore, the power series solution of Eq. (15) can be obtained in terms of λ_{+} is

$$u_{+}=\lambda_{+}+\left(\frac{1216}{21}+\frac{30\pi}{a_{0}}\right)\lambda_{+}^{3}-\frac{160}{3}\lambda_{+}^{4}+\cdots\,. \tag{16}$$

Now substituting the value of u_{+} from Eq. (16) into Eq. (14), the relation between ω_{+} and λ_{+} is determined and then the approximate angular frequency approximated as

$$\omega_{+}=\sqrt{2+\frac{8\,a_{0}}{3\pi}-\frac{8\,a_{0}}{5\pi}\lambda_{+}+\frac{72\,a_{0}}{35\pi}\lambda_{+}^{2}-48\lambda_{+}^{3}-\frac{9728\,a_{0}}{105\pi}\lambda_{+}^{3}+\cdots}\,. \tag{17}$$

Therefore, Eq. (9) represents the second approximate solution of Eq. (8) where ω_{+} and u_{+} are respectively given by Eq. (17) and (16). Now for the region $y<0$, the second approximate solution of Eq. (9) can be written as

$$y_{-}=b_{0}(\rho\cos(\omega_{-}t)+u_{-}\cos(3\omega_{-}t))\,, \tag{18}$$

where b_{0} is determined as

$$b_{0}=(3+2a_{0}-3\sqrt{1-4a_{0}(1+a_{0})/3})/4\,. \tag{19}$$

From Eq. (19), it has been shown that $a_{0}\le 0.5$ and $b_{0}\le 1$, (see [20] for details). Without repeating the solution process, u_{-} and ω_{-} are respectively obtained from Eqs. (16)-(17), replacing a_{0} by $-b_{0}$. Therefore, it becomes

$$u_{-}=\lambda_{-}+\left(\frac{1216}{21}-\frac{30\pi}{b_{0}}\right)\lambda_{-}^{3}-\frac{160}{3}\lambda_{-}^{4}+\cdots\;,\;\lambda_{-}=\frac{-7b_{0}}{210\pi-282b_{0}}, \tag{20}$$

and

$$\omega_{-}=\sqrt{2-\frac{8b_{0}}{3\pi}+\frac{8b_{0}}{5\pi}\lambda_{-}-\frac{72\,b_{0}}{35\pi}\lambda_{-}^{2}-48\lambda_{+}^{3}+\frac{9728\,b_{0}}{105\pi}\lambda_{-}^{3}+\cdots}\,. \tag{21}$$

4. Results and Discussions

The approximated solutions have been compared with the existing solutions obtained by homotopic perturbation method (HPM) and modified MHPM [21], and the available exact solutions which are illustrated in table 1 for initial oscillation amplitude $a_{0}=0.1$. The comparison between the results obtained by present study and the corresponding numerical solutions are shown in figure 1 to figure 2 for the initial oscillation amplitude $a_{0}=0.2$ and $a_{0}=0.3$ respectively. It is observed from all tables and figures that the results obtained by modified harmonic balance method (MHBM) are in good agreement with the corresponding numerical solutions compare to those obtained by Momani et. al. [21] applying homotopy perturbation method. Moreover, using the MHBM the angular frequency as well as the period of oscillatory problem can be calculated for different values of initial amplitudes.

5. Conclusion

In this paper, an analytical approximate technique based on the Harmonic Balance Method has been introduced to determine approximate angular frequencies and the corresponding periodic solutions of the oscillators with anti-symmetric quadratic nonlinearity. The solution procedure is straightforward and simple as compared with the existing methods. The approximated solutions show a good agreement with its exact ones and much better than the existing results. All of these allow us to conclude that the method presented in this article for solving the oscillators with anti-symmetric quadratic nonlinearity can be considered as an efficient alternative of the already published methods.

Table 1. The comparison the approximated results with existing solutions and the corresponding numerical solutions for the initial oscillation amplitude $a_0 = 0.1$.

t	$y_{[21]}^{HPM}$	$y_{[21]}^{MHPM}$	$y_{[Numerical]}$	$y_{[presented]}^{HBM}$
0	0.1	0.1	0.1	0.1
0.5	0.07492588	0.07492588	0.07492584	0.07606976
1	0.01261291	0.01262114	0.01260204	0.01574689
1.5	-0.05542130	-0.05564197	-0.05599434	-0.05208194
2	-0.08949363	-0.09641338	-0.09857369	-0.09488080
2.5	-0.02186309	-0.08941601	-0.09605590	-0.09218266
3	0.34944163	-0.03758556	-0.04954852	-0.04547820
3.5	1.63139818	0.03306331	0.02002292	0.02285490
4	5.37207116	0.08689346	0.07971480	0.08009427
4.5	14.98507704	0.09750473	0.09971271	0.09873981
5	37.20789062	0.05990474	0.06971159	0.07017312

Note: In table 1, $y_{[21]}^{HPM}$ and $y_{[21]}^{MHPM}$ respectively denote approximate solutions determined previously in Momani et. al. [21] by using Homotopy Perturbation Method and Modified Homotopy Perturbation Method. $y_{[presented]}^{HBM}$ represents the approximate solutions obtained in the presented technique. $y_{[Numerical]}$ indicates the numerical solutions (Runge-Kutta fourth order method) which has considered to be the exact solutions.

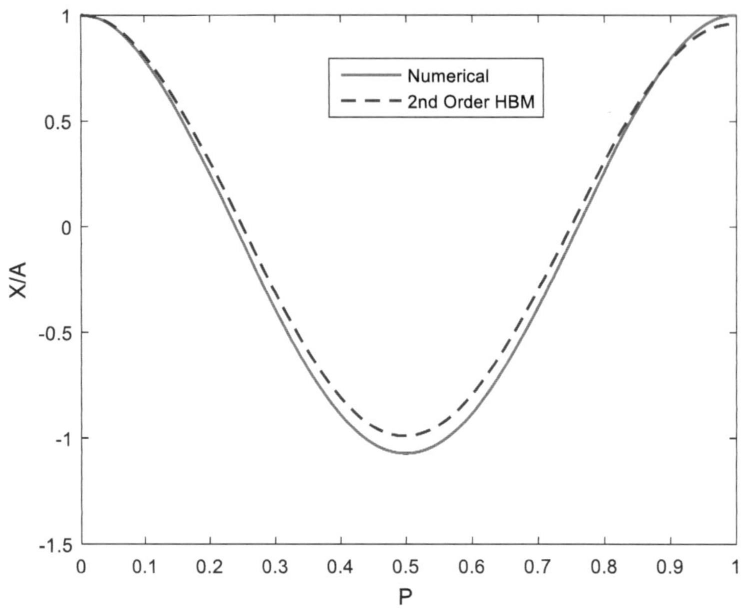

Figure 1. Comparison between the approximated solutions (represented by blue dashed line) and the corresponding numerical solutions (represented by red solid line) for the initial amplitude $a_0 = 0.2$.

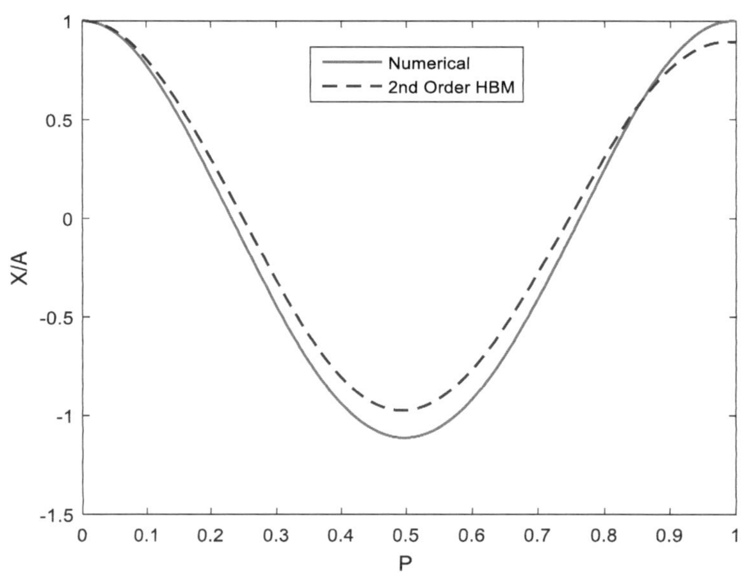

Figure 2. Comparison between the approximated solutions (represented by blue dashed line) and the corresponding numerical solutions (represented by red solid line) for the initial amplitude.

Acknowledgments

This research was funded by Ministry of Science, Technology and Innovation, Malaysia under Research Grant SF15- 016-0066. The work is also partially supported by the International Islamic University Malaysia (IIUM) and Ministry of Higher Education Malaysia through the research grant FRGS-14-143-0384.

6. References

[1] Porwal R and Vyas N S 2008 Damped quadratic and mixed-parity oscillator response using Krylov-Bogoliubov method and energy balance *J. Sound Vib.* **309** 877-86.
[2] Marion J B 1970 *Classical Dynamics of Particles and System* (San Diego, CA: Harcourt Brace Jovanovich).
[3] Krylov N N and Bogoliubov N N 1947 *Introduction to Nonlinear Mechanics* (Princeton University Press, New Jersey).
[4] Nayfeh A H 1973 *Perturbation Methods* (J. Wiley, New York).
[5] Azad A K, Hosen M A and Rahman M S 2012 A perturbation technique to compute initial amplitude and phase for the Krylov-Bogoliubov-Mitropolskii method *Tamkang J Math* **43(4)** 563-75.
[6] Ganji D D and Azimi M 2012 Application of max min approach and amplitude frequency formulation to nonlinear oscillation systems *UPB Scientific Bulletin, Series A: App. Math. Phy.* **74(3)** 131-40.
[7] Wang S Q and He J H 2008 Nonlinear oscillator with discontinuity by parameter-expansion method *Chaos, Solitons & Fractals* **35(4)** 688-91.
[8] Baghani M, Fattahi M and Amjadian A 2012 Application of the variational iteration method for nonlinear free vibration of conservative oscillators *Scientia Iranica* **19(3)** 513-18.
[9] Akbarzade M and Farshidianfar A 2014 Application of the amplitude-frequency formulation to a nonlinear vibration system typified by a mass attached to a stretched wire *Inter. Applied Mechanics* **50** 476-83.
[10] Hosen M A, Chowdhury M S H, Ali M Y and Ismail A F 2017 An analytical approximation technique for the duffing oscillator based on the energy balance method *Italian journal of pure and applied mathematics* **37** 455-66.
[11] Hosen M A, Chowdhury M S H, Ali M Y and Ismail A F 2016 A new analytical approximation technique for highly nonlinear oscillations based on the energy balance method *Results in Physics* 6 496-504.
[12] Akbarzade M and Farshidianfar A 2017 Nonlinear transversely vibrating beams by the improved energy balance method and the global residue harmonic balance method *Applied Mathematical Modelling* **45** 393-404.
[13] Khan Y and Mirzabeigy A 2014 Improved accuracy of He's energy balance method for analysis of conservative nonlinear oscillator *Neural Comput & Applic* **25** 889-95.
[14] Peijun J 2015 Global residue harmonic balance method for Helmholtz–Duffing oscillator *Applied Mathematical Modelling* **39(8)** 2172-79.
[15] Cveticanin L 2004 Vibrations of the nonlinear oscillator with quadratic nonlinearity *Physica A* **341** 123-35.
[16] Mickens R E and Mixon M 1992 Application of generalized harmonic balance method to an anti-symmetric quadratic non-linear oscillator *Journal of Sound and Vibrations* **159** 546-68.
[17] Chen S H, Yang X M and Cheung Y K 1998 Periodic solutions of strongly quadratic nonlinear oscillators by the elliptic perturbation method *Journal of Sound and Vibrations* **212** 771-80.
[18] Beléndez A, Pascual C, Beléndez T and Hernández A 2009 Solution of an anti-symmetric quadratic nonlinear oscillator by a modified He's homotopy perturbation method *Nonlinear Analysis: Real World Applications* **10** 416-27.
[19] Beléndez A, Gimeno E, Alvarez M L, Yebra M S and Méndez D I 2010 Analytical approximate solutions for conservative nonlinear oscillators by modified rational harmonic balance method *International Journal of Computer Mathematics* **87** 1497-1511.
[20] Hu H 2006 Solution of a quadratic nonlinear oscillator by the method of harmonic balance *J. Sound Vib.* **293** 462-68.
[21] Momani S, Erjaee G H and Alnasr M H 2009 The modified homotopy perturbation method for solving strongly nonlinear oscillators *Computers and Mathematics with Applications* **58** 2209-20.

[22] Cveticanin L, Zukovic M, Mester G, Biro I and Sarosi J 2016 Oscillators with symmetric and asymmetric quadratic nonlinearity *Acta Mechanica* **227(6)** 1727-42.
[23] Mondal M M H, Molla M H U, Razzak M A and Alam M S 2016 A new analytical approach for solving quadratic nonlinear oscillators *Alexandria Engineering Journal* (In Press).
[24] Haque B M I and Hossain M R 2016 An analytic investigation of the quadratic nonlinear oscillator by an iteration method *British Journal of Mathematics & Computer Science* **13(1)** 1-8.
[25] Mickens R E 2010 *Truly nonlinear oscillations* (World Scientific Publishing, Singapore).
[26] Mickens R E 1986 A generalization of the method of harmonic balance *J. Sound Vib.* **111** 515-18.
[27] Hosen M A, Rahman M S, Alam M S and Amin M R 2012 An analytical technique for solving a class of strongly nonlinear conservative systems *Applied Mathematics and Computation* **218** 5474-86.
[28] Hosen M A and Chowdhury M S H 2015 A new reliable analytical solution for strongly nonlinear oscillator with cubic and harmonic restoring force *Results in Physics* **5** 111-114.
[29] Razzak M A 2016 A simple harmonic balance method for solving strongly nonlinear oscillators *Journal of the Association of Arab Universities for Basic and Applied Sciences* 21 68-76.
[30] Hosen M A and Chowdhury M S H 2015 A new analytical technique based on harmonic balance method to determine approximate periods for *Duffing-harmonic* oscillator *Alexandria Engineering Journal* **54(2)** 233-39.
[31] Hosen M A 2014 Approximate solutions of the equation of motion's of the rigid rod which rocks on the circular surface without slipping *Ain Shams Engineering Journal* **5(3)** 895-99.
[32] Hosen M A 2015 Determination of approximate periods of *Duffing-harmonic* oscillator *Walailak Journal of Science and Technology* **12(10)** 923-31.

IOP Conf. Series: Journal of Physics: Conf. Series **949** (2017) 012007 doi:10.1088/1742-6596/949/1/012007

Impact of donor-acceptor morphology on the charge carrier generation in organic photovoltaic devices

M. L. Inche Ibrahim

Department of Science in Engineering, Faculty of Engineering, International Islamic University Malaysia, P.O. Box 10, 50728 Kuala Lumpur, Malaysia

E-mail: lukmanibrahim@iium.edu.my

Abstract. We investigate the impact of donor-acceptor morphology on the generation of charge carriers in organic photovoltaic cells and photodiodes based on a recently proposed model. It is found that for devices with typical values of permittivity and actual charge carrier mobilities, the donor-acceptor morphology plays an important role in the charge carrier generation, practically at all operating electric fields. Higher permittivity and actual mobilities increase the charge carrier generation, and also decrease the effect of donor-acceptor morphology on the charge carrier generation. Although the donor-acceptor morphology is important for charge carrier generation in general, but having permittivity and/or actual mobilities that are significantly higher than the typical values are more desirable in order to maximize the charge carrier generation.

1. Introduction

Organic photovoltaic (OPV) devices (a term used here to refer both organic solar cells and photodiodes) have distinctive and interesting properties that could lead to various applications. They are also relatively cheap to fabricate, which mean that organic solar cells have the potential to be a viable energy source.

When light is absorbed by the organic photoactive layer (PL), excitons (strongly bound electron-hole pairs) are produced. In order to efficiently generate free charge carriers, the PLs are made by combining (usually two) different organic semiconductors with different energy levels, where one semiconductor acts as the electron donating material while the other one acts as the electron accepting material. How free charge carriers are generated is a fundamental step in the working operation of OPV devices.

At the donor-acceptor (DA) interface, excitons are generally transformed into charge-transfer (CT) states. A CT state consists of an electron in the acceptor and a hole in the donor that are still bound to each other. Several models have been proposed to explain how a CT state can overcome its binding energy to produce a free electron and a free hole. Since CT states are produced with excess energies, it has been suggested that the energy required to overcome the binding energy is provided by the excess energies [1,2].

Another situation is that CT states with excess energies relax to the ground CT states before dissociating into free charge carriers. The dissociation via the relaxed CT states has been supported experimentally [3–7]. The Onsager–Braun (OB) model [8] is widely used to describe the dissociation of relaxed CT states. However, the OB model has several fundamental issues [9].

IOP Conf. Series: Journal of Physics: Conf. Series **949** (2017) 012007 doi:10.1088/1742-6596/949/1/012007

Very recently, Ref. [9] has made an improvement to the OB model by including the effect of DA morphology and by employing the correct charge carrier mobilities. However, the detailed investigation on the impact of DA morphology on the generation of charge carriers using the modified OB model [9] has not been done yet. A more efficient charge carrier generation of course would increase the potential efficiency of the device.

The impacts of the PL's properties such as the carrier mobilities and the permittivity on the performance of OPV devices have been investigated by many studies (for example by Refs. [10-12]) and well understood. However, the impact of the DA morphology of the PL on the charge carrier generation is still unclear. The aim of this paper is to investigate how the DA morphology impacts the generation of charge carriers based on the new model proposed by Ref. [9].

2. Model and procedures

The dissociation probability of a CT state is defined as [8]

$$P_d = \frac{k_d}{k_d + k_f} \tag{1}$$

where k_d is the CT state dissociation rate coefficient, and k_f is the CT state decay rate coefficient. According to the modified OB model proposed by Ref. [9], k_d is given by

$$k_d = \frac{3q\left(\mu_{n,i} + \mu_{p,i}\right)}{4\pi\varepsilon a^3} \exp\left(\frac{-E_b}{k_B T}\right) \frac{J_1\left(2\sqrt{-2b_{\text{eff}}}\right)}{\sqrt{-2b_{\text{eff}}}} \tag{2}$$

where q is the elementary charge, $\mu_{n,i}$ is the actual electron mobility at DA interface, $\mu_{p,i}$ is the actual hole mobility at DA interface, ε is the effective permittivity of the photoactive layer (PL), a is the separation distance between the electron and the hole of the CT state, $E_b = q^2/(4\pi a)$ is the binding energy of the CT state, k_B is the Boltzmann constant, T is the absolute temperature, J_1 is the Bessel function of the first kind of order 1, and b_{eff} is

$$b_{\text{eff}} = \frac{q^3 \lambda |F|}{8\pi\varepsilon k_B^2 T^2} \tag{3}$$

where $|F|$ is the magnitude of the electric field, and λ is a parameter that describes how the DA morphology influences the CT state dissociation. Note that $\mu_{n,i}$ and $\mu_{p,i}$ are generally not the same as (can be higher than) the conventional carrier mobilities that are used to calculate the electric current [9]. $\mu_{n,i}$ and $\mu_{p,i}$ are fixed for any given DA mixture (independent of the DA morphology and device architecture) unlike the conventional carrier mobilities [9]. The detailed explanation on the actual carrier mobilities can be obtained in Ref. [9].

A CT state's direction is defined as the direction from the electron to the hole of the CT state. A more positive value of λ means the effective direction of the CT states inside the PL is more aligned in the direction of the electric field inside the PL, while a more negative value of λ means the effective direction of the CT states inside the PL is more aligned in the direction opposite to the direction of the electric field. $\lambda = 1$ (the maximum value) means that all the CT states inside the PL are in the same direction as the direction of the electric field. $\lambda = -1$ (the minimum value) means that all the CT states inside the PL are exactly opposite to the direction of the electric field. $\lambda = 0$ means the effective direction of the CT states is perpendicular to the direction of the electric field. In other words, $\lambda = 0$ means that the electric field effectively does not contribute to the CT state dissociation. However, since $\lambda = 0$ gives an undefined value of k_d according to equation (2), we simply neglect

IOP Conf. Series: Journal of Physics: Conf. Series **949** (2017) 012007 doi:10.1088/1742-6596/949/1/012007

the term $J_1\left(2\sqrt{-2b_{\text{eff}}}\right)/\sqrt{-2b_{\text{eff}}}$ in equation (2) in order to represent k_d (and thus P_d) when $\lambda = 0$. As explained in Ref. [9], the value of λ can be controlled by properly designing the distributions (morphologies) and/or the architectures of the DA mixture.

The generation rate of free charge carriers is given by $P_d G_{CT}$, where P_d is CT state dissociation probability, and G_{CT} is the generation rate of CT states. Since P_d is very closely linked to the charge carrier generation, understanding of the impact of DA morphology on the charge carrier generation can be meaningfully achieved by understanding the impact of DA morphology on P_d. In fact, P_d may also be called the free charge carrier generation probability, which will be used from here on.

As can be seen from the equations above, the electric field also affects the charge carrier generation. If the electric field is uniform, $|F|$ inside the PL can be given by (see Ref. [13] for example)

$$|F| = \left|\frac{V_{bi} - V_a}{L}\right| \tag{4}$$

where V_{bi} is the built in voltage, V_a is the applied voltage, and L is the PL's thickness. The operating $|F|$ inside the PL of an OPV device can be up to several 10^7 Vm^{-1}. For example, an organic ultraviolet photodetector may have $V_{bi} = 3.5$ eV and may operate at $V_a = -1$ V (reverse bias) with $L = 100$ nm, which means $|F| = 4.5 \times 10^7$ Vm^{-1} inside the PL. Our analysis must obviously cover the range of the possible operating $|F|$ in OPV devices.

In order to achieve a comprehensive understanding on how the charge carrier generation is influenced by the DA morphology, we will calculate and analyze the charge carrier generation probabilities with respect to λ (which represents the DA morphology) together with other relevant parameters, which are the effective permittivity, the actual mobilities, and the electric field.

3. Results

Table 1 shows the values of the parameters used for calculations. The PL's effective permittivity ε is given by $\varepsilon = \varepsilon_0 \varepsilon_r$, where ε_0 is the vacuum permittivity, and ε_r is the effective relative permittivity of the PL.

IOP Conf. Series: Journal of Physics: Conf. Series **949** (2017) 012007 doi:10.1088/1742-6596/949/1/012007

Table 1. Values of the parameters.

Parameter (symbol)	Value
Actual electron mobility at DA interface ($\mu_{n,i}$)	See text
Actual hole mobility at DA interface ($\mu_{p,i}$)	See text
Relative permittivity (ε_r)	See text
CT state decay rate coefficient (k_f)	1.0×10^{8} s^{-1}
Electron-hole separation (a)	1.3×10^{-9} m
Temperature (T)	300 K

Figure 1 shows the charge carrier generation probabilities P_d's calculated using $\varepsilon_r = 3.5$, $\mu_{n,i} = \mu_{p,i} = 2\times10^{-6}$ m^{2}V^{-1}s^{-1}, and several different values of λ as functions of the magnitude of the electric field $|F|$. The values of ε_r, $\mu_{n,i}$, and $\mu_{p,i}$ used for figure 1 are the typical values for organic PLs. It can be seen that for the OPV device of figure 1, the value of λ (and thus the DA morphology) plays an important role in the charge carrier generation. In fact, it is important for the PL of the OPV device of figure 1 to have λ of at least around 0.75 in order to produce a relatively good P_d in comparison to P_d when $\lambda = 1$. At $|F| \approx 0$ Vm^{-1}, the impact of DA morphology on P_d is not important.

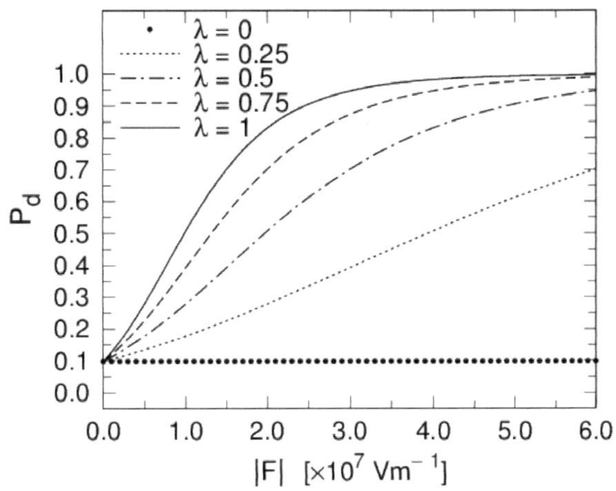

Figure 1. Charge carrier generation probabilities P_d 's for several values of λ as functions of the magnitude of the electric field $|F|$. The parameters as given in table 1 with $\varepsilon_r = 3.5$ and $\mu_{n,i} = \mu_{p,i} = 2\times10^{-6}\,\text{m}^2\text{V}^{-1}\text{s}^{-1}$ are used for calculations.

Figure 2 shows P_d 's for several different values of λ as functions of $|F|$ using the same $\mu_{n,i}$ and $\mu_{p,i}$ as used in figure 1 but with $\varepsilon_r = 5$. From figure 2, it is clear that increasing ε_r increases P_d at any given $|F|$. Increasing ε_r from 3.5 to 5 significantly increases P_d in general. For example, at $|F| = 0$, increasing ε_r from 3.5 to 5 increases P_d from 0.1 to 0.75. For the OPV device of figure 2, it is important for the PL to have λ of at least about 0.5 in order to produce a relatively good P_d in comparison to P_d when $\lambda = 1$. However, for the device of figure 2, even a very low (but positive) λ produce a quite high P_d at practically the whole range of the studied $|F|$.

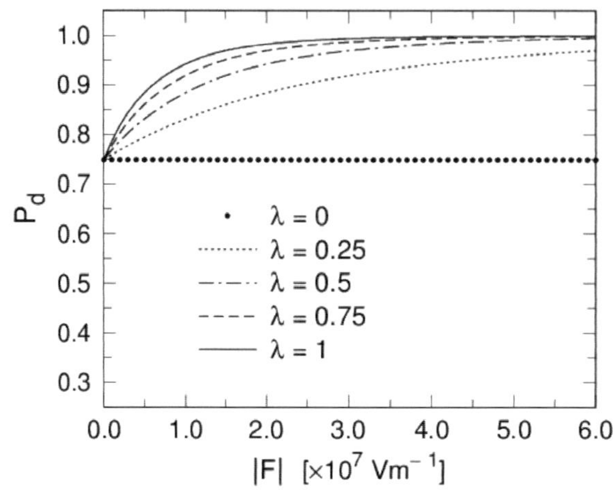

Figure 2. Charge carrier generation probabilities P_d 's for several values of λ as functions of the magnitude of the electric field $|F|$. The parameters as given in table 1 with $\varepsilon_r = 5$ and $\mu_{n,i} = \mu_{p,i} = 2\times10^{-6}$ m^{2}V^{-1}s^{-1} are used for calculations.

Figure 3 shows P_d 's for several different values of λ as functions of $|F|$ using the same $\mu_{n,i}$ and $\mu_{p,i}$ as used in figure 1 but with $\varepsilon_r = 6.5$. For the OPV device of figure 3, the role of λ (and thus the DA morphology) is generally not important for the generation of charge carriers. However, it is still quite important to at least have $\lambda > 0$ in general.

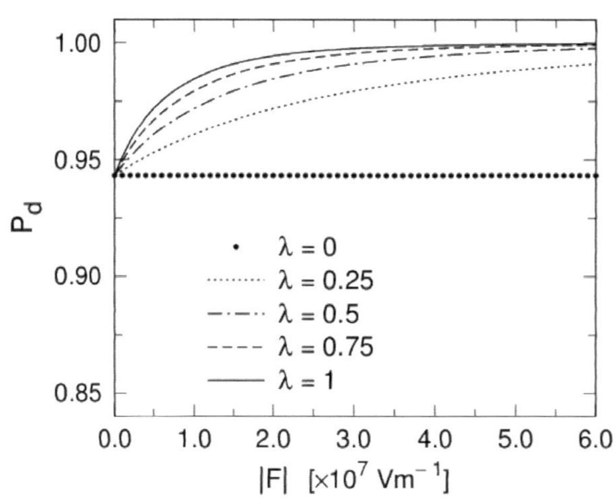

Figure 3. Charge carrier generation probabilities P_d 's for several values of λ as functions of the magnitude of the electric field $|F|$. The parameters as given in table 1 with $\varepsilon_r = 6.5$ and $\mu_{n,i} = \mu_{p,i} = 2\times10^{-6}$ m^{2}V^{-1}s^{-1} are used for calculations.

IOP Conf. Series: Journal of Physics: Conf. Series **949** (2017) 012007 doi:10.1088/1742-6596/949/1/012007

Figure 4 shows P_d's calculated using $\varepsilon_r = 3.5$, $\mu_{n,i} = \mu_{p,i} = 6 \times 10^{-5}$ m^{2}V^{-1}s^{-1}, and several different values of λ as functions of $|F|$. From figure 4, it can be seen that higher $\mu_{n,i} + \mu_{p,i}$ increases P_d, similar as the effect of a higher ε_r. A higher $\mu_{n,i} + \mu_{p,i}$ makes the role of λ (and thus the DA morphology) to be less significant in generating the charge carriers.

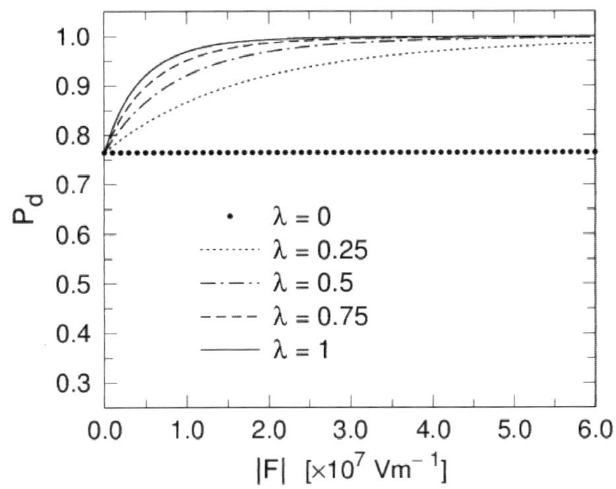

Figure 4. Charge carrier generation probabilities P_d's for several values of λ as functions of the magnitude of the electric field $|F|$. The parameters as given in table 1 with $\varepsilon_r = 3.5$ and $\mu_{n,i} = \mu_{p,i} = 6 \times 10^{-5}$ m^{2}V^{-1}s^{-1} are used for calculations.

It can be concluded that OPV devices with relatively low λ but with ε_r and/or $\mu_{n,i} + \mu_{p,i}$ that are significantly higher than the typical values are likely to have better charge carrier generation than OPV devices with very high λ but with typical values of ε_r and $\mu_{n,i} + \mu_{p,i}$. Hence, to maximize the charge carrier generation, devices with relatively low λ but having ε_r and/or $\mu_{n,i} + \mu_{p,i}$ that are significantly higher than the (current) typical values are more desirable than devices with very high λ but having typical values of ε_r and $\mu_{n,i} + \mu_{p,i}$. However, for devices with typical values of ε_r and $\mu_{n,i} + \mu_{p,i}$, having a very high λ is important to maximize the charge carrier generation.

4. Summary

Based on a recently proposed model, we have performed a study on the impact of DA morphology on the generation of charge carriers in OPV devices by calculating and analyzing the charge carrier generation probabilities P_d with respect to the parameter λ (which represents the DA morphology), the effective permittivity, the actual mobilities, and the electric field. It is found that for OPV devices with typical values of ε_r (or ε) and $\mu_{n,i} + \mu_{p,i}$, the DA morphology plays an important role in the charge carrier generation, practically at all possible operating electric fields. Increasing the values of ε_r and/or $\mu_{n,i} + \mu_{p,i}$ increase P_d and at the same time making the effect of DA morphology to be less significant. In order to maximize charge carrier generation, devices with relatively poorly-designed donor-acceptor morphology (low λ) but having ε_r and/or $\mu_{n,i} + \mu_{p,i}$ that are significantly higher than the (current) typical values are more desirable than devices with excellently-designed donor-acceptor morphology (very high λ) but having typical values of ε_r and $\mu_{n,i} + \mu_{p,i}$.

5. Acknowledgments

This work is financially supported by International Islamic University of Malaysia

References

[1] Ohkita H et al 2008 J. Am. Chem. Soc. **130** 3030
[2] Grancini G, Maiuri M, Fazzi D, Petrozza A, Egelhaaf H J, Brida D, Cerullo G and Lanzani G 2013 Nat. Mater. **12** 29
[3] Howard I A, Mauer R, Meister M and Laquai F 2010 J. Am. Chem. Soc. **132** 14866
[4] Lee J, Vandewal K, Yost S R, Bahlke M E, Goris L, Baldo M A, Manca J V and Voorhis T V 2010 J. Am. Chem. Soc. **132** 11878
[5] van der Hofstad T G J, Di Nuzzo D, van den Berg M, Janssen R A J and Meskers S C J 2012 Adv. Energy Mater. **2** 1095
[6] Vandewal K et al 2014 Nat. Mater. **13** 63
[7] Devizis A, De Jonghe-Risse J, Hany R, Nuesch F, Jenatsch S, Gulbinas V and Moser J E 2015 J. Am. Chem. Soc. **137** 8192
[8] Braun C L 1984 J. Chem. Phys. **80** 4157
[9] Inche Ibrahim M L 2017 J. Phys. D: Appl. Phys. **50** 065103
[10] Inche Ibrahim M L, Ahmad Z, K Sulaiman K, and Muniandy S V 2014 AIP Adv. **4** 057133
[11] Koster L J A, Shaheen S E, and Hummelen J C 2012 Adv. Energy Mater. **2** 1246.
[12] Shieh J T, Liu C H, Meng H F, Tseng S R, Chao Y C, and Horng S F 2010 J. Appl. Phys. **107** 084503
[13] Inche Ibrahim M L, Ahmad Z and Sulaiman K 2015 AIP Adv. **5** 027115

IOP Conf. Series: Journal of Physics: Conf. Series **949** (2017) 012008 doi:10.1088/1742-6596/949/1/012008

Insertion algorithms for network model database management systems

Abdurashid Mamadolimov[1], Khikmat Saburov[2]

[1,2]Mathematical Modelling Laboratory, MIMOS,
Technology Park Malaysia, Kuala Lumpur, Malaysia

[1]rashid.mdolimov@mimos.my, [2]khikmat.saburov@mimos.my

Abstract. The network model is a database model conceived as a flexible way of representing objects and their relationships. Its distinguishing feature is that the schema, viewed as a graph in which object types are nodes and relationship types are arcs, forms partial order. When a database is large and a query comparison is expensive then the efficiency requirement of managing algorithms is minimizing the number of query comparisons. We consider updating operation for network model database management systems. We develop a new sequential algorithm for updating operation. Also we suggest a distributed version of the algorithm.

1. Introduction

A database management system (DBMS) is a software package with computer programs that controls the creation, maintenance, and use of a database. One of the components of DBMS is a Data Manipulation Subsystem. Data manipulation subsystem helps the user to add, change, and delete information in a database and query it for valuable information. Software tools within the data manipulation subsystem are most often the primary interface between user and the information contained in a database. It allows the user to specify its logical information requirements.

DBMSs are designed to use one of five database models to provide simplistic access to information stored in databases. The most popular five database models are the followings:

(i) hierarchical model,
(ii) network model,
(iii) relational model,
(iv) multidimensional model, and
(v) object model.

The network model is a database model conceived as a flexible way of representing objects and their relationships. Its distinguishing feature is that the schema, viewed as a graph in which object types are nodes and relationship types are arcs, forms partial order, is not restricted to being a tree or lattice (Figure 1). When a database is large and a query comparison is expensive then the efficiency requirement of managing algorithms is minimizing the number of query comparisons i.e., the efficiency is measured by query complexity of an algorithm. For example, conceptual graphs form a network model structure and query comparison will be a graph homomorphism checking which a very expensive operation is [1]-[9].

In this paper we consider updating operation for network model database management systems. We develop a new sequential algorithm for updating operation. Also we suggest a distributed version of the algorithm.

One of the important components of updating operation is a searching. In [10], Mozes, Onak, and Weimann present a linear-time algorithm that finds the optimal strategy for searching a tree-like partially ordered set. Dereniowski [11] proves that finding an optimal search strategy for general posets is hard and gives a polynomial-time approximation algorithm with sub-logarithmic approximation ratio. In our early works [12], [13] we proposed new sequential and distributed algorithms for searching operation. Another component of updating operation is a finding parents. For finding parents operation Levinson [14] and Ellis [15] have proposed several modifications of breadth/depth first search technique using some simple properties of partially ordered sets. In this paper we used height of elements of poset for a good ordering of a dataset.

2. Partially ordered set (poset)

Definition. A *partial order* is a binary relation " $\leq$ " over a set P which is reflexive, antisymmetric, and transitive, that is, for all a, b and c in P, we have that:

(i) $a \leq a$ (reflexivity);
(ii) if $a \leq b$ and $b \leq a$ then $a = b$ (antisymmetry);
(iii) if $a \leq b$ and $b \leq c$ then $a \leq c$ (transitivity).

A set P with a partial order " $\leq$ " is called a *partially ordered set* or *poset* $(P, \leq)$. For any two elements a and b of a poset $(P, \leq)$ we use $a \geq b$ if $b \leq a$. Also we use $a < b$ $(a > b)$ if $a \leq b$ $(a \geq b)$ and $a \neq b$. Let $(P, \leq)$ be a poset. If $\top \in P$ such that $\forall x \in P, x \leq \top$, then $\top$ is called the *top* element of P (the *bottom* element $\bot$ is dually defined); a poset does not necessarily have a top (bottom) element, but if it has a top (bottom) element then it is unique. Any finite poset can be represented by directed acyclic graph and this representation is called Hasse diagram of a poset (Figure 1). Network model databases have a partially ordered data structure, i.e., data and relations between them form a poset.

Let $(P, \leq)$ be a poset and D be any non-empty, finite subset of P: $D \subseteq P$, $|D| = n$, where n is a positive integer. We note that $(D, \leq)$ is also a poset. We all times assume that the subset D has both top and bottom elements and that they are different: $\top \neq \bot$. We call the set D as a *dataset*. Let x be an element of P: $x \in P$. We call the element x a *query element*. The following definitions are useful for future description. $Ancestors(x, D) = \{a \in D: a > x\}$;
$Descendants(x, D) = \{d \in D: d < x\}$;
$Parents(x, D) = \{p \in Ancestors(x, D):$
$\nexists a \in Ancestors(x, D), a < p\}$;
$Children(x, D) = \{c \in Descendants(x, D):$
$\nexists d \in Descendants(x, D), d > c\}$;
$Indeg(x, D) = |Parents(x, D)|$;
$Outdeg(x, D) = |Children(x, D)|$.
We use $Ancestor(x, D), Descendant(x, D), Parent(x, D)$ and $Child(x, D)$ for an element of $Ancestors(x, D), Descendants(x, D), Parents(x, D)$ and $Children(x, D)$ respectively.

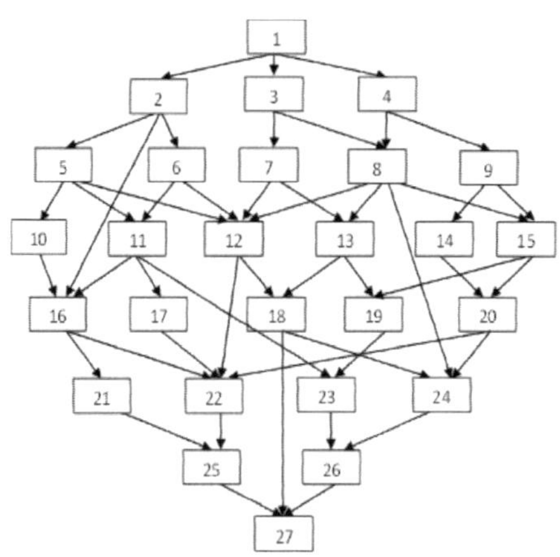

Figure 1. Example of a graph representation of a poset

Definition. A *chain* from an element a to an element b ($\neq a$) in a poset D is a sequence of different elements $a = x_0, x_1, \dots, x_l = b$ of a poset D such that each of its elements, except last one, is a parent for the next element in the sequence; l is called a *length* of the chain; A chain from element a to element b with maximum length is called *maximum chain* from a to b.

Definition. A length of maximum chain from top element T to an element x is called *height* of x in a poset D and denoted as $Height(x, D)$. Height of bottom element is called *height of poset* i.e., $Height(D) = Height(\perp, D)$.

We do not provide here the proofs of the Propositions 1-3.

Proposition 1. For any two elements x and y of a poset D, if $x > y$ then $Height(x, D) < Height(y, D)$.

Proposition 2. For any element x of a poset D,

$$Height(x, D) = \max_{y \in Parents(x,D)} Height(y, D) + 1.$$

Proposition 3. Let x, y, z be elements of some poset. If $x \geq y$ is false then for any z such that $z \leq x$, we have that $z \geq y$ is also false.

Also we use the next proposition in our algorithms.

Proposition 4. Let $(P, \leq)$ be a poset and D be any non-empty, finite subset of P. Let $q \in P \backslash D$. Let L_h be a set of elements of D with fixed height h i.e., $L_h = \{x \in D | \, Height(x, D) = h\}$. Let L_h^+ be a set of elements of D with height more than h i.e., $L_h^+ = \{x \in D | \, Height(x, D) > h\}$. If $\forall x \in L_h, \ x > q$ is false, then $\forall y \in L_h^+, y > q$ is also false.

Proof. Suppose that $\exists y \in L_h^+, y > q$. Let height of y be l i.e., $height(y) = l$. Then there exists a (maximum) chain $y_0 = \text{T}, y_1, y_2, \dots, y_l = y$ such that
$y_0 = \text{T} > y_1 > y_2 > \cdots > y_l = y$. Since $l > h$ we have a sequence $y_0, y_1, y_2, \dots, y_h, y_{h+1}, \dots, y_l$ and $y_0, y_1, y_2, \dots, y_h$ is a maximum chain from y_0 to y_h. Therefore $height(y_h) = h$ or $y_h \in L_h$. However $y_h > y$ and $y > q$ then $y_h > q$. This contradicts to the condition of Proposition 3. □

3. Representation of a poset: the extended descendants list data structure

For representation of a poset we use list of descendants or list of ancestors in special form. Let $c_1, \dots, c_n$ be elements of poset D and let c_1 be top element, i.e., $c_1 = \text{T}$. We define partial order over integers $D' = \{1,2, \dots, n\}$ in the following way: for any $x, y \in D'$, $x \leq y$ if and only if $c_x \leq c_y$. Two lists are associated with an element $c_i \in D$, $i = 1..n$: *list of descendants* **D**(i) and *list of ancestors* **A**(i). The list of descendants includes $Height(i, D)$ as a first component of the list, second component is $Outdeg(i, D')$, and first $Outdeg(i, D')$ descendants are $Children(i, D')$, and other descendants:

D$(i) = [Height(i, D');\ Outdeg(i, D');$
$Child_1(i, D'), \dots, Child_{Outdeg(i,D')}(i, D');$
$Descendant_{Outdeg(i,D')+1}(i, D'), \dots, \perp], \quad i = 1..n.$

We assume that elements $c_1, \dots, c_n$ of poset D are ordered such that $Height(i, D')$ is a non-decreasing function of i. Also we denote the number of elements of dataset D' with $Height(x, D') = h$ as n_h, $h = 0,..Height(D') = H$. It is clear that $\sum_{h=0}^{H} n_h = n$. The list of ancestors is dually defined. The following is the list of descendants of the poset given in Figure 1.

D(1) = [0; 3; 2,3,4; 5,6,7,8,9,10,11,12,13,14,15,16,17,18,19,20,21,22,23,24,25,26,27],
D(2) = [1; 3; 5,6,16; 10,11,12,16,17,18,21,22,23,24,25,26,27],
D(3) = [1; 2; 7,8; 12,13,15,18,19,20,22,23,24,25,26,27],
D(4) = [1; 2; 8,9; 12,13,14,15,18,19,20,22,23,24,25,26,27],
D(5) = [2; 3; 10,11,12; 16,17,18,21,22,23,24,25,26,27],

....

D(25) = [6; 1; 27],
D(26) = [6; 1; 27],
D(27) = [7; 0].

4. Updating operation

Updating can be done using *Inserting* and *Deleting* of a query element. The *Deleting* first uses *Searching* of a query element then deletes it and its links to parents and children. *Deleting* and *Searching* have the same query complexity. These two operations are almost same. Since in early works [12], [13] we developed algorithms for *Searching* operation, in this invention we do not consider the *Deleting* operation. *Inserting* can be done using *Finding Parents* and *Finding Children* operations. *Finding Parents* and *Finding Children* are dual operations, i.e., an algorithm for *Finding Parents* can be very easily adapted to *Finding Children* operation and vice versa. Thereby we consider *Finding Parents* operation in this paper. We assume that a query element is not in a database; otherwise again we have an operation which is equivalent to *Searching*. Since our original operation is *Inserting* this assumption is inartificial.

5. Algorithm for finding parents operation

Let $(P, \leq)$ be a poset and D be any dataset i.e., non-empty, finite subset of P: $D \subseteq P$, $|D| = n$, where n is a positive integer. Let a query element $q \in P$ be not in D i.e., $q \notin D$. Although the query element q is not in the dataset D, it has parents, children and ancestors, descendants in D i.e., query element has a virtual address in the dataset (Figure 2). The proposed algorithm takes list of descendants **D**(i), $i = 1..n$ of a given poset D, the numbers n_h, $h = 0,..Height(D') = H$ of elements of dataset D' with $Height(x, D') = h$ and a query element q ($\notin D$) as input and finds $Parents(q, D)$. Data representation way described in section 3 allows us to distribute the elements of dataset in the table (Table 1). The height of the table is $H + 1$, where H is height of dataset. Width of the table is equal to $\max\{n_0, n_1, \dots, n_H\}$. The top element $\top$ is only element on the 0-th row of the table, for $h = 1, \dots, H$, h-th row of the table contains elements of dataset from $\sum_{i=0}^{h-1} n_i + 1$ to $\sum_{i=0}^{h} n_i$. Proposed algorithm scans elements of this table from up to down and from left to right. Scanning is a comparison of current element c and query element q: if $c > q$ then YES status will be assigned to the element c, else NO. Regarding Proposition 3 if NO status is assigned to the element c then NO status will be assigned to all its descendants also. Algorithm scans not only current element c but all its children also. If all children of c have NO status then c is a parent of query element q. Regarding Proposition 4 algorithm ends if all elements of some row of the table have NO status.

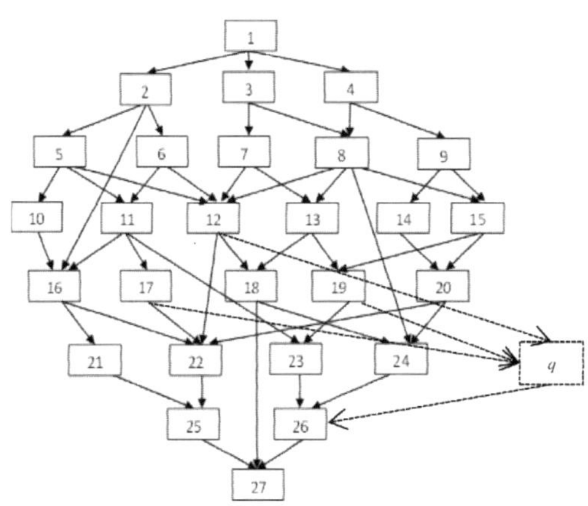

Figure 2. Example of a virtual address of a query element

In Figure 3 you can see the pseudo-code description of proposed algorithm.

```
Algorithm 1: FindingParents(Dataset, query element);
Input: Elements c_i, i = 1..n of dataset D and list of descendants D(i) of elements c_i, i = 1..n of dataset D,
    Numbers n_h, h = 0..H of elements of dataset D with Height(x, D') = h, query element q (∉ D);
Output: Parents(q, D);
     Status(1)=YES;
     for each x ∈ {2,...,n}, Status(x) ⟵ UNVISITED;
     Parents(q, D) ← ⊤;
     for h = 1 to H do
          for x = Σ_{i=0}^{h-1} n_i + 1 to Σ_{i=0}^{h} n_i do
              if Status(x) = YES then
                          go to lab;
              fi
              if Status(x) = UNVISITED then
                  if c_x > q then
                        Status(x) ← YES;
lab:                  for i = 1 to Outdeg(x, D') do
                           y ← Child_i(x, D');
                           if Status(y) = UNVISITED then
                                 if c_y > q then Status(y) ← YES;
                                        else Status(y) ← NO;
                                           for each Descendant(y, D') ∈ Descendants(y, D'): Status(Descendant(y, D')) ← NO;
                                 fi
                           fi
                      od
                      if for each Child(x, D') ∈ Children(x, D'), Status(Child(x, D')) = NO then Parents(q, D) ← Parents(q, D) ∪ {x};
                      fi
                      else Status(x) = NO;
                         for each Descendant(x, D') ∈ Descendants(x, D'), Status(Descendant(x, D')) ← NO;
                  fi
              fi
          od
          if for each x ∈ {Σ_{i=0}^{h-1} n_i + 1, ..., Σ_{i=0}^{h} n_i}, Status(x)=NO and h = 1 then
                                          return Parents(q, D);
          else if x ∈ {Σ_{i=0}^{h-1} n_i + 1, ..., Σ_{i=0}^{h} n_i}, Status(x)=NO then
                                          return Parents(q, D)\{⊤};
          fi
     od
```

Figure 3. Finding parents algorithm

```
Algorithm 2: ParallelFindingParents(Dataset, query element);
Input: Elements c_i, i = 1..n of dataset D and list of descendants D(i) of elements c_i, i = 1..n of dataset D,
    Numbers n_h, h = 0..H of elements of dataset D with Height(x, D') = h, query element q (∉ D);
Output: Parents(q, D);
    For processor P_1 do
        Status(1)=YES;
        for each x ∈ {2,...,n}, Status(x) ⟵ UNVISITED;
        Parents(q, D) ← T;
        for h = 1 to H do
    od
    For all processors P_s, s = 1..m do in parallel
            for x_s = Σ_{i=0}^{h-1} n_i + 1 + (s − 1)⌈n_h/m⌉ to Σ_{i=0}^{h-1} n_i + s⌈n_h/m⌉ do
              if Status(x_s) = YES then
                        go to lab;
              fi
              if Status(x_s) = UNVISITED then
                 if c_{x_s} > q then
                     Status(x_s) ← YES;
lab:                 for i_s = 1 to Outdeg(x_s, D') do
                         y_s ← Child_{i_s}(x_s, D');
                         if Status(y_s) = UNVISITED then
                               if c_{y_s} > q then Status(y_s) ← YES;
                                      else Status(y_s) ← NO;
                                        for each Descendant(y_s, D') ∈ Descendants(y_s, D'): Status(Descendant(y_s, D')) ← NO;
                               fi
                         fi
                     od
                     if for each Child(x_s, D') ∈ Children(x_s, D'), Status(Child(x_s, D')) = NO then Parents(q, D) ← Parents(q, D) ∪ {x_s};
                     fi
                     else Status(x_s) = NO;
                        for each Descendant(x_s, D') ∈ Descendants(x_s, D'), Status(Descendant(x_s, D')) ← NO;
                 fi
              fi
            od
    od
    For processor P_1 do
        if for each x ∈ {Σ_{i=0}^{h-1} n_i + 1, ..., Σ_{i=0}^{h} n_i}, Status(x)=NO and h = 1 then
                                            return Parents(q, D);
        else if x ∈ {Σ_{i=0}^{h-1} n_i + 1, ..., Σ_{i=0}^{h} n_i}, Status(x)=NO then
                                            return Parents(q, D)\{T};
        fi
    od
    od
```

Figure 4. Distributed algorithm

6. Distributed algorithm

Proposed algorithm can be adapted for distributed computing. For distributed algorithm we use parallel random-access machine (PRAM) model of computation with $P_1, P_2, \ldots, P_m,\ m \geq 2$ processors. In PRAM model of computation all processors can access common RAM in a single algorithm-step.

In the distributed algorithm one of the processors starts and ends to compute following Algorithm 1. Scanning of elements in each row will be done parallelly by m processors. Each row of Table 1 i.e., elements of dataset with fixed height, is divided to m equal intervals:

$$\{\textstyle\sum_{i=0}^{h-1} n_i + 1 + (s-1)\left\lceil \frac{n_h}{m} \right\rceil,\ \sum_{i=0}^{h-1} n_i + s\left\lceil \frac{n_h}{m} \right\rceil\},$$
$$s = 1..m,$$

where $\lceil a \rceil$ is the smallest integer not less than a.

Each interval is scanned by one processor. It is given the pseudo-code description of distributed algorithm in Figure 4. In Algorithm 2 two or more processors do not check the same query comparison if they do not come to common child of two elements with same height and from different intervals

Table 1. Distribution of the elements of a dataset by heights

0	1	-	…	-	-	…	-
1	2	3	…	$1+n_1$	-	…	-
	…	…	…	…	…	…	…
h	$\sum_{i=0}^{h-1} n_i + 1$	…	…	$\sum_{i=0}^{h} n_i$	-	…	-
	…	…	…	…	…	…	…
H	n	-	…	-	-	…	-

at exactly the same time. In other words, there is no overlapping of partition of scanning space up to query comparisons. We also note that the processors are almost uniformly tasked.

7. Conclusions

Updating consists of two sub-operations: (1) deleting of an existing object from a database; (2) inserting of a new object to a database. In both case we call of a deleting/inserting object as a query object (element). For deleting/inserting of query object it will be compared with objects of database. This comparison is called a query comparison. It is clear that any updating algorithm uses at worst n query comparisons, where n is a database size. If a query comparison is easily computable (for example, comparison of d-dimensional vectors) then even an algorithm with exactly n query comparisons is suitable; thereby to develop of an updating algorithm is very easy. However when a query comparison is not easily computable (for example, comparison of graphs) then efficiency of an updating algorithm is required to minimize the number of query comparisons. A *query complexity* of an algorithm is the number of query comparisons that is used in the algorithm. Our goal was to develop an updating algorithm(s) with a query complexity as less as possible.

We first analyzed an updating operation and picked out an operation *Finding Parents* as a most considerable sub-operation. We proposed an algorithm for *Finding Parents* operation. We used a new data structure for posets: it is an extended descendant list of a poset. The main extra parameter of this representation is a height/level of an element of a database. The extended descendant list of poset allows us to distribute the elements of a database in a rectangular table. Then algorithm starts to scan cells of the table. For minimizing of query complexity we have also used the properties of posets.

We also proposed a distributed version of the algorithm. Proposed distributed algorithm responds two main requirements for distributed algorithms: uniform distribution of computation among processors and no overlapping computation.

References

[1] Sowa J.F., *Conceptual Graphs*, IBM Journal of Research and Development, 20(4), 1976, pp336 - 357.

[2] Sowa J.F., *Conceptual Structures: Information Processing in Mind and Machine*. 1984, Addison- Wesley.

[3] Chein M., Mugnier M.-L., *Graph-based Knowledge Representation. Computational Foundations of Conceptual Graphs*. 2009, Springer.

[4] Chein M., Mugnier M.-L., *Conceptual Graphs: Fundamental Notions*, Revue d'Intelligence Artificielle, 6(4), 1992, pp365-406.

[5] Willems M., *Projection and unification for conceptual graphs*, In Proc. of ICCS'95, Vol. 954 of LNCS, pp 278-292, 1995, Springer.

[6] Mugnier M.-L, *On Generalization / Specialization for Conceptual Graphs*, Journal of Experimental and Theoretical Artificial Intelligence, 7, 1995, pp325-344.

[7] Baget J.-F., *Representer des connaissances et raisonner avec des hypergraphes: de la projection a la derivation sous contraintes*. PhD Thesis, 2001, University Montpellier II.

[8] Baget J.-F., *Simple Conceptual Graphs Revisited: Hypergraphs and Conjuctive Types for Efficient Projection Algorithms*, In Proc. of ICCS'03, vol. 2746 of LNAI. 2003, Springer.
[9] Croitoru M. and Compatangelo E., *A tree decomposition algorithm for conceptual graph Projection,* In Proc. of KR'06, pp271-276. 2006, AAAI Press.
[10] Mozes S., Onak K., Weimann O., *Finding an optimal tree searching strategy in linear time*, In: SODA'08: Proceedings of the nineteenth annual ACM-SIAM symposium on Discrete algorithms, Philadelphia, PA, USA, Society for Industrial and Applied Mathematics, 2008, pp1096-1105.
[11] Dereniowski D., *Edge ranking and searching in partial orders*, Discrete Appl. Math. 156(13), 2008, pp2493-2500.
[12] Mamadolimov A., *Search algorithms for conceptual graph databases*, International Journal of database Management Systems (IJDMS), Vol.5, No.1, 2013, pp 35-44.
[13] Mamadolimov A., *Searching Algorithms in Partially Ordered Set,* SIAM Conference on Discreet Mathematics, June 18-21, 2012, Dalhosie University, Halifax, Nova Scotia, Canada.
[14] Levinson R.A., *A self-organizing retrieval system for graphs*, In The 3rd National Conference of the American Association for Artificial Intelligence, 1984, pp 203-206, Austin, Texas, AAAI Press, Menlo Park.
[15] Ellis G.R., *Managing Complex Objects*. PhD Thesis. University of Queensland, 1995.

IOP Conf. Series: Journal of Physics: Conf. Series **949** (2017) 012009 doi:10.1088/1742-6596/949/1/012009

Multicollinearity and Regression Analysis

Jamal I. Daoud
Department of Science in Engineering, IIUM, 53100, Jalan Gombak, Selangor Darul Ehsan, Malaysia

E-mail: jamal58@iium.edu.my

Abstract: In regression analysis it is obvious to have a correlation between the response and predictor(s), but having correlation among predictors is something undesired. The number of predictors included in the regression model depends on many factors among which, historical data, experience, etc. At the end selection of most important predictors is something objective due to the researcher. Multicollinearity is a phenomena when two or more predictors are correlated, if this happens, the standard error of the coefficients will increase [8]. Increased standard errors means that the coefficients for some or all independent variables may be found to be significantly different from 0. In other words, by overinflating the standard errors, multicollinearity makes some variables statistically insignificant when they should be significant. In this paper we focus on the multicollinearity, reasons and consequences on the reliability of the regression model.

1. Introduction

In regression analysis there are many assumptions about the model, namely, multicollinearity, nonconsistant variance (non-homogeneity), linearity, and autocorrelation [6]. If one or more assumption is violated, then the model in hand is no more reliable and also is not acceptable in estimating the population parameters.

In this study we focus on multicollinearity as a violation of one of basic assumption for successful regression model assumptions of successful regression model. Multicollinearity appears when two or more independent variables in the regression model are correlated. a little bit of multicollinearity sometimes will cause big problem but when it is moderate of high then it will be a problem to be solved.

Multicollinearity, or near-linear dependence, is a statistical phenomenon in which two or more predictors variables in a multiple regression model are highly correlated. If there is no linear relationship between predictor variables, they are said to be orthogonal [2].

IOP Conf. Series: Journal of Physics: Conf. Series **949** (2017) 012009 doi:10.1088/1742-6596/949/1/012009

In most applications of regression, the predictor's variables are usually not orthogonal. Multicollinearity can be observed in the following cases

i) Large changes in the estimated coefficients when a variable is added or deleted.
ii) Large changes in the coefficients when a data point is altered or dropped.

Multicollinearity may be present if:

i) The algebraic signs of the estimated coefficients do not conform to the prior expectation; or
ii) Coefficients of variables that are expected to be important have large standard errors (small t-values).

In fact, the researcher has no tools to know the multicollinearity unless the data has been collected. There are two types of multicollinearity:

i) Data-based multicollinearity which occurs because of the researcher, when the experiment is poorly designed, or the collected data are purely observational
ii) Structural multicollinearity: it occurs when the researcher generates new independent variable from one or more existing variables, for example creating x^3 from x, it is in fact mathematical artifact which leads to multicollinearity.

Therefore, in this research we will focus on the impact of multicollinearity existence among predictor variables on hypotheses testing decision taken.

2. Correlation of predictors and the impact on regression model

What impact does the correlation between predictors have on the regression model and subsequent conclusions? Correlation can be high or otherwise, to illustrate the impact of correlation among predictors on the reliability of the model obtained we use two sets of data one set with low correlation among predictors and other set with high correlation between predictors. The analysis of regression for the first set of data yielded the following regression information.

We start by fitting simple models with one predictor variable each time, then by fitting multiple model containing both predictor variables. The multiple regression model found include both variables the correlation coefficients between two predictors was very low (-0.038). Results are shown in Table (1)

Table-1: Coefficients of models for first data set

Term	Coef	SECoef	t-value	P-value	VIF
X1	2.20	3.30	0.67	0.524	1.00
X2	0.354	0.638	0.55	0.594	1.00
X1	2.27	3.46	0.66	0.523	1.00
X2	0.371	0.662	0.56	0.593	1.00

From the Table (1), we can see that t-value for x_1 when included alone in the model is not far from the t-value when both predictors were included, same for x_2 the t-value if not different from its value when both predictors are included. The decision of the parameter test will be same. On the other hand the standard error of coefficients have not been changed dramatically, for x_1 from 3.3 to 3.46 for simple and multiple model, and for x_2 from 0.638 to 0.662. All this can be viewed as a result of low correlation between predictors

Now for the second set of data which is showing a very high correlation between predictors (0.996) the summary of the analysis are shown in Table (2).

Table-2: Coefficients of models for first data set

Term	Coef	SECoef	t-value	P-value	VIF
X11	-0.309	0.279	-1.11	0.299	1.00
X21	-0.704	0.579	-1.22	0.258	1.00
X11	2.71	2.96	0.92	0.390	113.67
X21	-6.39	6.24	-1.02	0.340	113.67

Also, from Table 2 we can see that, there is a huge change in the coefficients values for x_{11} from (-0.309) to (2.71) in simple and multiple models. In addition to that, we can notice that the standard errors for coefficients have been increased hugely as well, for x_1 from (0.279) to (2.96) and for x_{21} from (0.579) to (6.24) when compare simple and multiple models.

The relation between predictor variables for first and second sets of data are shown below.

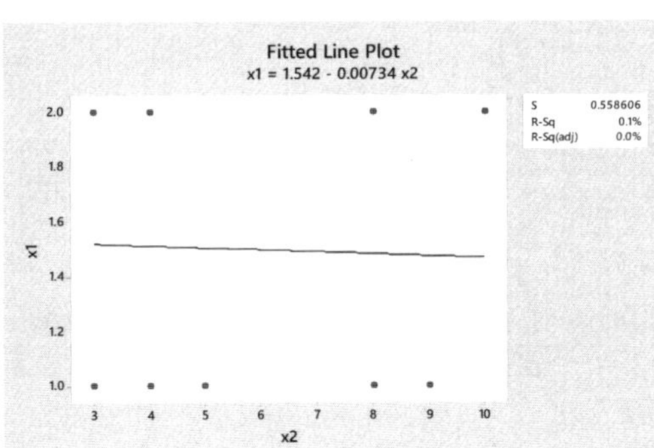

Graph-1: Predictors with low correlation

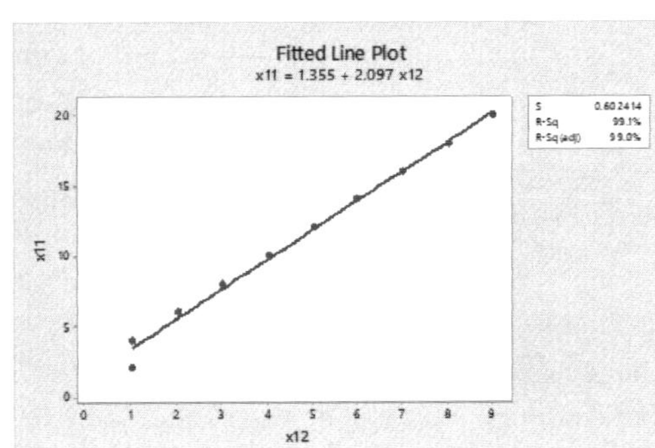

Graph-2 Predictors with high correlation

This comparison evident on the impact of correlation on the standard error of coefficients was huge in changing the standard errors of the coefficients for the second set of data, which will lead to wrong conclusions on the model. Some or all predictors will become insignificant when they should be significant because of inflation in standard error for predictors coefficients. As a summary having the high correlation among predictors will prevent the researcher capturing the most influential predictors for inclusion in the model.

IOP Conf. Series: Journal of Physics: Conf. Series **949** (2017) 012009 doi:10.1088/1742-6596/949/1/012009

3. Diagnostic of Multicollinearity

There are many signs in the analysis for the multicollinearity among which

- The correlation among predictors is large
- In case if the correlation is not calculated the following are signs of having the multicollinearity:
  i) When the predictor's coefficients vary from one to another model.
  ii) When applying t-test, the coefficient are not significant but put all together ($F - test$) for the whole model it is significant.

Relying only on correlation between pairs of predictors has limitation, the small or large value of correlation is something subjective depends on individual and also on the field of research that is why most of the time to detect the multicollinearity we use some indicator called variance inflation factors (VIF).

4. Variance Inflation Factors (VIF)

When correlation exists among predictor's the standard error of predictors coefficients will increase and consequently the variance of predictor's coefficients are inflated. The VIF is a tool to measure and quantify how much the variance is inflated. *VIFs* are usually calculated by the software as part of regression analysis and will appear in VIF column as part of the output. To interpret the value of VIF the following rule is used in the table below:

Table-3: *VIF* interpretation

VIF -value	conclusion
$VIF = 1$	Not correlated
$1 < VIF \leq 5$	Moderately correlated
$VIF > 5$	Highly correlated

In addition to the meaning of VIF itself in showing whether the predictors are correlated, the square roof of VIF indicates how much larger the standard error is, for example if $VIF = 9$ this means that the standard error for the coefficient of that predictor is 3 times as large as it would be if that predictor is uncorrelated with other predictors. VIF can be calculated using the formula:

$$VIF = \frac{1}{1 - R_i^2}$$

This VIF can be calculated for each predictor in the model, and the way is to regress the variable assume it is i^{th} variable against all other predictors. We obtain R_i^2 which can be used to find VIF , same thing can be applied to all other predictors.

Back to results of our data analysis, from Table-1, we can see that the value of variance inflation factor for x_1 was 1 for both simple and multiple models, same for x_2 unchanged and it was 1, this is due to a very low correlation for first set of data, while for second set of data, *VIF* for both variables were changed from 1 for simple model to 113.67 for multiple model. In the latter case we cannot commence with regression unless this problem is solved [3].

5. Problem solving

When two or more predictors are highly correlated, the relationship between the independent variables and the dependent variables is distorted by the very strong relationship between the independent variables, leading to the likelihood that our interpretation of relationships will be incorrect. In the worst case, if the variables are perfectly correlated, the regression cannot be computed [4].

Multicollinearity is detected by examining the tolerance for each independent variable. Tolerance is the amount of variability in one independent variable that is no explained by the other independent variables, and it is in fact $1 - R^2$ .Tolerance values less than 0.10 indicate collinearity.

If we discover collinearity in the regression output, we should reject the interpretation of the relationships as false until the issue is resolved [3].

Multicollinearity can be resolved by combining the highly correlated variables through principal component analysis, or omitting a variable from the analysis that associated with other variable(s) highly.

6. Conclusions

1. Multicollinearity is one of serious problems that should be resolved before starting the process of modeling the data.
2. It is highly recommended that all regression analysis assumption should be met as they are contributing to accurate conclusion and helps to make inference on the population.
3. Ignore and dismiss the model if the multicollinearity discovered after finding the model specially with high correlation as the model cannot be interpreted.

References

[1] Carl F. Mela* and Praveen K. Kopalle. The impact of collinearity on regression analysis: the asymmetric effect of negative and positive correlations. J. of Applied Economics, 2002, 43,667-677.
[2] D.R.Jensen and D.E. Ramirez. Variance Inflation in Regression, Advances in Decision Sciences, 2012, 2013, 1-15.
[3] Debbie J. Dupuis1 and Maria-Pia Victoria-Feser. Robust VIF regression with application to variable Selection in large data sets, The Annals of Applied Statistics, 2013, 7,319-341.
[4] George A. Milliken, Dallas E. Johnson (2002). Analysis of Messy data, Vol.3, Chapman & Hall/CRC
[5] Golberg, M. Introduction to regression analysis, Billerica, MA: Computational Mechanics Publications, 2004, 436.
[6] Jason W. Osborne and Elaine Waters (2002). Four Assumptions of Multiple Regressions that Researchers should always Test. J. of Practical Assessment, Research, and Evaluation. Vol.8, No.2, PP1-5.
[7] Kleinbaum, David G. Applied regression analysis and other multivariable methods, Australia;

Belmont, CA: Brooks/Cole, 2008,906.
[8] McClendon, McKee J. Multiple regression and causal analysis, Prospects Heights, Ill.: Waveland Press 2002,358.
[9] Seber, G. A. F. (George Arthur Frederick). Linear regression analysis, Hoboken, N.J.: Wiley-Interscience, 2003,557.

IOP Conf. Series: Journal of Physics: Conf. Series **949** (2017) 012010 doi:10.1088/1742-6596/949/1/012010

Mechanism of the free charge carrier generation in the dielectric breakdown

N A A Rahim[1], R Ranom[1] and H Zainuddin[1]

[1]Department of Electrical Engineering, Universiti Teknikal Malaysia Melaka 76100 UTeM Durian Tunggal, Melaka Bandaraya Bersejarah, MALAYSIA

E-mail: rahifa@utem.edu.my

Abstract. Many studies have been conducted to investigate the effect of environmental, mechanical and electrical stresses on insulator. However, studies on physical process of discharge phenomenon, leading to the breakdown of the insulator surface are lacking and difficult to comprehend. Therefore, this paper analysed charge carrier generation mechanism that can cause free charge carrier generation, leading toward surface discharge development. Besides, this paper developed a model of surface discharge based on the charge generation mechanism on the outdoor insulator. Nernst's Planck theory was used in order to model the behaviour of the charge carriers while Poisson's equation was used to determine the distribution of electric field on insulator surface. In the modelling of surface discharge on the outdoor insulator, electric field dependent molecular ionization was used as the charge generation mechanism. A mathematical model of the surface discharge was solved using method of line technique (MOL). The result from the mathematical model showed that the behaviour of net space charge density was correlated with the electric field distribution.

1. Introduction

Insulation is the most important part in the high voltage application to resist some mechanical, thermal, electrical and environmental stresses that act permanently or temporarily on insulators [1]. Thus, a precaution must be taken to reduce the factors of insulation failures caused by stresses. A good insulator must be able to avoid the flow of current to the undesired path.

Outdoor insulators in polluted environments are often exposed to pollution. This condition does not affect the performance of the insulators until the surface of the insulators become moisturized because of dew, rain or fog [2]. Due to the surface discharge activities, a conducting path arising from the drying out process will be formed on the surface of the insulator and it will allow the flow of leakage current from the high voltage electrode to the ground electrode [3]. Although the leakage current flow is quite small, the long term of leakage current flow due to the discharge may lead to the insulator breakdown. Thus, studies on physical discharge activities on insulator are essential to decrease system failure.

In order to understand the propagation of surface discharge on insulators, the mechanism for the charge carrier generation and transportation must be known. There are several types of mechanisms for the charge carrier generation and transportation that can lead to the breakdown of the insulators; impact ionization [4], field emission [5], secondary electron emission [6], ionic dissociation [7] and molecular ionization [8]. Besides, recombination of free charge carriers like electrons, positive ions, and negative ions with each other and the surrounding media also contributes in the charge carrier generation of the insulation. In addition to the recombination of the free charge, electrons are also combined with neutral molecules to form negative ion in the process of electron attachment [5].

The charge transport continuity equation which accounts for Nernst Planck theory is used to understand the charge carrier behavior on the insulator [9] and this equation is coupled with Poisson's equation for the field distribution equation. Therefore, studies on the mechanisms that can lead to the breakdown of the insulators are very important to prevent some failures in the high voltage system. Thus, this paper was to study the charge carrier generation mechanism that could cause the surface discharge activities. In order to investigate the behavior of the charge carrier, a model of surface discharge was developed.

2. Governing Equation

The charge transport continuity equation which accounts for Nernst Planck theory is used to model the behavior of the charge carriers like positive ions, negative ions and electrons in the insulator system. The Nernst-Planck theory has been used to model the ion transport in many physical and biological system such as protein channel of biological membranes [10], neuron cell membranes [11], and electrolytic solutions [12]. Generally, the charge transport continuity equation can be written as [5]:

$$\frac{\partial N_i}{\partial t} + \nabla . J_i = f(\vec{E}, \mu_i) \quad (1)$$

where the suffix $i = p, n, e$ indicates the species of the charge; positive ion, negative ion and electron in the dielectric system respectively. J_i $[mol\, m^{-2}\, s^{-1}]$ is a total current density for species i at distance x $[m]$ between the electrode, N_i $[mol\, m^{-3}]$ is the density of each charge carrier, and $f(\vec{E}, \mu_i)$ describes the charge generation and recombination and it is depending on the electric field $(\overrightarrow{E})$ and ion mobility (μ_i). The total current density for each charge carrier in equation (1) can be given as follows [13]:

$$J_i = -D_i \frac{\partial N_i}{\partial x} - z_i \mu_i N_i \vec{E} + N_i v \quad (2)$$

where D_i $[m^2\, s^{-1}]$, z_i and N_i $[mol\, m^{-3}]$ are the diffusion coefficient, the valence and the concentration for the species i respectively, v $[m\, s^{-1}]$ is the charge velocity, and $\frac{\partial N_i}{\partial x}$ is the concentration gradient. The first term on the right hand side of equation (2) represents Fick's law equation, the second term accounts for the migration of the species in an electric field and the last term represents the convection movement of the solution induced, density gradients or bulk movement of fluid [14]. The charge transport mechanism continuity equations are coupled with Poisson's equation to determine the electric field distribution. The equation of Poisson's equation can be expressed as follows:

$$\nabla . \left(-\varepsilon_0 \varepsilon_r \vec{E}\right) = \left(N_p - N_n - N_e\right) q N_A \quad (3)$$

where $\vec{E}$ $[V\, m^{-1}]$ is the electric field vector, ε_0 $[F\, m^{-1}]$ is permittivity of free space charge, ε_r is relative permittivity of material, $N_p, N_n, and\ N_e$ are the density of each charge carrier determined from the charge transport mechanism continuity equations (1), q $[C]$ is the elementary charge and N_A $[mol^{-1}]$ is the Avogadro's number.

3. Charge Generation

There are some mechanisms for the free charge carrier growth in insulator when it is exposed to the high electrical fields such as impact ionization, field emission, electric field dependent ionic dissociation and electric field dependent molecular ionization. Under normal condition, insulators have few free charge carriers and in normal condition. However, insulator will transform into conductors if there is a high electrical stress that may increase the free charge. Eventually, these free charge carriers will lead to the insulation breakdown. Therefore, identifying the mechanisms of the growth of free charge carriers in the insulator under high electrical stresses are important.

3.1 Impact ionization

In terms of impact ionization, electrons move away from the cathode because of the influence of the electric field. Through impact ionization, as the electron travel, they gain enough energy to liberate new electrons and positive ions. Positive ions due to the ionization drift towards cathode and then the bombardment of positive ions leads to the emission of electrons. A theory on impact ionization in insulator often comes from gas discharge theory. A new method in describing the evaluation of impact ionization coefficient α in insulators had been used by Solomo et. al., in their study of impact ionization for Silicon Dioxide SiO_2 [4]. The results showed the impact ionization and recombination as the dominant process in the breakdown events for Silicon Dioxide SiO_2. The model for ion impact ionization charge generation can be expressed as follows:

$$\alpha(\vec{E}) = A_t \exp\left(\frac{-B_t}{|\vec{E}|}\right) \quad (4)$$

where A_t is the pre-exponential coefficient with units in $[m^{-1}]$ and B_t is the exponential term in $[\frac{V}{m}]$.

Nonetheless, in the case like dense medium such as water, the value of ionization coefficients for gas cannot be used by other dielectric elements [15], causing difficulties in modelling the impact ionization for different dielectric elements such as liquid and solid. However, there are some studies in the dielectric liquids that use impact ionization as the charge generation. Haidar et. al. conducted a study on Cyclohexane and Propane liquids [16] and the results showed that the regular pulse regime of this two dielectric liquids are correspond to the electron multiplication. Obodovski et. al. showed the electron multiplication through impact ionization which is observed in xenon liquid by the occurring space charge [17].

During the early studies, breakdown in dielectric liquids concentrated on the bubble mechanism in the liquid. However, study by Atrazhev et. al. [15], showed that high mobility liquid breakdown can be defined by the electron ionization breakdown mechanism. They also concluded that the electron ionization mechanism defines upper limit of electrical strength under very short voltage pulses approximately several nanosecond (ns) compared to gas bubble formation that is similar or more than microsecond ($\geq \mu s$). In another study, Kumara et. al. investigated the surface charge effect on the impulse flashover of outdoor polymeric insulators and found, that the effective ionization coefficient depends on the electrical field strength [18].

3.2 Field emission

The field emission of electrons occurs at cathode electrodes. The field emission theory is referred as Fowler-Nordheim theory, [19] field emission of electron from metal to free space. The Fowler-Nordheim equation indicates that emission should be apparent at fields of about 10^7 V/cm or more [20]. Both Fowler and Nordheim relates the field current and the voltage gradient as shown in the equation (5). According to the physical process, the electron must have sufficient energy to escape above the top of the barrier. Therefore, the tunnelling of the electron increases exponentially with the electrical field. The basic equation of Fowler-Nordheim theory of field emission can be expressed as [19]:

$$J(\vec{E}) = \frac{e^3|\vec{E}|^2}{8\pi h\emptyset} \exp\left(-\frac{8\pi\sqrt{2m}\emptyset^{\frac{3}{2}}}{3he|\vec{E}|}\right) \quad (5)$$

where J, is the current density, e is the electron charge, $|\vec{E}|$is electric field, h is Planck's constant and $\emptyset$ is metal's work function. As described by Fowler and Nordheim [21], this field emission theory is suitable for electron emission from the metal to the free space. Thus, some modifications have been done for describing the field emission from metal to other medium like liquid [5]. A theory of electron emission from electrode into the liquid or free space is known as boundary effect theory. The source of charge generation can be accounted by setting boundary conditions and initial conditions for the charge continuity equation [5].

From the equation of Fowler-Nordheim theory, an increase in the electrical field will lead to an increase in the Fowler-Nordheim tunneling current [22]. This theory is supported by Chen who found that the mobility of charge carriers is a function of the electrical field [22]. However, Kao stated that the field emission equation of Fowler and Nordheim must have been modified to correlate with the experimental current-voltage characteristics of liquid [23]. This is because it is hard for electron to gain sufficient energy to ionize a molecule and the mean free path of an electron in a dielectric liquid in order of 10 A. Meanwhile, Nakano et. al., investigated the mechanism of surface discharge in vacuum by the charge activities on solid insulators. According to this study, explosive electron emission (EEE) on cathode is the mechanism that causes the impulse surface flashover due to extreme field emission [24].

EEE can occurs for several times due to the erosion and regeneration of the electron. However, the field emission of electrons only occurs at the negative electrode and this mechanism is only suitable for the negative streamer growth. Therefore, according to O'Sullivan, this mechanism is unimportant for the streamer development along the dielectric surface [5].

3.3 Electric field dependent ionic dissociation

Other mechanism of dielectric breakdown is electric field dependent ionic dissociation. According to Onsager theory, electric field dependent ionic dissociation is defined as the formation of ions by dissociation of neutral molecules in the presence of electric field. This theory is confirmed by the deviations from Ohm's law for solutions of weak electrolytic of liquid such as water and solid such as glass [7]. From this theory, it shows that ionic dissociation depends on the strong electric field. In addition, this theory concerns about the rates of dissociation and recombination of the ions. According to this theory, dissociation rates are dependent on the electric field because dissociation increases by the factor of $F(|\vec{E}|)$ while the recombination rates are independent to the electric field because the derivation of recombination is equivalent to the Langevin equation. The electric field dependence of dissociation constant can be expressed as:

$$K_D(\vec{E}) = K_D^0 F(\vec{E}) \tag{6}$$

where K_D^0 is the dissociation constant when electric field is zero, and $F(E)$ is constant's electric field. $F(\vec{E})$ can be expressed as:

$$F(\vec{E}) = f(b) = \frac{I_1 4(b)}{2b} \text{ where } b = \sqrt{\frac{q^3 \vec{E}}{16\pi\varepsilon k^2 T^2}} \tag{7}$$

where I_1 is the modified Bessel function, q is the electron charge, E is the magnitude of electric field, ε is the permittivity of dielectric liquid, k is Boltzmann's constant and T is the temperature. The relative concentrations of neutral ion-pairs and free ions in thermal equilibrium and no applied electric field can be expressed as:

$$cK_D^0 = n_{\pm 0}^2 K_R \tag{8}$$

Because the value of n_{+0} is equal to n_{-0}, the zero field free ion concentration can be expressed as:

$$n_{\pm 0} = \frac{\sigma}{q(\mu_+ + \mu_-)} \tag{9}$$

where σ is the conductivity of dielectric liquid. Therefore, charge generation rate based on Onsager's theory can be expressed as follows:

$$K_D(\vec{E}) = \left[\frac{\sigma}{q(\mu_+ + \mu_-)}\right]^2 K_R F(\vec{E}) \tag{10}$$

There are many studies that involve electric field dependent ionic dissociation as a charge generation mechanism. Castellanos et. al. generalized the Thomson model to include field enhanced dissociation in the bulk and field dependent unipolar charge injection [25]. Gafvert et. al., presented a physical drift-diffusion model that uses electric field dependent ionic dissociation to predict the behavior of the electric field in oil-pressboard structures [26]. In this study, the predicted behavior of the electric field is in agreement with Kerr's measurements compared to the RC-model that is strongly deviated from the measured fields. However, according to O'Sullivan et. al, the model for electric field dependent ionic dissociation does not produce the same results as the results of electrical breakdown in transformer oil [27]. This is because the electric field levels necessary for dissociation to occur can only be found in the region close to the tip of needle electrode. In addition, the low mobility of dissociated ions only leads to the formation of current density at the needle tip. Thus, a very limited thermal enhancement occurring in the dielectric liquid.

3.4 Electric field dependent molecular ionization

Another mechanism that might also be contributing to the breakdown of the dielectrics in the presence of high electric field is a molecular ionization process. Electric field dependent molecular ionization is a direct ionization mechanism that extracts electrons from neutral molecule which then generates free electrons and positive ions. Thus, Zener theory [8] of tunneling electron in solid that is correlated with molecular ionization mechanism in the presence of high electric field is used in the modelling of dielectric breakdown. The ionization rate γ of Zener theory is given as:

$$\gamma|\vec{E}| = \frac{ea|\vec{E}|}{h}\exp\left(-\frac{\pi^2 m^* a\Delta^2}{eh^2|\vec{E}|}\right) \quad (11)$$

where e is the electronic charge, a is the molecular separation, $|\vec{E}|$ is the magnitude of electric field, h is the Planck's constant, m^* is the effective electron mass, Δ is the molecular ionization potential. In order to give a charge density rate source term for inclusion in the positive ion and electron, electric field dependent molecular ionization rate must be multiplied by the ionisable charge density, N_0 [5]. Hence, the ionization rate can be expressed as:

$$G_i|\vec{E}| = \frac{e^2 N_0 a|\vec{E}|}{h}\exp\left(-\frac{\pi^2 m^* a\Delta^2}{eh^2|\vec{E}|}\right) \quad (12)$$

Charge generation mechanism by molecular ionization in the presence of high electric field has been analyzed by few researchers. Some researchers use this theory in the modelling of streamer in dielectric liquid such as in transformer oil [5][28][29]. However, this mechanism is not suitable to measure negative sharp points as stated by Becerra et. al. in the study of charge carrier generation in Cyclohexane [30]. According to this study, although this equation is suitable for the current-voltage (IV) characteristics in positive point, it is fails to predict the currents for the negative points.

3.5 Comparison of the charge generation mechanism

By studying different types of charge carrier generation, the modelling of surface discharge on the insulator surface is developed. Not all type of the charge generation mechanism has the same effect for the real cases. The summary of the charge carrier generation is shown in Table 1.

Table1: The comparison of charge generation mechanism

Charge generation	Basic equation	Theory	Description
Impact Ionization	$\alpha(\vec{E}) = A_t \exp\left(\frac{-B_t}{\lvert\vec{E}\rvert}\right)$ [4]	Gas discharge theory of breakdown. The process of bombardment of positive ions that leads to the electron emission	Not suitable for the charge generation in dense medium.
Field Emission	$J(\vec{E}) = \frac{e^3\lvert\vec{E}\rvert^2}{8\pi h\emptyset}\exp\left(-\frac{8\pi\sqrt{2m}\emptyset^{\frac{3}{2}}}{3he\lvert\vec{E}\rvert}\right)$ [19] (Boundary effect generation)	The emission of electron from the metal surface ooccurred with the high electric field greater than 5 x 10^8 V/m.	Only occurred at the negatively electrode and suitable for negative streamer growth.
Electric Field Dependent Ionic Dissociation	$K_D(\vec{E}) = \left[\frac{\sigma}{q(\mu_+ + \mu_-)}\right]^2 K_R F(\vec{E})$ [7]	Formation of positive and negative ions by dissociation of neutral molecules in the presence of high electric field.	The low mobility of dissociated ions only leads to the formation of current density at the high voltage electrode.

Electric Field Dependent Molecular Ionization	$G_i\left\|\vec{E}\right\| = \frac{e^2 N_0 a\left\|\vec{E}\right\|}{h}\exp\left(-\frac{\pi^2 m^* a\Delta^2}{eh^2\left\|\vec{E}\right\|}\right)$ [8]	Zener theory extracting the electron from neutral molecule and generating free electron and positive ions.	The fast mobility of electron leads to the electric field wave propagation from the high voltage electrode to the ground electrode.

The ionization coefficients for gasses cannot be used by another dense insulator like liquid and solid [15]. Hence, impact ionization is not suitable for the modelling of surface discharge on the outdoor insulator that involves the flow of electrolyte. Meanwhile, field emission of electrons always occurs on the negative electrode in which the development of negative streamer propagation can take place [5]. Thus, the field emission of electrons is not suitable for the modelling of surface discharge along the surface of outdoor insulator.

Electric field dependent ionic dissociation is also one of the charge carrier generation mechanisms that has been discussed earlier. From a study by Sullivan et. al., electric field dependent ionic dissociation mechanism occurs on the region near the high voltage electrode [27]. Therefore, this mechanism is not applicable to model the surface discharge formation that occurs along the surface of the outdoor insulator that involves the flowing of the electrolyte from the high voltage electrode to ground electrode. However, the electric field dependent molecular ionization mechanism has been used by some researchers in the modelling of positive streamer in the liquid insulator [5][28][29]. In their study, at good correlation between simulation result and experimental result for the streamer formation in the liquid insulator has been observed. Thus, the results showed the distribution of charge carrier and electric field along the insulator.

3.6 Recombination

The concentrations of free charge carriers can exist during the formation of surface discharge on the insulator surface and is generated by the generation mechanism discussed previously. The combination of free charge carriers like electrons, positive ions, and negative ions with each other and the surrounding media may lead toward of electron/positive ion recombination, positive/negative ion recombination and electron attachment. For the charge recombination, Langevin's relation for recombination coefficient has been used [5] [31]. In Langevin's relation the recombination coefficient is based on the diffusion equation of charge carriers in the electric field [32] and the Langevin' theory which can be expressed as follows:

$$K_{rpn} = \frac{q(\mu_p+\mu_n)}{\varepsilon\varepsilon_0} \quad (13)$$

$$K_{rpe} = \frac{q(\mu_p+\mu_e)}{\varepsilon\varepsilon_0} \quad (14)$$

where, $\mu_p, \mu_n, and\ \mu_e$ are the mobility of positive ion, negative ion and electron respectively, $K_{rpn}\ and\ K_{rpe}$ are recombination coefficient of positive/negative ion and positive ion/ electron recombination, $q, \varepsilon_0\ and\ \varepsilon$ are elementary charge, relative permittivity and permittivity of free space charge respectively. However, in the dense media like gases and non-polar liquid that have higher electron mobility, the value of recombination coefficient is deviated from the diffusion controlled rate constant [33]. In the dielectric liquid studies, some authors have used Langevin's recombination coefficient for ion-ion recombination to model ion-electron recombination [5][28]. This method has been effectively compensating the electron-ion recombination by reducing the electron mobility.

3.7 Electron attachment

In addition to the recombination theory, electrons are also combined with neutral molecules to form negative ion. This process of electron attachment, τ_a is modelled by an attachment time constant that

IOP Conf. Series: Journal of Physics: Conf. Series **949** (2017) 012010 doi:10.1088/1742-6596/949/1/012010

describes the lifetime of energetic electron in a dielectric. The electron attachment time constant is calculated simply by the electron attenuation length, electron mobility and electric field strength and can be described as follows:

$$\tau_a = \frac{\lambda_a}{\mu_e|\vec{E}|} \quad (15)$$

where, τ_a is the electron attachment time constant, λ_a is the electron attenuation length, μ_e is the electron mobility and $|\vec{E}|$is the magnitude of electric field. Becerra et. al., showed that field-dependent attachment should be considered in the estimation of electric current in the liquid phase [30]. Meanwhile, according to the studies on the streamer growth in dielectric liquid [5][28] , few authors stated that the impact of electron attachment during the streamer growth in dielectric liquids is reasonably small due to the attachment processes which have a longer time scale. Therefore, the electron attachment gives small impact on the charge carrier generated.

4. Toy Model of Surface Discharge on the Outdoor Insulator Surface

Outdoor insulators in polluted areas are often exposed to pollution. This condition does not affect the performance of the insulator until the surface of the insulator becomes moisturized because of dew, rain or fog [34]. Due to the surface discharge activities, a conducting path due to the drying out process will be formed on the surface of the insulator and it will allow the flow of leakage current from the high voltage electrode to the ground electrode [35]. Although the leakage current flow is quite small, the long term of leakage current flow due to the discharge may lead to the insulator breakdown. Thus, studies on the physical discharge activities on the insulator are quite important to decrease the system failure.

In summary, electric field dependent molecular ionization is chosen in the modelling of surface discharge on the outdoor insulator. This is due to the surface discharge activities which occur along the insulator surface when there is a flow of wet contaminant and leads to the flow of leakage current. In order to solve this problem, a toy model of surface discharge is developed by taking into account the simplification of all the parameters involved. The purpose of the toy model development is to investigate whether this model is suitable for the exact problem. For the development of surface discharge model, charge transport continuity equation accounted for the Nernst Planck theory is used to investigate the charge behaviour and Poisson's equation is used to determine the electric field distribution. For the equation of the current density in equation (2), only diffusion and migration terms are considered. This is because in the surface discharge development high electric field is given and the charge carrier is diffused from high concentration to the low concentration in the electrolyte. Thus, the diffusion and migration terms are taken as a dominant role and the convection term is neglected because no bulk movement of fluid is involved [14]. Thus, the migration-diffusion equation for the charge carrier (positive ion, negative ion and electron) generation includes electric field dependent molecular ionization, recombination, electron attachment and charge transport which can be expressed as follows:

Positive ion:

$$\frac{\partial N_p}{\partial t} + \nabla.\left(-D\nabla N_i + N_i\mu_i\vec{E}\right) = \frac{qN_0 a|\vec{E}|}{h}\exp\left(\frac{-\pi^2 m^* a\Delta^2}{qh^2|\vec{E}|}\right) - N_pN_nK_{rpn} - N_pN_eK_{rpe} \quad (16)$$

Negative ion:

$$\frac{\partial N_n}{\partial t} + \nabla.\left(-D\nabla N_i - N_i\mu_i\vec{E}\right) = \frac{N_e}{\tau_a} - N_pN_nK_{rpn} \quad (17)$$

Electron:

$$\frac{\partial N_e}{\partial t} + \nabla.\left(-D\nabla N_i - N_i\mu_i\vec{E}\right) = \frac{qN_0 a|\vec{E}|}{h}\exp\left(\frac{-\pi^2 m^* a\Delta^2}{qh^2|\vec{E}|}\right) - N_pN_eK_{rpe} - \frac{N_e}{\tau_a} \quad (18)$$

where F, R, and T are Faraday's constant $(\frac{C}{mol})$, relative gas constant $(\frac{J}{mol.K})$, and temperature (K) respectively. The charge generation term in equation (1) can be seen only in equation (16) and equation (18) because by electric field dependent molecular ionization as discussed previously, only positive ion and electron are extracted from the neutral molecule. Meanwhile, the generation term for negative ion in equation (17) is the electron attachment. The Poisson's equation can be expressed as follows:

$$\nabla.\left(-\varepsilon_0\varepsilon_r\vec{E}\right) = \left(N_p - N_n - N_e\right)qN_A \quad (19)$$

IOP Conf. Series: Journal of Physics: Conf. Series **949** (2017) 012010 doi:10.1088/1742-6596/949/1/012010

The boundary conditions that were applied for this model can be expressed as follows:

- Charge Transport Continuity Equations: The boundary equation at the high voltage electrode and ground electrode is set to zero normal flux as follows:

$$\hat{n}(\nabla N_i) = 0 \tag{20}$$

- Poisson's Equation: The high voltage electrode was set to a supply voltage V_0 at boundary of $x = 0$. This boundary was supplied with DC voltage. Meanwhile, for the ground voltage at $x = d$, it was set to zero normal electric field component.

$$x = 0; \quad \hat{n}(V) = V_0 \tag{21}$$
$$x = d; \quad \hat{n}(V_i) = 0 \tag{22}$$

In order to solve the mathematical modelling, method of line technique (MOL) is used and can be solved using the finite difference method. Thus, this finite difference method was solved using ODE15s in MATLAB software.

From the simulation of the mathematical model, the net space charge density is shown in Figure 1. Meanwhile, the electric field distribution due to the charge generation is shown in Figure 2. In the simple model, the scaling of the charge was about 10^{-2} and the scaling of the distance was order 1. The resulting charge generation from the molecular ionization and attachment, tended to move to the opposite direction of their charge. The slow mobility of the charge was essentially remaining stationary. Thus, this condition led to the formation of the net space charge density.

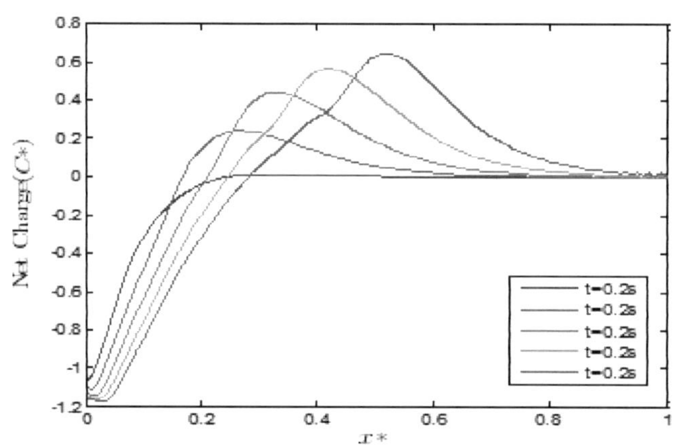

Figure 1: Net Space Charge Density

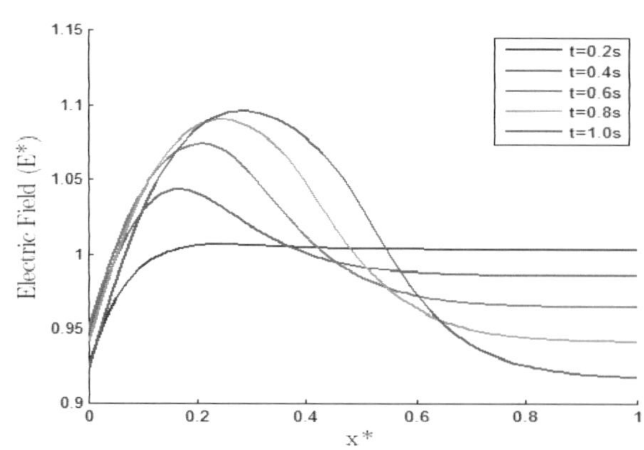

Figure 2: Electric Field Distribution

Figure 1 and Figure 2 shows, the relationship between the charge density and the electric field distribution. Similar to equation (19), the charge density influenced the electric field distribution [5]. The position of the region of net space charge density indicated the location of the peak of the electric field distribution and ionization during the surface discharge on the outdoor insulator. In summary, this simple model can be used to predict the behavior of the charge density during surface discharge.

5. Conclusion

Studies on the physical process of charge generation mechanism is important to prevent the system failure in the high voltage application. This paper provides fundamental study of the charge carrier generation mechanism that can cause surface discharge and then lead to the breakdown of the insulator. There are several types of charge generation mechanism and each mechanism is depending upon the parameters of the insulator, surrounding media and the electric field applied. Similar to previous studies, the total net space charge in the insulator influences the amount of electric field distribution on the insulator. Based on the simple model presented in this paper, the results show that the total net space charge density is influencing the electric field distribution of this model. In the next study, this model will be used to investigate the charge behavior using exact parameter values.

Acknowledgments
We wish to thank the Faculty of Electrical Engineering and Universiti Teknikal Malaysia Melaka (UTeM) for the help and support in carrying this research under the grant vote number PJP/2015/FKE(4C)/S01402.

References

[1] C. Rusu-Zagar, P. V. Notingher, and C. Stancu, 'Ageing and Degradation of Electrical Machines Insulation', *J. Int. Sci. Publ. Mater. Methods Technol.*, vol. 8, pp. 526–546, 2007.
[2] Y. Li, H. Yang, Q. Zhang, X. Yang, X. Yu, and J. Zhou, 'Pollution flashover calculation model based on characteristics of AC partial arc on top and bottom wet-polluted dielectric surfaces', *IEEE Trans. Dielectr. Electr. Insul.*, vol. 21, no. 4, pp. 1735–1746, 2014.
[3] A. Syakur, H. Berahim, Tumiran, and Rochmadi, 'Hydrophobic Contact Angle and Surface Degradation of Epoxy Resin Compound with Silicon Rubber and Silica', *Electr. Electron. Eng.*, vol. 2, no. 5, pp. 284–291, 2012.
[4] P. Solomon and N. Klein, 'Impact Ionization in Silicon Dioxide at Fields in the Breakdown Range', *Solid State Commun.*, vol. 17, pp. 1397–1400, 1975.
[5] F. M. O. Sullivan, 'A Model for the Initiation and Propagation of Electrical Streamers in Transformer Oil and Transformer Oil Based Nanofluids', Massachusetts Institute of Technology, 2007.
[6] E. M. J. Niessen, 'Numerical simulation of secondary electron emission charging at insulator surfaces', *Int. Symp. Discharges Electr. Insul. Vacuum-Eindhove*, vol. 1, no. 18, pp. 162–165, 1998.
[7] L. Onsager, 'Deviations from Ohm's Law in Weak Electrolytes', *J. Chem. Phys.*, vol. 2, pp. 599–615.
[8] C. Zener, 'A Theory of the Electrical Breakdown of Solid Dielectrics', *Proc. R. Soc.*, vol. Volume 109, no. Society, The Royal Society, Royal Sciences, Physical, pp. 523–529, 1926.
[9] H. Zainuddin and P.L. Lewin, 'Modeling of Degradation Mechanism at the Oil-Pressboard Interface due to Surface Discharge', in *COMSOL Confference*, 2015, pp. 1–22.
[10] A. Singer, 'A poisson-nernst-planck model for biological ion channels — an asymptotic analysis in a 3-d narrow funnel', *SIAM J. Appl. Math.*, vol. 70, no. 3, pp. 27–29, 2009.
[11] P. M. Nanninga, 'A computational neuron model based on Poisson – Nernst – Planck theory', *ANZIAM J. 50*, vol. 50, pp. 46–59, 2008.
[12] E. Samson and J. Marchand, 'Numerical Solution of the Extended Nernst-Planck Model', *J.*

Colloid Interface Sci., vol. 8, pp. 1–8, 1999.
[13] R. Rahifa, 'Mathematical Modelling of Lithium Ion Batteries', University of Southampton, 2014.
[14] C. G. Zoski, *Handbook of electrochemistry*, First edit., vol. 53, no. 9. New Mexico, USA: Elsivier, 2007.
[15] V. M. Atrazhev, E. G. Dmitriev, and I. T. Iakubov, 'The impact ionization and electrical breakdown strength for atomic and molecular liquids', *IEEE Trans. Electr. Insul.*, vol. 26, no. 4, pp. 586–591, 1991.
[16] M. Haidara and A. Denat, 'Electron multiplication in liquid cyclohexane and propane: An estimation of the ionization coefficcient', *IEEE Trans. Electr. Insul.*, vol. 26, no. 4, pp. 592–597, 1991.
[17] I. Obodovski, M. Kirsanov, and S. Pokachalov, 'Electron Avalanches in Liquid Xenon Induced by Space Charge', *Int. Conf. Conduct. Break. Dielectr. Liq.*, pp. 81–84, 1996.
[18] S. Kumara, Y. Serdyuk, and S. Gubanski, 'Simulation of surface charge effect on impulse flashover characteristics of outdoor polymeric insulators', *IEEE Trans. Dielectr. Electr. Insul.*, vol. 17, no. 6, pp. 1754–1763, 2010.
[19] B. R. H Fowler and L. Nordheim, 'Electron Emission in Intense Electric Fields', *Proc. R. Soc. London. Ser. A, Contain. Pap. Math. Phys. Character*, vol. 119, no. 781, pp. 173–181, 1928.
[20] J. M. Meek and J. D. Graggs, *Electrical Breakdown of Gases*, vol. 24, no. 10. London, 1953.
[21] L. W. Nordheim, 'The Effect of the Image Force on the Emission and Reflection of Electrons by Metals', *Proc. R. Soc. London*, pp. 626–639, 1928.
[22] G. Chen, 'Anomalous phenomena in solid dielectrics under high electric fields', *Proc. IEEE Int. Conf. Prop. Appl. Dielectr. Mater.*, pp. 954–960, 2009.
[23] K. C. Kao, 'Theory of High-Field Electric Conduction and Breakdown in Dielectric Liquids', *IEEE Trans. Electr. Insul.*, vol. EI-11, no. 4, pp. 121–128, 1976.
[24] Y. Nakano, H. Kojima, K. Tsuchiya, and N. Hayakawa, 'Impulse surface flashover development associated with transient charging by explosive electron emission in vacuum', *IEEE Trans. Dielectr. Electr. Insul.*, vol. 22, no. 4, pp. 2390–2397, 2015.
[25] A. Castellanos and F. Pontiga, 'Generalised Thomson-Onsager model for charge injection into dielectric liquids', *Proc. 1995 Conf. Electr. Insul. Dielectr. Phenom.*, no. 4, pp. 616–620, 1995.
[26] U. Gäfvert, O. Hjortstam, Y. Serdyuk, C. Törnkvist, and L. Walfridsson, 'Modeling and Measurements of Electric Fields in Composite Oil / Cellulose Insulation', in *Annual Report Conference on Electrical Insulation and Dielectric Phenomena*, 2006, pp. 154–157.
[27] F. O'Sullivan, L. Se-Hee, M. Zahn, L. Pettersson, R. Liu, O. Hjortstam, T. Auletta, and U. Gafvert, 'Modeling the Effect of Ionic Dissociation on Charge Transport in Transformer Oil', in *Annual Report Conference on Electrical Insulation and Dielectric Phenomena*, 2006, pp. 756–759.
[28] J.-W. G. Hwang, 'Elucidating the mechanisms behind pre-breakdown phenomena in transformer oil systems', Massachusetts Institute of Technology, 2010.
[29] N. Davari, P. O. Åstrand, M. Unge, L. E. Lundgaard, and D. Linhjell, 'Field-dependent molecular ionization and excitation energies: Implications for electrically insulating liquids', *AIP Adv.*, vol. 4, no. 3, pp. 0–13, 2014.
[30] M. Becerra and H. Frid, 'On the Modeling of the Production and Drift of Carriers in Cyclohexane', *Annu. Rep. Conf. Electr. Insul. Dielectr. Phenom.*, pp. 905–908, 2013.
[31] B. Gross, 'Charge Storage and Transport in Solid Dielectrics (The Case of Irradiated Polymers)', *Conf. Electr. Insul. Dielectr. Phenom. - Annu. Rep.*, pp. 55–70, 1978.
[32] H. Huang and J. Huang, *Organic and Hybrid Solar Cells*. New York: Springer, 2014.
[33] K. Shinsaka, T. Odaka, K. Isoda, H. Yamada, M. Ukai, N. Kouchi, and Y. Hatano, 'Electron Mobilities and Electron-ion Recombination Rate Constants in Nonpolar Solvents', *JAERI-Conference*, pp. 69–75, 1995.
[34] A. Ahmed, H. Singer, and P. K. Mukherjee, 'Numerical Model Using Surface Charges for the Calculation of Electric Fields and Leakage Currents on Polluted Insulator Surfaces', in *Annual*

Report Conference on Electrical Insulation and Dielectric Phenomena, 1998, vol. 1, pp. 116–119.

[35] A. Syakur, H. Berahim, Tumiran, and Rochmadi, 'Electrical Tracking Formation on Silane Epoxy Resin under Various Contaminants', *TELKOMNIKA (Telecommunication Comput. Electron. Control.*, vol. 11, no. 1, pp. 17–28, 2013.

IOP Conf. Series: Journal of Physics: Conf. Series **949** (2017) 012011 doi:10.1088/1742-6596/949/1/012011

An Empirical Estimation of Underground Thermal Performance for Malaysian Climate

Azfarizal Mukhtar, Mohd Zamri Yusoff, and Ng Khai Ching

Center for Fluid Dynamics, College of Engineering, Universiti Tenaga Nasional (UNITEN), Jalan IKRAM-UNITEN 43000 Kajang, Selangor.

Email: azfarizal.mukhtar@gmail.com

Abstract. In this study, the soil temperature profile was computed based on the harmonic heat transfer equations at various depths. The meteorological data ranging from January, 1st 2016 to December, 31st 2016 measured by local weather stations were employed. The findings indicted that as the soil depth increases, the temperature changes are negligible and the soil temperature is nearly equal to the mean annual air temperature. Likewise, the results have been compared with those reported by other researchers. Overall, the predicted soil temperature can be readily adopted in various engineering applications in Malaysia.

1. Introduction

Soil is not a good insulator for underground structure: the resistance of heat flow of dry earth is less than 15% of that of moisture and decreases rapidly when the moisture content increases. However, the soil is a terrific moderator for temperature fluctuations [1] in which it does not responds to daily temperature variations and only reacts gradually to seasonal changes. Due to the high thermal inertia (51% of the energy from the sun absorbed by the soil as shown in Figure 1) and the great heat capacity of the soil structure, it allows a damping of surface temperature fluctuations at a rate that varies exponentially with respect to the soil depth [2]. In other words, at a sufficient depth, the soil temperature is lower than the surface air temperature during daytime (the trend is opposite during night-time). Therefore, soil temperature is seemingly a key factor in affecting the heat transfer of an underground building and the thermal storage for new sources of renewable energy [3–7]. Therefore, the investigation of soil temperature at different depths is necessary to calculate the heat flux through the building surface. Meanwhile, the soil temperature data can be used for designing underground building in Malaysia.

Soil temperature has been investigated for different geographical locations [8], geothermal assessments [3,9] and soil depths [9–11]. Also, research work on Ground Heat Exchanger (GHE) implementation [12,13], underground building [5,6,14] and underground wine cellar [2,15] have been performed as well. According to the papers above, the research of soil temperature variation within a tropical climate region is still incipient. Alam et al. [16] applied the soil temperature variation in designing an energy-efficient green building in Malaysia. However, the climate data in China was implemented directly in Malaysia environment. Yusof et al. [4] investigated the potential of ground cooling in Malaysia by using weather data measured in Kuantan, Pahang. However, the data provided by Yusof et al. [4] was limited to the soil depth of up to 6m only, which is not readily applicable in designing underground building in Malaysia. In the present work, an equation was established to determine the soil temperature for Malaysian climate. The analysis included the investigations of high and low dry bulb temperatures, average dry bulb temperature, amplitude of surface temperature and mean annual air temperature. In addition, the phase constant (the minimum surface temperature throughout the year) was investigated as well. All analyses were performed by using the climate data provided by the Malaysian Meteorological Department for station 48684. It is worth to mention here that our proposed correlation considered the soil depth of up to 15m.

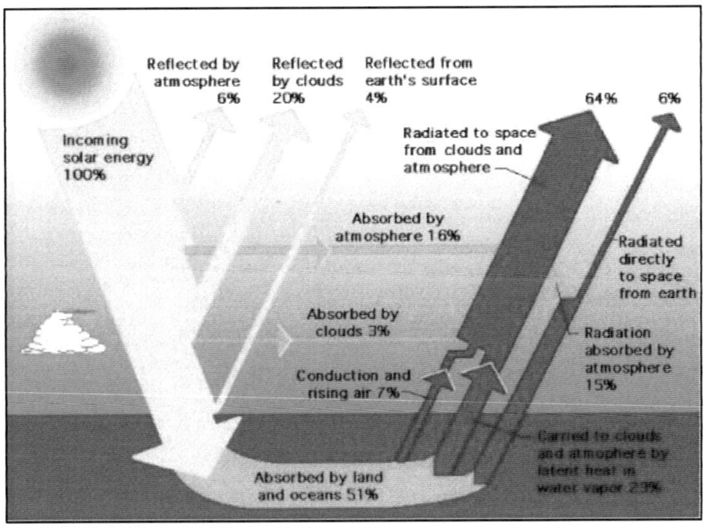

Figure 1. Earth energy balance

2. Method

2.1. Study area

Malaysia is a maritime country close to the equator which lies between 1° and 7° North latitude and 100° and 119° East longitude. It consists of west and east regions separated by the South China Sea. As a tropical country, Malaysia experiences high temperature and relative humidity, light and variable wind conditions and long hour of sunshine with rainfall occurred throughout the year. The average sunshine is around 6 hours. The daily temperature in Malaysia ranges from 24°C to 38°C. The lowest temperature is usually recorded during night-time. The mean daily humidity can be as low as 42% to as high as 94%. The annual evaporation rate in Malaysia is about 4mm–5mm per day depending on the cloud cover and the air temperature. Due to its hot and humid climate, cooler days are generally exhibiting lower evaporation rate and higher relative humidity as compared to warmer days [17]. In the present study, the soil temperature in Petaling Jaya, Selangor, Malaysia was investigated.

2.2. Numerical simulation

The soil temperature can be affected by variables such as topography, meteorology and sub-surface. Basically, the energies from meteorology elements such as solar, atmospheric agents and air are continuously transmitted to the soil surface (see Figure 1). Solar energy is the most dominant factor. In addition, daily and seasonal changes in solar energy will inflict a cyclical variation in air and soil temperatures. Other factors such as wind and rain could cause local variations as well. The typical annual temperature variation at the surface (or close to the soil surface) follows the pattern of simple harmonic equation. This harmonic equation can be modelled based on the Kasuda model [8], where the soil temperature is a function of time (day), soil depth and air temperature:

$$T(z,t) = T_{mean} - \left[\left(T_{amp} e^{-z\left(\frac{\pi}{365\alpha}\right)^{1/2}}\right)\cos\left\{\left(\frac{2\pi}{365}\right)\left((t - t_0) - \left(\frac{z}{2}\right)\left(\frac{365}{2\alpha}\right)^{1/2}\right)\right\}\right] \tag{1}$$

IOP Conf. Series: Journal of Physics: Conf. Series **949** (2017) 012011 doi:10.1088/1742-6596/949/1/012011

Here, $T(z,t)$ is the undisturbed soil temperature at time t(day) and depth z(m), T_{mean} is the mean surface temperature, T_{amp} is the amplitude of surface temperature, α is the thermal diffusivity of the soil (m^2/day) and t_0 is the day of the year with minimum surface temperature. T_{mean}, T_{amp} and t_0 can be determined from the meteorology data. However, α should be determined based on the physical soil characteristic such as soil type and moisture content.

2.3. Data Analyses

In order to apply Eq.(1) for Malaysia climate, the data provided by the Malaysian Meteorological Department was used. The data has been generated hourly starting from 0001H 1st January 2016 until 2359H 31st December 2016. Table 1 shows the monthly average temperature (derived from the hourly temperature).

Table 1. Climate Data for Petaling Jaya

Month	Jan	Feb	Mar	Apr	May	June	Jul	Aug	Sept	Oct	Nov	Dec	Yearly
Record High Dry Bulb Temp (°C)	35.90	35.70	36.20	36.90	36.20	35.70	35.80	34.80	34.80	34.20	34.10	35.10	36.90
Record Low Dry Bulb Temp (°C)	23.80	25.00	23.70	24.60	23.40	23.40	27.41	23.80	28.01	23.60	23.30	22.50	22.50
Ave Dry Bulb Temp (°C)	29.13	29.19	29.99	30.13	29.23	28.96	28.59	29.51	28.84	28.70	27.51	27.67	28.95
Amplitude of Surface Temp (°C)	5.99	6.18	5.82	7.20	6.39	7.49	7.92	4.92	6.56	6.10	5.64	7.50	6.48

From Table 1, the lowest and highest temperatures are 22.50°C and 36.90°C, respectively. The temperature amplitude is 6.48°C and t_0 is taken from the basis of daily average temperature as shown in Figure 2. Meanwhile, the mean annual air temperature is 28.95°C. The thermal diffusivity, α is 0.06 m^2/day with R^2 value of 0.9999 [4]. Substituting the above-mentioned values into Eq. (1), the soil temperature can be predicted as:

$$T(z,t) = 28.95 - 6.48\, e^{\left(-z \times \left[\frac{\pi}{366\alpha}\right]^{0.5}\right)} \cos\left[\frac{2\pi}{366}(t-169) - z \times \left[\frac{\pi}{366\alpha}\right]^{0.5}\right] \quad (2)$$

The soil temperature, which is generated from MATLAB software, is applicable up to soil depth of 15m. The results are shown in Figure 3.

3. Result and Discussion

3.1. Soil temperature measurement

Figure 3 shows the time history of temperature distribution for various depths calculated from Eq. (2). It seems that the soil surface temperature remains in phase with that of the daily temperature in Figure 2. However, at the soil depth of approximately 1m, the temperature fluctuation, which is previously observed near the soil surface, is diminishing. At the soil depth of 5m, the maximum soil temperature condition occurs during the first 6 months of year 2016. However, the temperature is almost constant for soil depth of 10 m and beyond, and the soil temperature is nearly equal to the mean annual air temperature of 28.95°C. It is believed that soil which is high in thermal inertia would induce temperature fluctuation. Meanwhile, factors such as time lag and temperature fluctuation between the surface and the soil could cause the soil to become isothermal after it reaches certain depth. These findings are consistent with those reported by previous researchers [3–6].

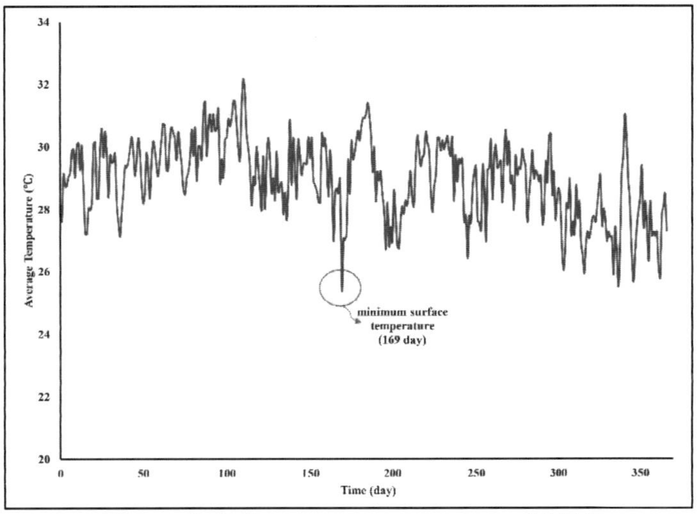

Figure 2. Daily average temperature of 2016

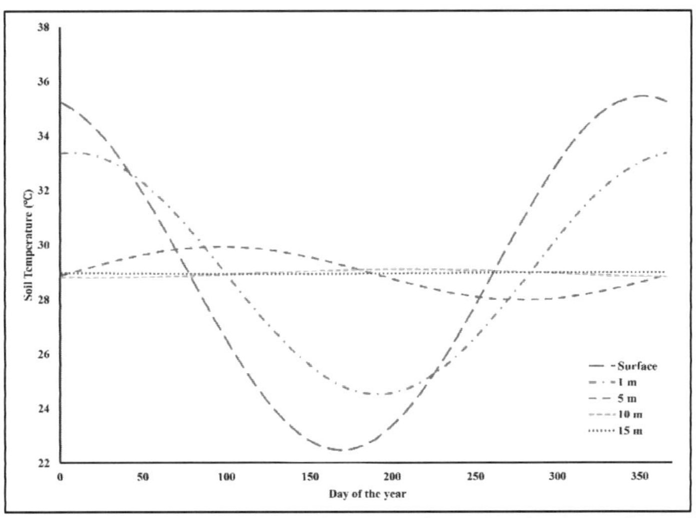

Figure 3. Soil temperature distribution at various depths throughout the year

3.2. Comparative analysis of other literature

The result has been compared with those reported by [4,18]. The meteorological data was analysed according to their respective local weather conditions. The soil depth and the thermal diffusivity were 4 m and 0.05 m^2/day, respectively. Table 2 indicates the values for all variables in Eq. (2). Figure 4 compares the results obtained from the present study and other references. The analysis was conducted using MATLAB software. From Table 2, the present mean annual air temperature is slightly higher than the others. Ref [18] reports a greater value of temperature amplitude. As shown in Figure 4, the harmonic pattern predicted from the present study is completely different from those reported by [4,18], even though the local weather conditions reported in Ref. [4] are quite similar to those considered in the present study. The only discrepancy between Ref. [4] and the present study was the measurement location of weather data. However, all results show that as soil depth increases, the soil temperature is converging to the annual mean air temperature. To the best of the authors' knowledge, it is believed that the phase constant plays a major role in producing the harmonic pattern of heat transfer at various depths. In addition, thermal diffusivity also plays an important role in the annual soil temperature variation.

Table 2. Variables data for each case

Variables	Al-Ajmi et al. [18]	Yusof et al. [4]	Present Study
T_{mean} (°C)	27	26.9	28.95
T_{amp} (°C)	13.3	5.7	6.48
t_0 (day)	16	365	169
α (m^2/day)	0.05	0.05	0.05
Location	Kuwait	Malaysia	Malaysia

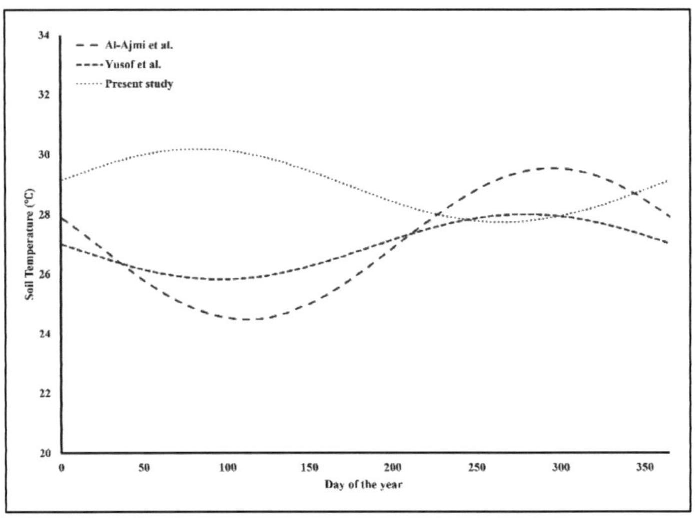

Figure 4. Comparison analysis of soil temperature

4. Conclusion

In this study, the soil temperature has been determined from the Malaysia meteorology data. An empirical equation has been proposed. At a depth of 10 m and beyond, the soil temperature is constant which is equal to the mean annual air temperature. The present results can then be used to calculate the heat flux through the underground building structure in Malaysia. Subsequently, the thermal performance in an underground building can then be determined accurately.

Acknowledgment

The author gratefully acknowledges the financial support given by Yayasan Tenaga Nasional (YTN).

References

[1] Anselm AJ. Earth Shelters; A Review of Energy Conservation Properties in Earth Sheltered Housing. Energy Conserv., InTech; 2012.

[2] Porras-amores C, Mazarron FR, Cid-falceto J, Cañas I. Energy efficient underground construction: natural ventilation during hot periods. Int. Conf. Pet. Sustain. Dev. IACSIT Press, vol. 26, 2011, p. 32–6.

[3] Márquez JMA, Bohórquez MÁM, Melgar SG. Ground Thermal Diffusivity Calculation by Direct Soil Temperature Measurement. Application to very Low Enthalpy Geothermal Energy Systems. Sensors 2016;16. doi:10.3390/s16030306.
[4] Yusof TM, Anuar S, Ibrahim H. Numerical Investigation of Ground Cooling Potential for Malaysian Climate. Int J Automot Mech Eng 2014;10:2081–90. doi:10.15282/ijame.10.2014.24.0175.
[5] Hazbei M, Nematollahi O, Behnia M, Adib Z. Reduction of Energy Consumption using Passive Architecture in Hot and Humid Climates. Tunn Undergr Sp Technol 2015;47:16–27. doi:10.1016/j.tust.2014.12.001.
[6] Moradi H, Eskandari H. An experimental and numerical investigation of Shovadan heating and cooling operation. Renew Energy 2012;48:364–8. doi: 10.1016/j.renene.2012.05.016.
[7] Mukhtar A, Ng KC, Yusoff MZ. Optimal Design of Opening Ventilation Shaft by Kriging Metamodel Assisted Multi-objective Genetic Algorithm. Int J Model Optim 2017;7:92–7. doi:10.7763/IJMO.2017.V7.565.
[8] Kusuda T, Achenbach PR. Earth Temperatures and Thermal Diffusivity at Selected Stations in the United States. ASHRAE Trans 1965:61–74.
[9] Florides G, Kalogirou S. Measurements of Ground Temperature at Various Depths. Proc. 3rd Int. Conf. Sustain. Energy Technol. Nottingham, UK, 2004.
[10] Florides G, Kalogirou S. Annual ground temperature measurements at various depths. 8th REHVA World Congr., 2005.
[11] Chow TT, Long H, Mok HY, Li KW. Estimation of soil temperature profile in Hong Kong from climatic variables. Energy Build 2011;43:3568–75. doi:10.1016/j.enbuild.2011.09.026.
[12] Mihalakakou G, Santamouris M, Asimakopoulos D. Modelling the thermal performance of earth-to-air heat exchangers. Sol Energy 1994;53:301–5. doi:10.1016/0038-092X(94)90636-X.
[13] Pfafferott J. Evaluation of earth-to-air heat exchangers with a standardised method to calculate energy efficiency. Energy Build 2003;35:971–83. doi:10.1016/S0378-7788(03)00055-0.
[14] Mukhtar A, Ng KC, Yusoff MZ. Design optimization for ventilation shafts of naturally-ventilated underground shelters for improvement of ventilation rate and thermal comfort. Renew Energy 2017. doi:10.1016/j.renene.2017.08.051.
[15] Mazarrón FR, Cañas I. Seasonal analysis of the thermal behaviour of traditional underground wine cellars in Spain. Renew Energy 2009;34:2484–92. doi:10.1016/j.renene.2009.03.002.
[16] Alam MR, Zain MFM, Kaish ABMA. Energy Efficient Green Building Based on Geo Cooling System in Sustainable Construction of Malaysia. Int J Sustain Constr Eng Technol 2012;3:96–105.
[17] Dahlan ND, Jones PJ, Alexander DK, Salleh E, Dixon D. Field Measurement and Subjects' Votes Assessment on Thermal Comfort in High-rise Hostels in Malaysia. Indoor Built Environ 2008;17:334–45. doi:10.1177/1420326X08094585.
[18] Al-Ajmi F, Loveday DL, Hanby VI. The cooling potential of earth-air heat exchangers for domestic buildings in a desert climate. Build Environ 2006;41:235–44. doi:10.1016/j.buildenv.2005.01.027.

IOP Conf. Series: Journal of Physics: Conf. Series **949** (2017) 012012 doi:10.1088/1742-6596/949/1/012012

A Response Surface Methodology for Mitigating Hot Gasses in Enclosed Car Park

Ahmad Faiz Tharima, Mohd Zamri Yusoff, Md Mujibur Rahman

College of Engineering, Universiti Tenaga Nasional (UNITEN), Putrajaya Campus, Jalan IKRAM-UNITEN, 43000 Kajang, Selangor, Malaysia.

Email: pait.afz@gmail.com

Abstract A hot gas rise towards ceiling due to fire buoyancy will cause severe damage to the building structure. The temperature rises need to be controlled as among the elements of compliance in performance-based design. The channel flow between beams has used in this study to mitigate hot gases out of the enclosure by mean of response surface methodology. Fire Dynamic Simulator was employed as a simulation tool while the result was statistically examined using analysis of variance via Minitab application. It was found that the result was linear with predicted R^2 (93.25%) and within the permissible R^2 (98.13%). The ceiling height has been identified not affect in controlling hot gases while four control parameters which are beam spacing, transversal beam, extraction rate and longitudinal beam with p-values of 0.00, 0.000, 0.023 and 0.000 respectively, have been found to have the significant effect on the smoke temperature control. This study contributes a good input to the fire safety community in providing the initial design of enclosed car park with better condition.

1 Introduction

The temperatures rise need to be controlled as among the elements of compliance in performance-based design as well as in perceptive code. A hot gas rise towards ceiling due to fire buoyancy will reduce oxygen concentration eventually create harmful to an occupants [1] and cause severe damage to the building structure such as enclosed car park [2], subway station [3] corridor -like structures [4], tunnel-like corridor [5] and underground shelter [6] as well as others building.

The fire temperatures measurement is essential to predict ignition of the object, the onset of flashover and structure damage [7]. Apart of that, it is also beneficial for predict smoke layer descent towards the floor [8] and arranges for a smoke detection [9–11] as well as sprinkler activation [12]. Following that, previous researchers had developed various ratio from experimental and numerical simulation such as beam depth to the ceiling height [13–15], radial distance to the ceiling height [16], beam spacing to ceiling clearance [15]. Most of it is to investigate the effect of the ceiling beam against flow visualization, temperature and velocity profile. In addition, it also can identify subcritical flow or density jump before and after ceiling jet intersect the ceiling beam.

However, according to the literature [11–15], most studies were conducted by means of trial and error based. These arrangements have not yet statistically proofed and which parameters show the significant effect of the temperature. In addition, the parameters that were investigated such as heat release rate (HRR) and the wind were categorized as non-control factors. Indeed, a few research papers related to the hot smoke control with the presence of the beam has been reported by [13–15,19,22–27] but still lacking regarding cost operational reducing (i.e in term of horizontal ventilation numbers and volume flow rate). Therefore, this study aims to channel the hot gasses by means statistical analysis based on established control parameters and resulting in the optimum operational cost of the ventilation fan.

2 Method

The research procedures were designed in the following steps [29]:

(a) identify the significant controllable factors
(b) performing the Design of Experiment (DOE)
(c) performing Computational Fluid Dynamics (CFD) simulation
(d) conducting the reliability test of DOE
(e) performing statistical analysis

2.1 Controllable Factors

The key factors that influence the temperature was identified which are ceiling height, beam span length, transversal beam depth, longitudinal beam depth and extraction fan rate [13–16]. The range of these factors is reported in Table 1. According to a general rule of Reynolds number, the ceiling height range from the literature review was changed to 0.3m due to the smallest height that supports turbulent flow is 0.3m [30,31]. The other amendment made within this range is beam span length; 0.556m to 0.57m in accordance to actual geometry size in Simulator Building at Fire and Rescue Malaysia Academy.

Table 1. Factor and response parameters

Parameters	Name	Coded Factor	Lower	Upper
Factors	Ceiling Height,	X	0.3m	0.442m
	Beam Span Length,	X_1	0.213m	0.57m
	Transversal Beam Depth,	X_2	0.02m	0.06m
	Extraction Fan Rate	X_3	$0.18m^3/s$	$0.31m^3/s$
	Longitudinal Beam Depth	X_4	0.02m	0.061m
Response	Temperature	Y		

2.2 Design of Experiment

The selected DOE was Central Composite Design (CCD) because it incorporates better design points. Correspondingly, the Face Central Design was employed to obtain the 32 design points. In order to maintain a hierarchical model at each step, terms were added during the process by using the stepwise procedure. The design points and their corresponding results are reported in Table 2.

IOP Conf. Series: Journal of Physics: Conf. Series **949** (2017) 012012 doi:10.1088/1742-6596/949/1/012012

Table 2. Numerical simulation design and results

Run	Factors					Temperature
	X	X_1	X_2	X_3	X_4	Y
1	0.442	0.213	0.02	0.31	0.061	315.97
2	0.371	0.3915	0.04	0.245	0.02	185.51
3	0.3	0.57	0.06	0.18	0.061	163.45
4	0.3	0.57	0.06	0.31	0.02	115.67
5	0.3	0.213	0.02	0.31	0.02	253.63
6	0.371	0.3915	0.04	0.245	0.0405	215.69
7	0.371	0.3915	0.04	0.245	0.0405	203.70
8	0.3	0.57	0.02	0.18	0.02	181.76
9	0.442	0.213	0.06	0.18	0.061	216.17
10	0.371	0.3915	0.04	0.18	0.0405	212.89
11	0.371	0.213	0.04	0.245	0.0405	231.64
12	0.371	0.3915	0.04	0.245	0.061	225.11
13	0.442	0.57	0.06	0.18	0.02	162.44
14	0.3	0.3915	0.04	0.245	0.0405	215.46
15	0.371	0.3915	0.04	0.245	0.0405	215.61
16	0.371	0.3915	0.04	0.31	0.0405	201.17
17	0.442	0.213	0.02	0.18	0.02	223.52
18	0.371	0.57	0.04	0.245	0.0405	179.82
19	0.371	0.3915	0.02	0.245	0.0405	262.6
20	0.3	0.213	0.06	0.31	0.061	265.12
21	0.3	0.213	0.02	0.18	0.061	398.07
22	0.371	0.3915	0.04	0.245	0.0405	197.79
23	0.371	0.3915	0.04	0.245	0.0405	211.11
24	0.442	0.213	0.06	0.31	0.02	181.86
25	0.442	0.57	0.06	0.31	0.061	185.51
26	0.3	0.213	0.06	0.18	0.02	125.71
27	0.442	0.57	0.02	0.31	0.02	206.29
28	0.371	0.3915	0.04	0.245	0.0405	205.38
29	0.442	0.57	0.02	0.18	0.061	223.92
30	0.371	0.3915	0.06	0.245	0.0405	193.45
31	0.3	0.57	0.02	0.31	0.061	278.27
32	0.442	0.3915	0.04	0.245	0.0405	209.38

2.3 CFD Simulation

The computational fluid dynamics (CFD) simulation was performed by using FDS software, which is specialized software in modelling fire-driven fluid flow. Flow turbulence was modelled via Large Eddy Simulation. Table 3 shows the numerical settings of the simulation.

IOP Conf. Series: Journal of Physics: Conf. Series **949** (2017) 012012 doi:10.1088/1742-6596/949/1/012012

Table 3. Numerical setting for the simulation

Parameter	Description of Car Park
Geometry dimension	4m x 1.6m x 0.3m
Mesh size	0.0094m
HRRPUA	2842.7kW/m^2
Fuel	Propane (C_3H_8)
CO yield	0.005
Soot yield	0.024
Fire source area	0.11684m x 0.0762m

2.3.1 Boundary conditions

- The ceiling, floor and side walls were prescribed as inert boundaries.
- The surrounding environment was not modelled. The ambient temperature was simply prescribed as 30°C.
- The wind effects were not considered.
- Two longitudinal beams placed at the center of car park were supported by transversal beams and columns of different sizes.
- The smoke extraction rate was specified at the downstream opening which was positioned below the transversal beam depth.
- The size of fuel source area was 0.11684m x 0.0762m with height of 0.02667 m.

2.3.2 Conservation Equation

The mass (1) momentum (2) and energy (3) conservation equations can be written as:

$$\frac{\partial \rho}{\partial t} + \nabla . \rho \mathrm{u} = \dot{\mathrm{m}}_{\mathrm{b}}''' \quad (1)$$

$$\frac{\partial}{\partial t}(\rho \mathrm{u}) + \nabla \cdot \rho \mathrm{uu} + \nabla p = \rho g + \mathrm{f_b} + \nabla \cdot \tau_{\mathrm{ij}} \quad (2)$$

$$\frac{\partial}{\partial t}(\rho \mathrm{h}) + \nabla \cdot \rho \mathrm{hu} = \frac{\mathrm{Dp}}{\mathrm{Dt}} + \dot{\mathrm{q}}''' - \dot{\mathrm{q}}_{\mathrm{b}}''' - \nabla \cdot \mathrm{q}''' \quad (3)$$

In these equations, ρ is density, t is time, **u** is velocity vector, $\dot{\mathrm{m}}_{\mathrm{b}}'''$ is net heat flux from thermal conduction and radiation, p is pressure, $\boldsymbol{g}$ is gravity vector, $\mathbf{f}_{\mathrm{b}}$ is external force vector, τ_{ij} is viscous stress tensor, $\mathrm{q}^{i'''}$ is heat release rate per unit volume from a chemical reaction, $\dot{\mathrm{q}}_{\mathrm{b}}'''$ is energy transferred to the evaporating droplets, $\dot{\mathbf{q}}''$ is conductive and radiative heat flux and ε is dissipation rate.

2.4 Reliability of DOE

The reliability of the DOE model was performed using the replication procedure which requires six models. As shown in Table 4, the results are ranging from 0.52% to 4.44%, indicating that the model is reliable.

IOP Conf. Series: Journal of Physics: Conf. Series **949** (2017) 012012 doi:10.1088/1742-6596/949/1/012012

Table 4. Replication of an enclosed car park model

Factors					Response	Error
Ceiling Height	Beam Span Length	Transversal Beam Depth	Extraction Rate	Longitudinal Beam Depth	Temperature	
0.371	0.3915	0.04	0.245	0.0405	205.11	0.65
0.371	0.3915	0.04	0.245	0.0405	203.7	1.33
0.371	0.3915	0.04	0.245	0.0405	215.61	4.44
0.371	0.3915	0.04	0.245	0.0405	197.79	4.19
0.371	0.3915	0.04	0.245	0.0405	211.11	2.26
0.371	0.3915	0.04	0.245	0.0405	205.38	0.52

2.5 Statistical Analysis

The form of mathematical model is as follows:

$$y = \propto_0 + \sum_{i=1}^{3} \propto_i x_i + \sum_{i=1}^{3} \propto_{ii} x_i^2 + \sum_{i=1}^{2} \sum_{j=i+1}^{3} \propto_{ij} x_i x_j \tag{4}$$

Where y is the predicted response (hot gases temperature); x_i and x_j are the uncoded independent variables and $\propto_0, \propto_i, \propto_{ii}$ *and* $\propto_{ij}$ are intercept, linear, quadratic and interaction constant coefficient respectively. Minitab was used for regression analysis and analysis of variance (ANOVA).

3 Results

3.1 Mesh independence Test and Geometry Validation

This model has been verified using a grid independence test and validated using a relative error percentage [32]. The grid independence study showed the finer mesh count was sufficient to conduct a further simulation. Moreover, the finest mesh was resulting in the lowest error with only 4.33%. The readers are referred to [33] for a complete description of the numerical experiment. Besides, a geometric validation also was performed between an enclosed car park and corridor-like structures experimented by Ji et al., (2015). According to Figure 1, it shows a similar pattern and indicating the enclosed car park design is good agreement with corridor-like structures simulation.

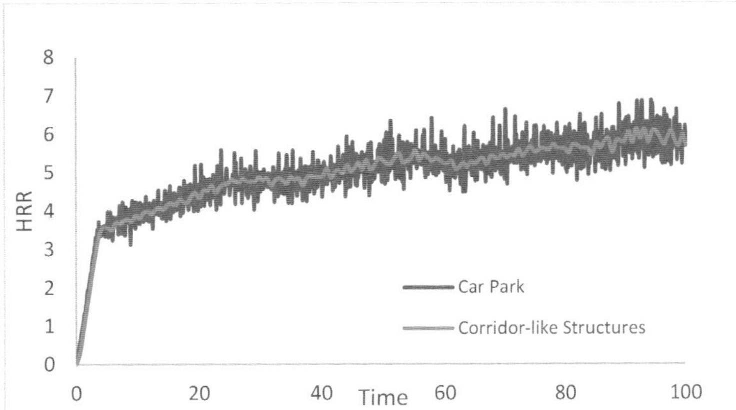

Figure 1. HRR for corridor-like structures fire and car park

3.2 Regression model

The relationship between temperature and five controllable factors (namely ceiling height X, beam span length X_1, transversal beam depth X_2, longitudinal beam depth X_3 and extraction rate X_4) was studied. The simulation result based on CCD model has developed a full quadratic equation as follows: -

$$y = 360.9 - 265X - 365.0X_1 - 8401X_2 + 81.7X_3 + 8032X_4 + 31285X_2^2 + 710XX_1 + 9598XX_2 - 11194XX_4 + 2436X_1X_2 - 3584X_1X_4 - 16229X_2X_4 \quad (5)$$

Based on statistical analysis, the result was linear with predicted R^2 (93.25%) and within the permissible R^2 (98.13%). It was found that the result was linear and good agreement can be seen between actual and predicted values.

3.3 ANOVA

Table 5 shows the linear, interaction and quadratic variables for the coded coefficient. In ANOVA analysis, the terms that found statistically significant only will be included in the model. Each variable with P-value less than 0.01 is considered highly significant, and between 0.01 and 0.05 is significant. The variable with P-value more than 0.05 is considered non-significant [34]. In the present work, it is observed that most of the variables have a highly significant on the linear effect, interaction and second order form. Only the ceiling height term in the linear effect was not significant. This illustrates the importance of employing the significant variables in design lower temperature in the enclosed car park.

Table 5. Coded Coefficients for Transformed Response

Coded Factor	Effect	Coef	SE Coef	T-Value	P-Value	Degree of Importance
X	-8.01	-4.00	2.15	-1.87	0.078	Non- significant
X_1	-57.17	-28.59	2.15	-13.32	0.000	High significant
X_2	-81.63	-40.81	2.15	-19.01	0.000	High significant
X_3	10.62	5.31	2.15	2.47	0.023	significant
X_4	70.58	35.29	2.15	16.44	0.000	High significant
X_2^2	25.03	12.51	3.25	3.86	0.001	Significant
XX_1	18.00	9.00	2.28	3.95	0.001	High significant
XX_2	27.26	13.63	2.28	5.99	0.000	High significant
XX_4	-32.58	-16.29	2.28	-7.16	0.000	High significant
X_1X_2	17.40	8.70	2.28	3.82	0.001	High significant
X_1X_4	-28.20	-14.10	2.28	-6.19	0.000	High significant
X_2X_4	-13.31	-6.65	2.28	- 2.92	0.009	High significant

Finally, based on P-value discussed above and significant effect, only the ceiling height term in the linear effect was not significant and should be removed. Thus, the model can be summarized as the following equation:

$$y = 360.9 - 365.0X_1 - 8401X_2 + 81.7X_3 + 8032X_4 + 31285X_2^2 + 710XX_1 + 9598XX_2 - 11194XX_4 + 2436X_1X_2 - 3584X_1X_4 - 16229X_2X_4 \quad (6)$$

4 Conclusion

In this research, the influence hot gasses temperature is investigated with identified factors such as ceiling height, beam span length, transversal and longitudinal beam depth, as well as extraction rate. For that purposes, the Response Surface Methodology (RSM) and Fire Dynamic Simulator (FDS) were used as a tool. The reliability result showed RSM can be used to investigate the controllable probability factors that influenced the response. In accordance to that, four factors have confirmed effect temperature in an enclosed car park. With only important parameters were included in this study, it is considered novelty in yielding lower temperature for the overall of car park geometry. The engineers afterwards could only have considered the high significant factors mentioned above in their design as compared to others factors studied in the previous study.

Acknowledgements:
This study work is supported by the scholarship provided by Malaysia Public Service Department to the first author.

References

[1] X.G. Zhang, Y.C. Guo, C.K. Chan, W.Y. Lin, Numerical simulations on fire spread and smoke movement in an underground car park, Build. Environ. 42 (2007) 3466–3475. doi:10.1016/j.buildenv.2006.11.002.
[2] M. Zahirasri, Multiple Vehicle Design Fire Scenarios in Car Parking Buildings by, (2015).
[3] N. Meng, L. Hu, S. Zhu, L. Yang, Effect of smoke screen height on smoke flow temperature profile beneath platform ceiling of subway station: An experimental investigation and scaling correlation, Tunn. Undergr. Sp. Technol. 43 (2014) 204–212. doi:10.1016/j.tust.2014.05.008.
[4] J. Ji, Y.Y. Fu, C.G. Fan, Z.H. Gao, K.Y. Li, An experimental investigation on thermal characteristics of sidewall fires in corridor-like structures with varying width, Int. J. Heat Mass Transf. 84 (2015) 562–570. doi:10.1016/j.ijheatmasstransfer.2015.01.020.
[5] T. Beji, S. Ukleja, J. Zhang., M.A. Delichatsios, Fire behaviour and external flames in corridor and tunnel-like enclosures, FIRE Mater. (2012). doi:10.1002/fam.
[6] A. Mukhtar, N.K. Ching, M.Z. Yusoff, Optimal Design of Opening Ventilation Shaft by

Kriging Metamodel Assisted Multi-objective Genetic Algorithm, 7 (2017). doi:10.7763/IJMO.2017.V7.565.
[7] W.D.W. and P.H. Thomas, Estimating Temperatures in Compartment Fires, in: SFPE Handb. Fire Prot. Eng., 2nd ed., 1995.
[8] E.K. Budnick, D.D. Evans, H.E. Nelson, Simplified Fire Growth Calculations, in: Fire Prot. Handb., 19th ed., 2003.
[9] M. He, Y. Jiang, J. Vythoulkas, T. AI-Farra, Smoke Detector Performance For Level Ceiling With Deep Beams and Deep Beam Pocket Configuration, Igarss 2014. (2014) 1–5. doi:10.1007/s13398-014-0173-7.2.
[10] F. J, R. S, G. D, Evaluating Smoke Detector Spacing Requirements for Parallel Beamed Hallways and Sloped Ceilings, Igarss 2014. (2014) 1–5. doi:10.1007/s13398-014-0173-7.2.
[11] F. G.P, D. W.D, K. J.H, Simulating the Effect of Beamed Ceilings on Smoke Flow , Part I . Comparison of Numerical and Experimental Results, (1992).
[12] R. Vettori, Effect of a Beamed, Sloped, and Sloped Beamed Ceilings on the Activation Time of a Residential Sprinkler, (2003).
[13] C.C. Koslowski, V. Motevalli, Effect of Beams on Ceiling Jet Behavior and Heat Detector Operation, J. Fire Prot. Eng. 5 (1993) 97–111. doi:10.1177/104239159300500302.
[14] C. Koslowski, V. Motevalli, Behavior of a 2-Dimensional Ceiling Jet Flow : A Beamed Ceiling Configuration, (1994) 469–480.
[15] Delichatsios, The Flow of Fire Gases Under a Beamed Ceiling, 10 (1981).
[16] N. Johansson, J. Wahlqvist, P. Van Hees, Numerical experiments in fire science : a study of ceiling jets, (2014). doi:10.1002/fam.
[17] N. Johansson, Numerical experiments and compartment fires, Fire Sci. Rev. 3 (2014) 2. doi:10.1186/s40038-014-0002-2.
[18] X. Deckers, S. Haga, N. Tilley, B. Merci, Smoke control in case of fire in a large car park: CFD simulations of full-scale configurations, Fire Saf. J. 57 (2013) 22–34. doi:10.1016/j.firesaf.2012.02.005.
[19] X. Deckers, S. Haga, B. Sette, B. Merci, Smoke control in case of fire in a large car park : Full-scale experiments, Fire Saf. J. 57 (2013) 11–21. doi:10.1016/j.firesaf.2012.10.017.
[20] C.-H. Hwang, a. Lock, M. Bundy, E. Johnsson, Studies on Fire Characteristics in Over- and Underventilated Full-scale Compartments, J. Fire Sci. 28 (2010) 459–486. doi:10.1177/0734904110363106.
[21] L.H. Hu, L.F. Chen, W. Tang, A global model on temperature profile of buoyant ceiling gas flow in a channel with combining mass and heat loss due to ceiling extraction and longitudinal forced air flow, Int. J. Heat Mass Transf. 79 (2014) 885–892. doi:10.1016/j.ijheatmasstransfer.2014.08.045.
[22] Deckers, Haga, Tilley, Merci, Smoke control in case of fire in a large car park: CFD simulations of full-scale configurations, Fire Saf. J. 57 (2013) 22–34. doi:10.1016/j.firesaf.2012.02.005.
[23] M. Deckers, Haga, Tilley, Smoke control in case of fire in a large car park: Full-scale experiments, Fire Saf. J. 57 (2013) 22–34. doi:10.1016/j.firesaf.2012.02.005.
[24] V. Motevalli, Z.P. Yuan, Steady State Ceiling Jet Behavior under an Unconfined Ceiling with Beams, Fire Technol. 44 (2008) 97–112. doi:10.1007/s10694-007-0027-3.
[25] Hinkley, Smoke & Heat Venting, in: SFPE Handb. Fire Prot. Eng., 2nd ed., 1995.
[26] J.H. Klote, J.A. Milke, Principles of smoke management, in: Princ. Smoke Manag., 2002.
[27] J.H. Klote, The ASHRAE design manual for smoke control, Fire Saf. J. 7 (1984) 93–98. doi:10.1016/0379-7112(84)90012-2.
[28] F.D. Simulations, Guide Guide for the Verification and Validation of Computational Fluid Dynamics Simulations, 1998 (2002).
[29] V. Gunaraj, N. Murugan, Application of response surface methodology for predicting weld bead quality in submerged arc welding of pipes, J. Mater. Process. Technol. 88 (1999) 266–

275. doi:10.1016/S0924-0136(98)00405-1.
[30] J.G. Quintiere, M.E. Dillon, Scale Model Reconstruction of Fire in an Atrium, Prog. Scale Model. (2008) 109–120.
[31] J.H. Klote, Scaling Modeling, in: Handb. Smoke Control Eng., 2012: p. xxiii, 484 .
[32] A.P. Smardz, Validation of Fire Dynamics Simulator (FDS) for forced and natural convection flows, 2006 (2006).
[33] A. Faiz Tharima, M. Mujibur Rahman, M. Zamri Yusoff, Verification and Validation of Fire Dynamic Simulator for Enclosed Car Park, Int. J. Model. Optim. 7 (2017) 105–108. doi:10.7763/IJMO.2017.V7.567.
[34] A.M. Doust, M. Rahimi, M. Feyzi, An Optimization Study by Response Surface Methodology (RSM) on Viscosity Reduction of Residue Fuel Oil Exposed Ultrasonic Waves and Solvent Injection, Iran. J. Chem. Eng. 13 (2016) 3–19.

IOP Conf. Series: Journal of Physics: Conf. Series **949** (2017) 012013 doi:10.1088/1742-6596/949/1/012013

Strongly localized dark modes in binary discrete media with cubic-quintic nonlinearity within the anti-continuum limit

L Abdul Taib[1], M S Abdul Hadi[1], B A Umarov[2]

[1] Department of Computational and Theoretical Sciences, Faculty of Science, International Islamic University Malaysia, 25200 Kuantan, Pahang Darul Makmur, Malaysia
[2] Department of Physics, Faculty of Science, International Islamic University Malaysia

E-mail: lukhman.taib@gmail.com

Abstract. The existence of dark strongly localized modes of binary discrete media with cubic-quintic nonlinearity is numerically demonstrated by solving the relevant discrete nonlinear Schrödinger equations. In the model, the coupling coefficients between adjacent sites are set to be relatively small representing the anti-continuum limit. In addition, approximated analytical solutions for vectorial solitons with various topologies are derived. Stability analysis of the localized states was performed using the standard linearized eigenfrequency problem. The prediction from the stability analysis are furthermore verified by direct numerical integrations.

1. Introduction

The study of localized solutions in discrete system has gain central attraction for nearly three decades, and this was motivated by its profound impacts on various branches of science, mainly solid-state physics and nonlinear optics [3, 6, 10–13]. The stability analysis and evolution of localized states frequently described by discrete nonlinear Schrödinger equation (DNLSE). Perhaps this was first initiated by experimental work on the fabricated AlGaAs waveguide arrays reported in [4], followed by the mathematical model presented in [5]. The constitution of fundamental dark discrete soliton is found in [9] on the anomalous diffraction region of AlGaAs waveguide arrays related with Kerr-type cubic nonlinearity. This was known as the effect of defocusing nonlinearity in DNLSE system. DNLSE also describing other areas in physics, i.e. Bose-Einstein condensates (BEC) and optically induced lattice [7].

In systems with more than one component, extra interaction mechanism that rise up can strongly affecting the conditions of the existence of solitons. Localized vectorial modes form a species of composite solutions that consist of two or more mutually self-trap components in a nonlinear medium. Significantly, if not all components exist at the same time, such composite systems could not be observed. In addition, the existence of dark solitons requires modulational stability of the background. Strongly localized, dark vectorial modes in discrete cubic media described by DNLSE's where shown to exist, along with stability region [8]. As far as discrete binary cubic-quintic (CQ) media being concerned, only modulational instability (MI) of the plane-wave solutions has been reported, where for the same set of initial parameters i.e. same amplitudes and inter-component coupling constants; quintic nonlinear yields fundamentally

IOP Conf. Series: Journal of Physics: Conf. Series **949** (2017) 012013 doi:10.1088/1742-6596/949/1/012013

stronger effect than cubic nonlinearity [2], which provides some information on the stability region of dark solitons but their existence is not yet verified. Thus, this paper examines the existence of dark strongly localized modes (SLM's) in binary discrete media with CQ nonlinearity.

This paper is organized as follows: in Section 2, we introduced the mathematical models representing binary discrete media with CQ nonlinearity with its conserved quantities, in Section 3, the existence and linear stability analysis of the solutions are discussed. Analytic expressions for vectorial dark SLM's of different topologies are derived in Section 3 too.

2. The Mathematical Model

Binary discrete media with cubic-quintic nonlinearity can be described by two-component CQDNLSE:

$$\begin{aligned} i\frac{dA_n}{dz} =& - c_a\left(A_{n+1}+A_{n-1}\right) - \lambda\left(|A_n|^2+\beta|B_n|^2\right)A_n \\ &- \gamma\left(|A_n|^4+2\alpha|A_n|^2|B_n|^2+\alpha|B_n|^4\right)A_n, \end{aligned} \tag{1}$$

$$\begin{aligned} i\frac{dB_n}{dz} =& - c_b\left(B_{n+1}+B_{n-1}\right) - \lambda\left(|B_n|^2+\beta|A_n|^2\right)B_n \\ &- \gamma\left(|B_n|^4+2\alpha|B_n|^2|A_n|^2+\alpha|A_n|^4\right)B_n, \end{aligned} \tag{2}$$

where $A_n(z)$, $B_n(z)$ are the complex-valued wave functions at site n; c_a, c_b denote the coupling constants between adjacent sites; variable z denotes propagation coordinate in optics or time in BEC application; λ and γ are the cubic and quintic nonlinear coefficients respectively; and β and α denote the cubic and quintic inter-component nonlinear coupling or cross-phase modulation (XPM) coefficients respectively. Here, the self-phase modulation (SPM) coefficients are rescaled to one. Although λ and γ also can be rescaled to one, these coefficients are preserved for convenience to examine the effects of only cubic ($\lambda \neq 0$, $\gamma = 0$), only quintic ($\lambda = 0$, $\gamma \neq 0$), or weighted contribution of both coefficients ($\lambda \neq 0$, $\gamma \neq 0$) to overall MI [2]. Setting $\gamma = 0$ and $\beta = 1/\lambda$ transforms Eqs. (1)-(2) into the coupled DNLSE studied in [8], while by posing $\beta = \alpha = 0$ Eqs. (1)-(2) become uncoupled and turned into two one-dimensional CQDNLSE's reported in [1, 14].

Eqs. (1)-(2) are derived from the Hamiltonian

$$H = H_a + H_b + H_{int} \tag{3}$$

through

$$i\frac{dA_n}{dz} = -\frac{\delta H}{\delta A_n^*}, \quad i\frac{dB_n}{dz} = -\frac{\delta H}{\delta B_n^*}, \tag{4}$$

where

$$H_a = \sum_n \left[c_a\left(A_n^* A_{n+1} + A_n A_{n+1}^*\right) + \frac{\lambda}{2}|A_n|^4 + \frac{\gamma}{3}|A_n|^6\right], \tag{5}$$

$$H_b = \sum_n \left[c_b\left(B_n^* B_{n+1} + B_n B_{n+1}^*\right) + \frac{\lambda}{2}|B_n|^4 + \frac{\gamma}{3}|B_n|^6\right], \tag{6}$$

$$H_{int} = \sum_n \left[\beta\lambda|A_n|^2|B_n|^2 + \alpha\gamma\left(|A_n|^4|B_n|^2 + |A_n|^2|B_n|^4\right)\right]. \tag{7}$$

Eqs. (1)-(2) have two conserved quantities: the Hamiltonian H and the total excitation norm $N = \sum_n\left(|A_n|^2 + |B_n|^2\right)$.

IOP Conf. Series: Journal of Physics: Conf. Series **949** (2017) 012013 doi:10.1088/1742-6596/949/1/012013

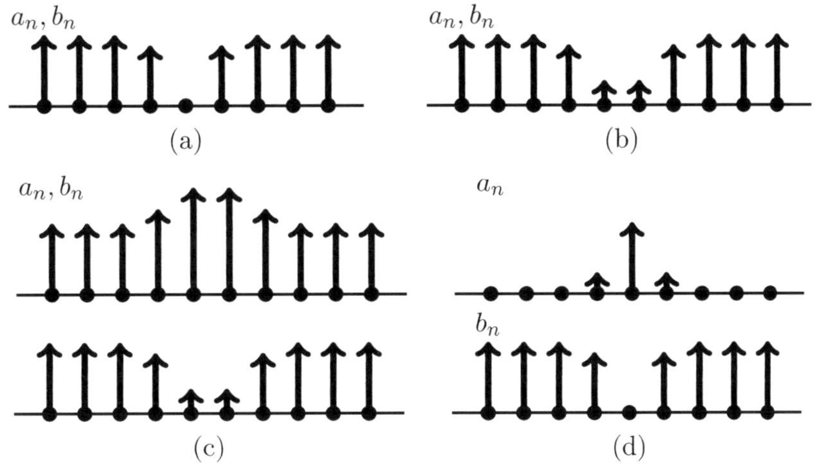

Figure 1: Schematic representation of different kinds of dark modes. (a) odd SLM; (b) even SLM; (c) and (d) symbiotic anti-dark–dark and bright-dark SLM's.

3. The Existence and Stability of Dark Strongly Localized Modes

The existence of dark SLM's can be investigated by setting $A_n = \alpha_n \exp(ik_a z), B_n = \beta_n \exp(ik_b z)$ and substitute into Eqs. (1)-(2) one obtains the steady-state equations for real amplitudes α_n and β_n

$$-k_a\alpha_n + c_a\left(\alpha_{n+1} + \alpha_{n-1}\right) + \lambda\left(\alpha_n^2 + \beta\beta_n^2\right)\alpha_n + \gamma\left(\alpha_n^4 + 2\alpha\alpha_n^2\beta_n^2 + \alpha\beta_n^4\right)\alpha_n = 0, \quad (8)$$

$$-k_b\beta_n + c_b\left(\beta_{n+1} + \beta_{n-1}\right) + \lambda\left(\beta_n^2 + \beta\alpha_n^2\right)\beta_n + \gamma\left(\beta_n^4 + 2\alpha\beta_n^2\alpha_n^2 + \alpha\alpha_n^4\right)\beta_n = 0. \quad (9)$$

The exact solutions of steady-state equations (8)-(9) give rise stationary excitations in each site of a lattice that yet constitute dark SLM's of different topologies. Topologies considered in this paper are restricted to the kink-like kinds, as what have been done in [8] for discrete cubic media.

Numerical stability analysis was performed by linearization using the ansatz

$$A_n = \exp(ik_a z)\{\alpha_n + \delta\left[c_n \exp(-i\omega z) + d_n \exp(i\omega^* z)\right]\},$$
$$B_n = \exp(ik_b z)\{\beta_n + \delta\left[f_n \exp(-i\omega z) + g_n \exp(i\omega^* z)\right]\},$$

where ω's denote the eigenfrequencies and $\delta \ll 1$. Substituting the ansatz into Eqs. (1)-(2) yields the (linearization) eigenfrequency problem at $O(\delta)$:

$$\omega \begin{pmatrix} c_k \\ d_k^* \\ f_k \\ g_k^* \end{pmatrix} = \begin{pmatrix} \frac{\partial F_{a,i}}{\partial \alpha_j} & \frac{\partial F_{a,i}}{\partial \alpha_j^*} & \frac{\partial F_{a,i}}{\partial \beta_j} & \frac{\partial F_{a,i}}{\partial \beta_j^*} \\ -\frac{\partial F_{a,i}^*}{\partial \alpha_j} & -\frac{\partial F_{a,i}^*}{\partial \alpha_j^*} & -\frac{\partial F_{a,i}^*}{\partial \beta_j} & -\frac{\partial F_{a,i}^*}{\partial \beta_j^*} \\ \frac{\partial F_{b,i}}{\partial \alpha_j} & \frac{\partial F_{b,i}}{\partial \alpha_j^*} & \frac{\partial F_{b,i}}{\partial \beta_j} & \frac{\partial F_{b,i}}{\partial \beta_j^*} \\ -\frac{\partial F_{b,i}^*}{\partial \alpha_j} & -\frac{\partial F_{b,i}^*}{\partial \alpha_j^*} & -\frac{\partial F_{b,i}^*}{\partial \beta_j} & -\frac{\partial F_{b,i}^*}{\partial \beta_j^*} \end{pmatrix} \begin{pmatrix} c_k \\ d_k^* \\ f_k \\ g_k^* \end{pmatrix}$$

where

$$F_{a,i} = c_a\left(\alpha_{i+1} + \alpha_{i-1}\right) + \lambda\left(|\alpha_i|^2 + \beta|\beta_i|^2\right)\alpha_i + \gamma\left(|\alpha_i|^4 + 2\alpha|\alpha_i|^2|\beta_i|^2 + \alpha|\beta_i|^4\right)\alpha_i - k_a\alpha_i,$$
$$F_{b,i} = c_b\left(\beta_{i+1} + \beta_{i-1}\right) + \lambda\left(|\beta_i|^2 + \beta|\alpha_i|^2\right)\beta_i + \gamma\left(|\beta_i|^4 + 2\alpha|\beta_i|^2|\alpha_i|^2 + \alpha|\alpha_i|^4\right)\beta_i - k_b\beta_i.$$

IOP Conf. Series: Journal of Physics: Conf. Series **949** (2017) 012013 doi:10.1088/1742-6596/949/1/012013

The existence of imaginary parts of the eigenfrequencies ω's in the eigenfrequency spectrum denotes the instability of the solutions.

The space-time evolution of the solutions can be achieved by direct numerical integrations of Eqs. (1)-(2). The initial conditions of the model are taken from the exact solutions of Eqs. (8)-(9) with additional random, uniformly distributed perturbation of amplitude 10^{-4}. We utilize the fourth-order Runge-Kutta procedure to perform the direct numerical integration.

3.1. *Odd dark SLM's*

These modes are defined as every site has nonzero excitation except the central site (see Fig. 1a), and have the solution in the form $\alpha_n = A(\ldots,1,1,1,a_1,0,a_1,1,1,1,\ldots)$ and $\beta_n = B(\ldots,1,1,1,b_1,0,b_1,1,1,1,\ldots)$. Substituting the above ansatz into Eqs. (8)-(9) straightforwardly we obtain the expressions for the dispersion relation

$$k_a = 2c_a + \lambda(A^2 + \beta B^2) + \gamma(A^4 + 2\alpha A^2 B^2 + \alpha B^4), \tag{10}$$

$$k_b = 2c_b + \lambda(B^2 + \beta A^2) + \gamma(B^4 + 2\alpha B^2 A^2 + \alpha A^4), \tag{11}$$

and the secondary excitations

$$a_1 = 1 - \zeta, \qquad b_1 = 1 - \xi, \qquad \zeta, \xi \ll 1 \tag{12}$$

where

$$\begin{aligned} \zeta &= \frac{\lambda\,(c_a - \beta c_b) + 2\gamma\left[\alpha A^2\,(c_a - c_b) - B^2\,(\alpha c_a - c_b)\right]}{2A^2\sigma}, \\ \xi &= \frac{\lambda\,(c_b - \beta c_a) + 2\gamma\left[\alpha B^2\,(c_a - c_b) - A^2\,(\alpha c_a - c_b)\right]}{2B^2\sigma}, \\ \sigma &= \lambda^2\left(\beta^2 - 1\right) + 4\gamma^2\,(\alpha - 1)\left[\alpha\left(A^2 + B^2\right) + \left(\alpha^2 + 1\right)A^2B^2\right] \\ &\quad + 2\lambda\gamma\left[\alpha\,(2\beta - 1) - 1\right]\left(A^2 + B^2\right). \end{aligned}$$

It is worth to note that the choice of the value of peak amplitudes A and B is arbitrary. The approximation $\zeta, \xi \ll 1$ is concordant with the subject of this paper, i.e. *strongly* localized modes. These derivations are restricted to the first-order approximations primarily because of the interest on the physical aspects of the problem, thus deviations from the background of the order ζ^2, ξ^2 and smaller are neglected. Examples of odd dark modes are displayed in Fig. 2.

In Fig. 2a, the absence of imaginary parts of the eigenfrequencies ω in the eigenfrequency spectrum (second top left panel) denotes that the solutions are stable at $c = -0.1$ (from hereby, we fix $c_a = c_b \equiv c$ to minimize the number of parameters). Meanwhile, the numerical stability analysis (second top right panel) showing that the solutions are said to be linearly stable for all c up to $c = -0.1$, where the dependence of the lowest squared eigenfrequencies ω^2 (solid line) is reported, as a function of c, to be always non-negative. Results from the direct numerical integrations of Eqs. (1)-(2) are found to be in full agreement with the stability analysis above. The deviation of the solutions from the exact ones are very small.

In contrast, in Fig. 2b, the eigenfrequency spectrum exhibits the presence of imaginary part of the eigenfrequencies ω, while the numerical stability analysis shows that the lowest squared eigenfrequencies ω^2 is always negative up to $c = -0.1$. These results indicate the instability of the solutions. Direct numerical integrations of the model depicts that the instability of the solutions gives rise to the localized states after transient, and the deviations of the solutions from the exacts' are large.

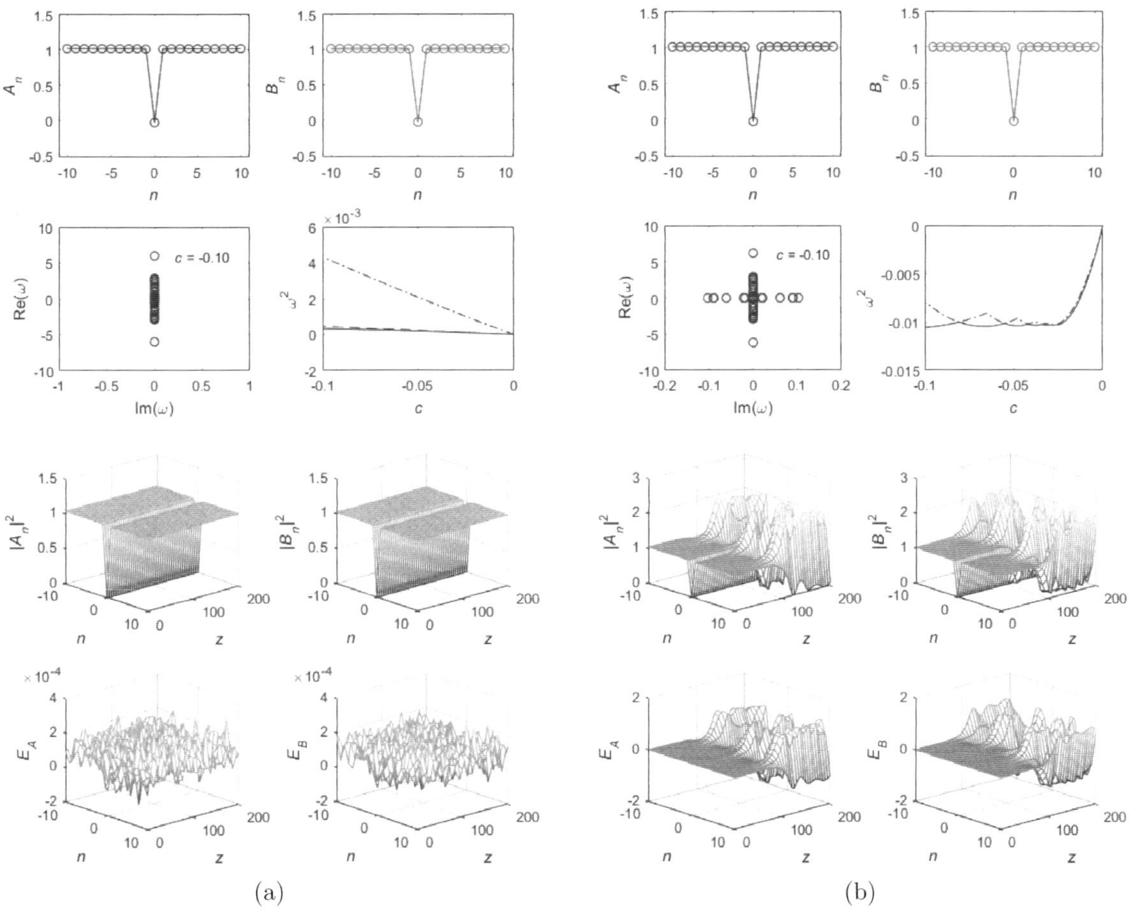

Figure 2: *First row panels.* Exact odd dark SLM's of Eqs. (1)-(2) of components A_n and B_n. *Second row panels.* Eigenfrequency spectrum (left panel) and the numerical stability analysis as functions of c (right panel) corresponds to the SLM's depicted in the top panels. *Third row panels.* Space-time evolution of $|A_n|^2$ and $|B_n|^2$, respectively. *Bottom row panels* Deviation of respective components from their exact solutions. Parameters: $A = 1.01$, $B = 1$, $c_a = c_b \equiv c = -0.1$, $\lambda = \gamma = 1$, $\alpha = 1$, and (a) $\beta = 0.9$, (b) $\beta = 1.1$.

3.2. Even dark SLM's

These modes are centered between two adjacent lattice sites (refer Fig. 1b), and have the ansatz of the form $\alpha_n = A(\ldots,1,1,1,a_2,a_1,a_1,a_2,1,1,1,\ldots)$ and $\beta_n = B(\ldots,1,1,1,b_2,b_1,b_1,b_2,1,1,1,\ldots)$ where a_2 and b_2 are assumed to be near to unity as a result of the strong localization. Substituting the above ansatz into Eqs. (8)-(9) one can obtain the dispersion relations same as in Eqs. (10)-(11) and the excitations a_2, b_2

$$a_2 = 1 - \frac{c_b\left[\lambda\beta - 2\gamma\alpha(B^2+A^2)\right](1-b_1) - c_a\left[\lambda - 4\gamma(A^2+\alpha B^2)\right](1-a_1)}{2A^2\sigma}, \quad (13)$$

$$b_2 = 1 - \frac{c_a\left[\lambda\beta - 2\gamma\alpha(A^2+B^2)\right](1-a_1) - c_b\left[\lambda - 4\gamma(\alpha A^2+B^2)\right](1-b_1)}{2B^2\sigma}, \quad (14)$$

where

$$\sigma = 8\lambda^2A^2B^2(\alpha^2+2) + 4\lambda\gamma\beta\alpha B^2 - 4\gamma^2\alpha(\alpha-4)(A^4+B^4) - 8\lambda\gamma(\alpha+1)(A^2+B^2) - \lambda^2\beta^2.$$

Here, a_1 and b_1 can be determined by solving the following system of nonlinear equations

$$\begin{aligned}\{A^2\left[\lambda+\gamma A^2(a_1^2+1)\right](a_1^2-1)+B^2\left[\lambda\beta+\gamma\alpha B^2(b_1^2+1)\right](b_1^2-1)\\ +2\gamma\alpha A^2B^2(a_1^2b_1^2-1)\}a_1+c_a(1-3a_1)=0,\end{aligned} \tag{15}$$

$$\begin{aligned}\{B^2\left[\lambda+\gamma B^2(b_1^2+1)\right](b_1^2-1)+A^2\left[\lambda\beta+\gamma\alpha A^2(a_1^2+1)\right](a_1^2-1)\\ +2\gamma\alpha B^2A^2(b_1^2a_1^2-1)\}b_1+c_b(1-3b_1)=0.\end{aligned} \tag{16}$$

For some particular cases Eqs. (15)-(16) can be substantially reduced. Suppose that $|a_1|, |b_1| \ll 1$, from Eqs. (13)-(16) one can obtain

$$a_1=\frac{c_a}{k_a-2c_a},\quad b_1=\frac{c_b}{k_b-2c_b},$$

$$a_2=1-\frac{c_b\left[\lambda\beta-2\gamma\alpha(B^2+A^2)\right]-c_a\left[\lambda-4\gamma(A^2+\alpha B^2)\right]}{2A^2\sigma},$$

$$b_2=1-\frac{c_a\left[\lambda\beta-2\gamma\alpha(A^2+B^2)\right]-c_b\left[\lambda-4\gamma(B^2+\alpha A^2)\right]}{2B^2\sigma}.$$

The evolutions of stable and unstable propagations are displayed in Fig. 3. The solutions exhibit the same behaviours discussed in the odd dark modes case. Numerous simulations on odd and even dark SLM's are found in excellent agreement with the background stability studied in [2].

3.3. Symbiotic dark–anti-dark SLM's

Beside the coupled CDNLSE, our model (1)-(2) also exhibit the special attribute of supporting symbiotic localized modes, i.e. modes with the combination of different topologies. A combination of dark and anti-dark localized modes (refer Fig. 1c) is described by Eqs. (15)-(16). This particular form of SLM's is entirely caused by the vectorial nature of the interaction between components which does not present in the continuum limit or in the scalar discrete system [8]. Nevertheless, if the solutions of Eqs. (15)-(16) correspond to the structure of this mode are found, it is not a guarantee that dark–anti-dark modes are exist across the continuation from anti-continuum limit.

However, the modulational stability of the background — although it is essential — is insufficient to probe the existence of stable solitons. In Fig. 4, the parameters used are the ones with stable background in [2]. In Fig. 4a, linear stability analysis predicts that the solutions are stable up to $c=-0.1$, and direct numerical integrations confirms the prediction. Meanwhile in Fig. 4b, the presence of two pairs of imaginary parts of the eigenfrequencies ω in the eigenfrequency spectrum illustrating the weak oscillatory instability of the solutions at $c=-0.1$, which is verified by direct numerical integrations of the model.

3.4. Symbiotic bright–dark SLM's

Another kind of symbiotic modes is the combination of a bright A-component and a dark B-component. The ansatz $\alpha_n = A(\ldots,0,0,0,a_1,1,a_1,0,0,0,\ldots)$ and $\beta_n = B(\ldots,1,1,1,b_1,0,b_1,1,1,1,\ldots)$ allow us to recognize an odd feature of bright–dark modes sketched in Fig. 1d. The secondary excitations a_1 and b_1 can be found from the related set of algebraic equations as

$$a_1=\frac{c_a}{\lambda(A^2-\beta B^2)+\gamma(A^4-\alpha B^4)}, \tag{17}$$

$$b_1=1-\xi,\quad \xi=\frac{-c_b}{2(\lambda-2\gamma B^2)B^2}\ll 1, \tag{18}$$

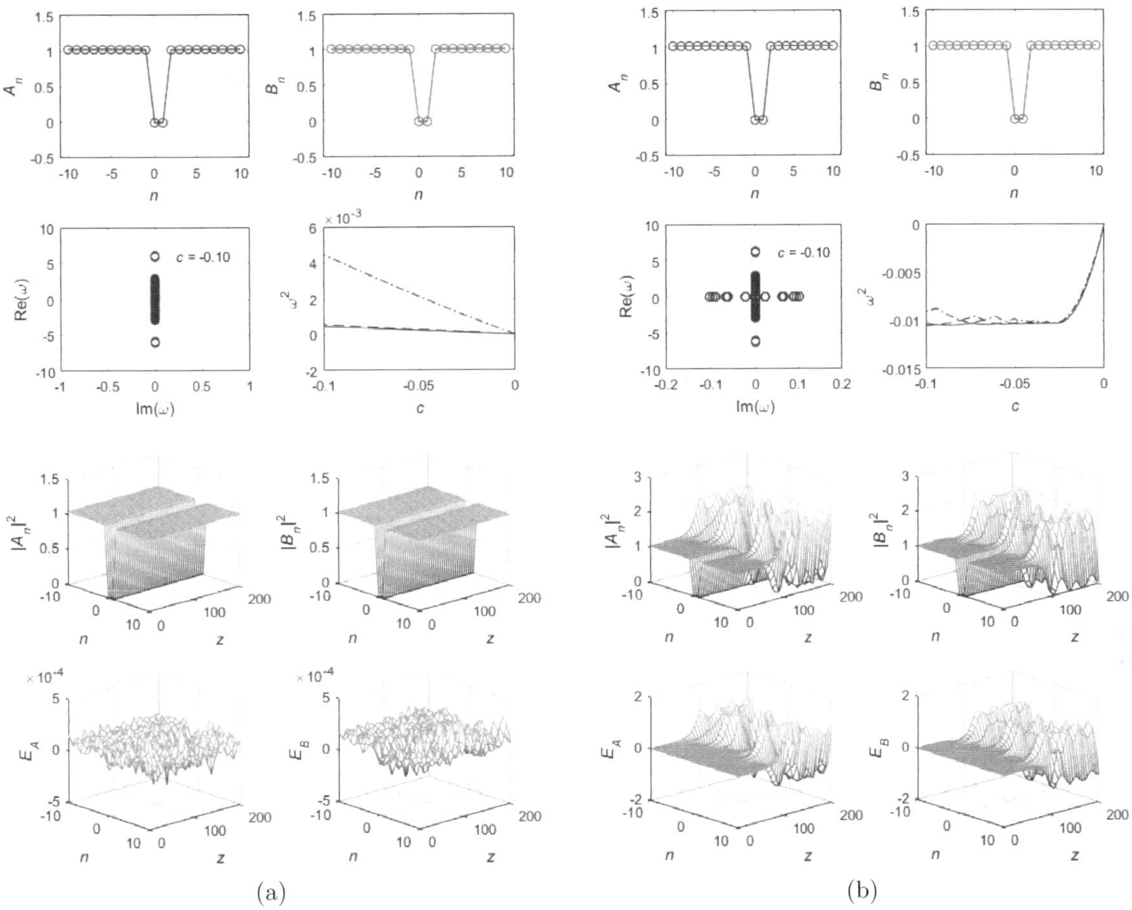

Figure 3: Same as in Fig. 2 but for even dark modes. All parameters are fixed as in Fig. 2.

and the corresponding dispersion relations as

$$k_a = (\lambda + \gamma A^2)A^2, \quad k_b = 2c_b + (\lambda + \gamma B^2)B^2. \tag{19}$$

In Fig. 5a, eigenfrequency spectrum shows purely real parts of eigenfrequencies ω, while the numerical linear analysis indicating that the solutions should be stable up to $c = -0.1$. Direct numerical integrations illustrate that the solutions propagate with stability across transient.

The same kind of solutions can arise in the even case $\alpha_n = A(\dots, 0, 0, 0, a, 1, s_a, s_a a, 0, 0, 0, \dots)$ and $\beta_n = B(\dots, 1, 1, 1, b_2, b_1, b_1, b_2, 1, 1, 1, \dots)$, where $s_a = \pm 1$, with dispersion relations

$$k_a = c_a s_a + (\lambda + \gamma A^2)A^2, \quad k_b = 2c_b + (\lambda + \gamma A^2)B^2 \tag{20}$$

and the excitations

$$a = \frac{c_a}{(\lambda + \gamma A^2)A^2 - (\lambda\beta + \gamma\alpha B^2)B^2}, \quad b_1 = \frac{c_b}{(\lambda + \gamma B^2)B^2 - (\lambda\beta + \gamma\alpha A^2)A^2}, \tag{21}$$

$$b_2 = 1 - \xi, \quad \xi = -\frac{c_b(1 - b_1)}{2(\lambda + 2\gamma B^2)B^2} \ll 1. \tag{22}$$

In Fig. 5b, a pair of imaginary parts of eigenfrequencies ω arise in the eigenfrequency spectrum due to the phase invariance, denoting strong exponential instability of the solutions. Negative

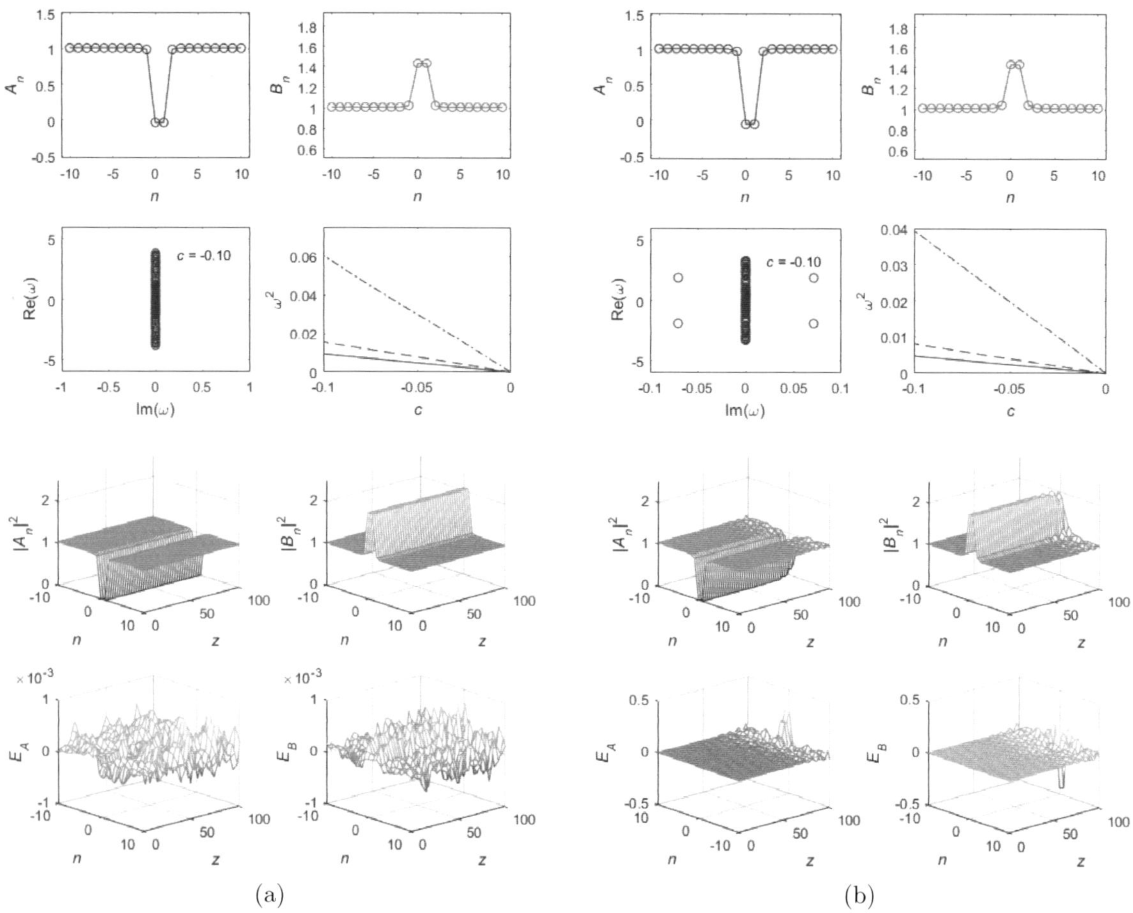

Figure 4: Same as in Fig. 2 but for symbiotic dark–anti-dark modes. Parameters: $A = 1.01$, $B = 1$, $c_a = c_b \equiv c = -0.1$, $\lambda = \gamma = 1$, and (a) $\beta = -5$, $\alpha = 3$ (b) $\beta = -2$, $\alpha = 2$.

values of the lowest squared eigenfrequencies ω^2 illustrate the instability of the solution up to $c = -0.1$. These predictions are verified by direct numerical integrations, where the effect of instability is stronger on bright A-component than dark B-component. The background stability is again inadequate condition for the existence of stable localized modes.

4. Conclusion

In this paper we have demonstrated the existence of dark strongly localized modes in binary discrete media with cubic and quintic nonlinearities modelled by the coupled CQDNLSE. The stability of these modes has been probed both by linear spectral analysis and direct numerical integrations. We found that stability of odd and even dark modes confirms the results on modulational stability region identified in [2], while the symbiotic modes are not restrained by it. For odd and even dark modes, instability gives rise to localized states. Meanwhile, the existence of symbiotic dark–anti-dark SLM's are restricted to a small range of parameters. For the symbiotic bright-dark SLM's, even bright component are found to be unstable earlier that dark component.

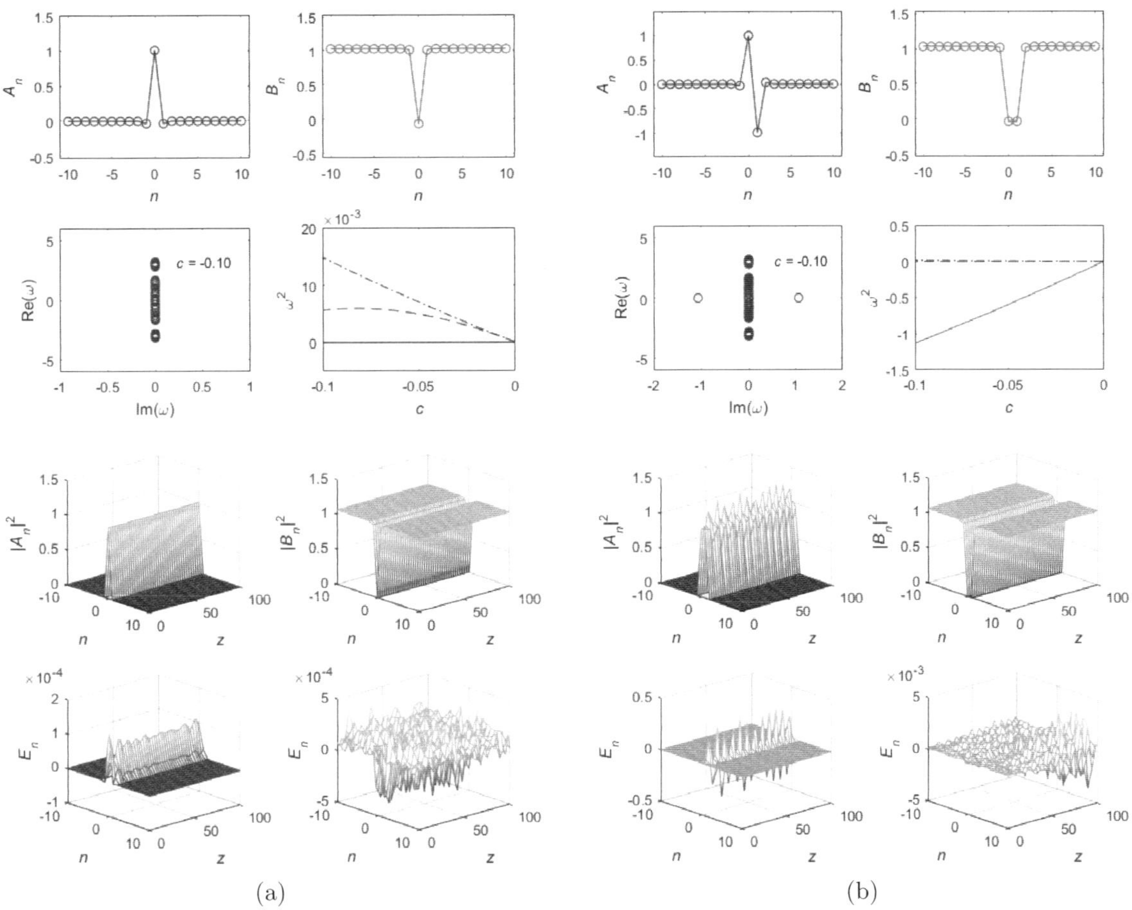

Figure 5: Same as in Fig. 2 but for symbiotic bright-dark modes. Parameters: $A = 1.01$, $B = 1$, $c_a = c_b \equiv c = -0.1$, $\lambda = \gamma = 1$, and $\beta = -3$, $\alpha = 2$, (a) odd modes and (b) even modes with $s_a = -1$.

Acknowledgment
The authors would like to acknowledge supports from the Ministry of Higher Education, Malaysia under research grant no. FRGS 144870-169605.

References

[1] Abdullaev F.Kh., Bouketir A, Messikh A and Umarov B A 2007 *Physica D* **232** 54
[2] Baizakov B B, Bouketir A, Messikh A and Umarov B A 2009 *Phys. Rev. E* **79** 046605
[3] Christodoulides D N and Joseph R I 1988 *Opt. Lett.* **13** 794
[4] Eisenberg H S, Silberberg Y, Morandotti R, Boyd A R and Aitchison J S 1998 *Phys. Rev. Lett.* **81** 3383
[5] Eisenberg H S, Silberberg Y, Morandotti R and Aitchison J S 2000 *Phys. Rev. Lett.* **83** 1863
[6] Eilbeck J C, Lomdahl P S and Scott A C 1985 *Physica D* **16** 318
[7] Kevrekidis P G 2009 *The Discrete Nonlinear Schrödinger Equation* (Berlin: Springer-Verlag)
[8] Kobyakov A, Darmanyan S, Lederer F and Schmidt E 1998 *Opt. Quantum Electron.* **30** 795
[9] Morandotti R, Eisenberg H S, Silberberg Y, Sorel M and Aitchison J S 2001 *Phys. Rev. Lett.* **86** 3296
[10] Page J B 1990 *Phys. Rev. B* **41** 7835
[11] Schmidt-Hattenberger C, Muschall R, Lederer F and Trutschel U 1993 *JOSA B* **10** 1592
[12] Scott A C and MacNeil L 1983 *Phys. Lett. A* **98** 87
[13] Sievers A J and Takeno S 1988 *Phys. Rev. Lett.* **61** 970
[14] Zhang A X and Xue J K 2008 *Phys. Lett. A* **372** 1147

IOP Conf. Series: Journal of Physics: Conf. Series **949** (2017) 012014 doi:10.1088/1742-6596/949/1/012014

Artificial Neural Network versus Linear Models Forecasting Doha Stock Market

Adil Yousif[1] and Faiz Elfaki[2]

[1, 2] Department of Mathematics, Statistics and Physics, College of Arts and Science, Qatar University, P.O. Box 2713, Doha, Qatar.

E-mail : [1] aeyousif@qu.edu.qa ; [2] felfaki@qu.edu.qa

Abstract. The purpose of this study is to determine the instability of Doha stock market and develop forecasting models. Linear time series models are used and compared with a nonlinear Artificial Neural Network (ANN) namely Multilayer Perceptron (MLP) Technique. It aims to establish the best useful model based on daily and monthly data which are collected from Qatar exchange for the period starting from January 2007 to January 2015. Proposed models are for the general index of Qatar stock exchange and also for the usages in other several sectors. With the help of these models, Doha stock market index and other various sectors were predicted. The study was conducted by using various time series techniques to study and analyze data trend in producing appropriate results. After applying several models, such as: Quadratic trend model, double exponential smoothing model, and ARIMA, it was concluded that ARIMA (2,2) was the most suitable linear model for the daily general index. However, ANN model was found to be more accurate than time series models.

1. Introduction

The stock market is instrumental in measuring the strength or weakness of a country's economy. In Arab countries, Egyptians took the initiative to set up a stock market in the18th century, which is one of the oldest stock markets in the world. In gulf region, the oldest stock market was established in Kuwait in 1962 [1]. Over the course of the next few decades, other Gulf countries also progressed economically resulting in the establishment of stock markets in countries like the United Arab Emirates, Oman, and Qatar.

For over a decade, Qatar worked hard to overcome the challenges, and over the next 15 years, it was able to build a competitive Liquefied Natural Gas (LNG) value chain that was able to reach the global market [2]. It was also successful in scaling up, integrating downstream, and built its reputation as a reliable and flexible partner and supplier. By the end of the last decade, Qatar directed its attention to the opportunities in converting lean gas into clean-burning gas-to-liquid fuels (GTLs) and it succeeded tremendously. Qatar is now the largest exporter of LNG and GTLs in the world with a supply chain that spans the entire globe [2].

The Qatari government adopted a policy that aims to diverse income resources and evolving economic infrastructures. Specifically, the government extended the exploration projects in oil and gas sectors and offered many incentives that appeal to foreign investors to perform similar projects. With its strong and open economy, the government of Qatar is determined to establish international standards and practices [3]. Besides infrastructure, there are many opportunities in the financial sector as Qatar aims to be a regional hub for foreign financial institutions. Local and foreign investors can avail the option of investing in Qatar Exchange, the country's stock market, which keeps a list of brokerage companies that facilitate business in Qatar for those looking to trade directly, thereby Qatar is currently experiencing in benefitting from the economic boom.

In Qatar, the stock market was established in 1995 under law no 14. After two years, in 1997, trade started with 17 listed companies and five brokerage firms. Work began with manual management of the trading floor. In August 1998, it was shifted to the central registration system. Within the next two years, the state of Qatar became the first country in the Gulf area to have the linkage with companies through the internet. At the same time, all manual trading systems were cancelled and trading became electronic. In 2002, it established its own website in the Internet and allowed investors to sell shares on the same day. After one year, the market was moved to a new building. By 2005, it allowed non-Qataris to invest in shares with a rate not exceeding 25% of the shares traded in the market. Then in 2006, both of the regular and irregular markets became integrated. In 2007, Qatar Exchange took membership of the World Federation of Exchanges (WFE). By 2009, Qatar witnessed a massive transformation into a world-class stock market when it signed a strategic partnership agreement with New York Stock Exchange (NYSE) Euronext.

As of now, 43 companies are listed for trading on Qatar exchange which are divided by seven sectors. The biggest sector is Bank Sector, which includes 12 joint-stock companies. Other sectors are: Service sector and consumer goods with eight joint-stock company, Industry includes 8 joint-stock companies, Insurance sector with 5 joint-stock companies, Transport sector includes 3 joint-stock companies, Real estate sector contains 4 joint-stock companies, and the smallest sector is telecommunications that includes 2 joint-stock companies.

Today, Qatar stock market has its own policy and legal power on this Exchange. The underlying vision of Qatar exchange is to support Qatar's economy by allowing the investors with a chance to trade fairly and efficiently. Qatar exchange is a part of Qatar's national strategy and to be global, it has signed an agreement with NYSE Euronext and renamed the market as Qatar exchange. It helps Qatar to be an important financial center in the gulf region. Finally, this initiative makes Qatar an attractive market, locally and regionally, for investors.

2. Neural Network

Predecting the stock market returns accurately can be a very challenging task in no small part because of the highly nonlinear nature of the financial time series [4]. Numerous studies have proven the existence of non-linearity in the financial markets. One very affective method to uncover this relationship is neural networks. Unfortunately, most of these studies failed to consider any alternative forecasting techniques. The also failed to determine the relevance of input variables, or consider the performance of the models under different trading strategies [5]. During the last few years, a number of hybrid models and neural network models were proposed for obtaining accurate prediction of the results in order to outperform the traditional linear and nonlinear approaches. This paper will evaluate the efficiency of neural network models which are known to be dynamic and effective in stock-market predictions [6]. Artificial Neural Networks have become the subject of interest of researches and studies in the last few years and are being applied successfully across a wide range of problem domains, such as: engineering, finance, geology, physics, and medicine. There have been many attempts to define neural networks formally. Artificial Neural Network can offer a valid approximation of a vast class of nonlinear process [7]. Researchers also added that, "*the most significant advantage of the ANN models over the classes of linear models is that ANNs are universal approximators that can approximate a large class of functions with high degree of accuracy*". In a recent research, it is stated that ANN models either control or at least resist the limitations of traditional forecasting models, including misspecification, biased outliers, and assumption of linearity [8].

IOP Conf. Series: Journal of Physics: Conf. Series **949** (2017) 012014 doi:10.1088/1742-6596/949/1/012014

Multilayer Perceptron (MLP) procedure yields a predictive model for one or more dependent (target) variables based on the values of the predictor variables. MLP is a widely used approach in regression-type problems.An MLP network consists of three layers: input layer, hidden layer, and output layer as shown in Figure 2. Neuron sums up the values of inputs parameters according to their assigned weights, and adds a bias. By applying the transfer function, values of the outputs would be determined. The number of neurons in the input layer is determined based on the number of input parameters [9]. MLP network is a function of one or more predictors (also referred to as independent variables or inputs) that minimizes the prediction error of one or more target variables (also referred to as outputs).

3. Data of the Study

Data for this study were obtained from Qatar exchange from January 2007 up to January 2015. Data were divided into two parts; the first part is used to identify a suitable time series model (January 2007 – December 2014) and the second part is used for model validation (January 2015).

Date: Date format for daily index is: Year, Month, Day; YYYY/MM/DD

Index Closing Value: The closing price represents the most up-to-date valuation of a security until trading resumes again the next trading day.

Figure 1 shows an increasing trend which reached its high peak at index 360 and dropped to the lowest point in between the year of 2007 and the beginning of 2009. This happened because of the global financial crisis which affected the stock markets all over the world. After that, the graph seems to be increasing with linear trend.

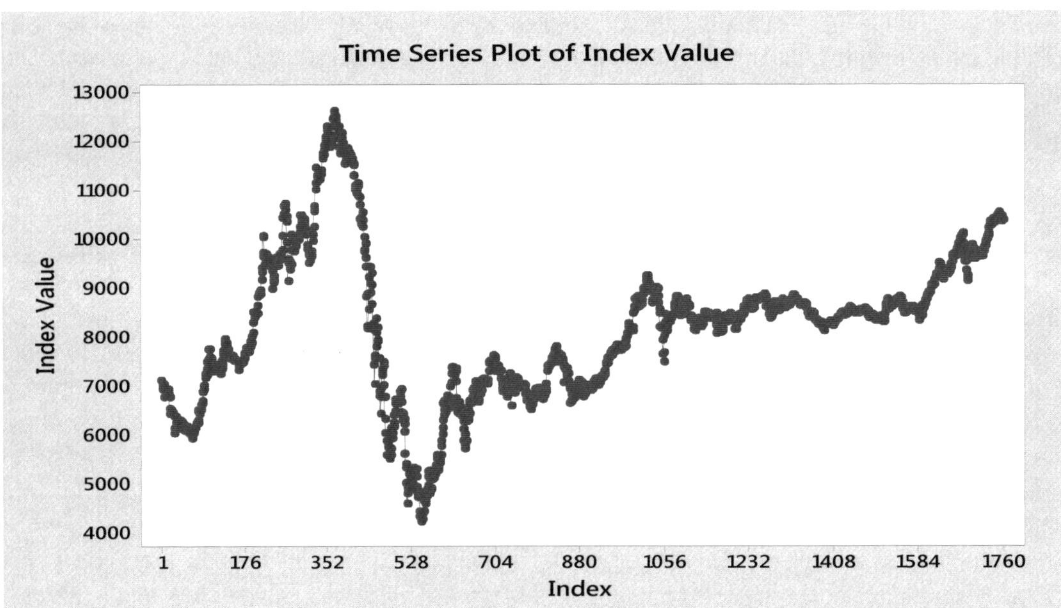

Figure 1. Daily Index Closing Value

4. Methodology

Two main methods were used for building a predictive model for the stock market data in the State of Qatar. Three time series models were utilized, namely: Trend model, Holt's Trend corrected models, and Box and Jenkins Autoregressive Moving Average (ARIMA) model. On the other hand, Artificial Neural Network which is one of none linear techniques was used.

4.1 Time Series Techniques
As the time series plot shows an increasing trend, several time series models were used. The Quadratic Trend technique was used, presented in equation (1), to investigate the trend of data.

$$Y_t = \beta_0 + \beta_1 t + \beta_2 t^2 \quad (1)$$

On the other hand the plot shows breakthrough in the level, which indicates parameters are not constant. So, Holt's Trend Exponential Smoothing Technique was also used for forecasting, as presented in equation (2) and (3).

$$L_t = \alpha \frac{Y_t}{S_{t-s}} + (1-\alpha)(L_{t-1} + b_{t-1}) \quad (2)$$

$$b_t = \beta(L_t + L_{t-1}) + (1-\beta)b_{t-1} \quad (3)$$

Equation (4) presents None-seasonal Autoregressive Moving Average (ARIMA) Technique which was also used as a mean linear stochastic model for forecasting.

$$Zt = \delta + \varphi_1 z_{t-1} + \varphi_2 z_{t-2} + \cdots + \varphi_p z_{t-p} + a_t - \theta_1 a_{t-1} - \theta_2 a_{t-2} - \cdots - \theta_q a_{t-q} \quad (4)$$

4.2 Artifical Neural Network
Network model, which is used in this study, has a single hidden layer that contains n nodes. The model has feed-forward functionality. Linear neuron activation function of the network model is presented in equation (5).

$$y_{t+h} = \alpha_0 + \sum_{j=1}^{n} w_j g\left(\alpha_0 + \sum_{i=1}^{p} w_i y_{t+h}\right) + \sum_{i=1}^{p} \beta_i y_{t-i} + \epsilon_{t+h} \quad (5)$$

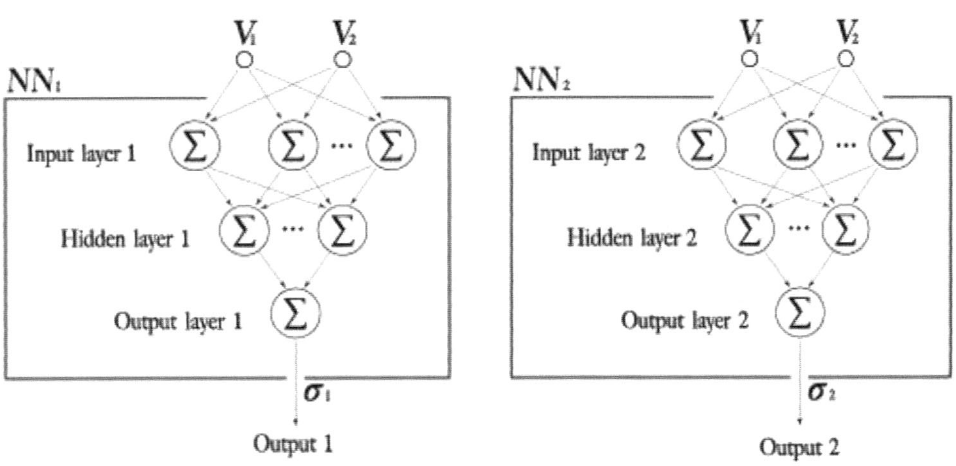

Figure 2. The Multilayer Perceptron (MLP) procedure.

5. Data Analysis and Discussion
Analysis of data for this study was performed via three linear time series models, which are: Trend model, Holt's corrected trend model, and Nonseasonal Box and Jenkins model (ARIMA). As a nonlinear model, Multilayer Perceptron (MLP) from Artificial Neural Network was also used. Among the linear models, ARIMA of order (2, 2) was identified as the most appropriate model with root mean square error (RMSE) as 287.97 as shown in Table 1. The Ljung Box Chi square statistic, as well as the

residual autocorrelation and residual partial autocorrelation were used to test the adequacy of ARIMA model.

Table 1. The Root Mean Square Error (RMSE) of the four models.

Model	RMSE
Quadratic Trend Model	42.19
Holt's Trend Model	26.31
ARIMA (2,2)	16.97
ANNW	3.01

On the other hand, when MLP was implemented using the hyperbolic tangent as a hidden layer activation function, the mean squares error was found as 82.3, the average overall relative error was obtained as 0.668, and the relative error for scale dependents was generated as 0.824. Table 2 displays the actual observation results for ten days which was observed after the period of the collected data and their corresponding point forecasts for ARIMA and ANN models.

Table 2. Comparison of the Forecasts for ARIMA and ANN models.

Day	Observed	ARIMA Model Forecast	ANNW Forecast
1	10526.95	10394.69	10458.17
2	10475.39	10383.96	10399.35
3	10468.59	10373.23	10406.84
4	10402.77	10362.51	10408.9
5	10453.15	10351.78	10477.88
6	10498.45	10341.05	10467.25
7	10462.54	10330.33	10409.01
8	10391.17	10319.6	10412.93
9	10360.02	10208.87	10317.48
10	10526.95	10394.69	10500.17

As it can be seen clearly in figure 3, Artificial Neural Network provided a better forecasting than linear time series models. On the other hand, ARIMA model appears to be the most suitable time series model followed by Holt's Trend model.

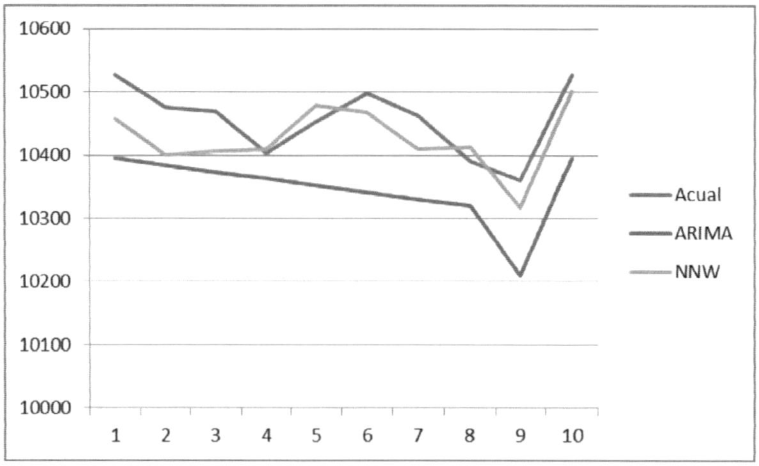

Figure 3. Comparison between Actual, ARIMA, and ANN models.

6. Summary and Conclusion

Qatar's stock market is a rapidly growing market which is due to the tremendous economic growth in the country [10]. As more and more companies are joining to the stock market, the number of investors increases as well. Therefore, it is essential that the investors should have an idea about the future of their investments. This future prediction can be done by modeling the data through several techniques.

This study analyzed the general closing index for the Qatar stock market from January 2007 to December 2016, as there was a drop in the time series because of the world's financial crisis in 2008-2009. Therefore, it was suggested by various similar studies, such as: studies on Saudi stock market and Johannesburg's stock market, to use data which are available after the crisis.

As the stock market is a highly fluctuating market, it is more appropriate to study the general closing index by analyzing data in daily basis. Based on time series plots, share prices for Doha stock market is continuously increasing and it was found that the data acquired a quadratic trend as it had the lowest mean absolute deviation (MAD) and mean square deviation (MSD). Holts' trend model, i.e. Double exponential smoothing model and ARIMA were applied and it was concluded that ARIMA (2,2) was the most suitable time series model for the daily general index. However, when Artificial Neural Network was applied, it shows a better forecasting than linear time series models.

References

[1] Al-Loughani, N., & Chappell, D. (2001). Modelling the day-of-the-week effect in the Kuwait Stock Exchange: a nonlinear GARCH representation. *Applied Financial Economics*, *11*(4), 353-359.

[2] Ibrahim, I., & Harrigan, F. (2012). Qatar's economy: Past, present and future. *QScience Connect*, (2012), 9.

[3] http://qatarhelplinegroup.com/business-opportunities-in-qatar/

[4] Qiu, M., Song, Y., & Akagi, F. (2016). Application of artificial neural network for the prediction of stock market returns: The case of the Japanese stock market. *Chaos, Solitons & Fractals*, *85*, 1-7.

[5] Enke, D., & Thawornwong, S. (2005). The use of data mining and neural networks for forecasting stock market returns. *Expert Systems with applications*, *29*(4), 927-940.

[6] Guresen, E., Kayakutlu, G., & Daim, T. U. (2011). Using artificial neural network models in stock market index prediction. *Expert Systems with Applications*, *38*(8), 10389-10397.

[7] Babikir, A., & Mwambi, H. (2014, July). A factor—Artificial neural network model for time series forecasting: The case of South Africa. In *Neural Networks (IJCNN), 2014 International Joint Conference on* (pp. 838-844). IEEE.

[8] Hill, T., O'Connor, M., & Remus, W. (1996). Neural network models for time series forecasts. *Management science*, *42*(7), 1082-1092.

[9] Moghaddam, A. H., Moghaddam, M. H., & Esfandyari, M. (2016). Stock market index prediction using artificial neural network. *Journal of Economics, Finance and Administrative Science*, *21*(41), 89-93.

[10] Sullivan, A., & Sheffrin, S. M. (2003). Economics: Principles in action. Upper Saddle River, New Jersey 07458: Pearson Prentice Hall.

IOP Conf. Series: Journal of Physics: Conf. Series **949** (2017) 012015 doi:10.1088/1742-6596/949/1/012015

Finite volume method for a Keller-Segel problem

C Messikh[1], M. S. H. Chowdhury[2,*], A Guesmia[3], A Berkane[1]

[1]University Badji Mokhtar, Faculty of sciences, Department of mathematics, Bp 12 El Hadjer Annaba, Algeria.

[2]Department of Science in Engineering, Faculty of Engineering, International Islamic University Malaysia, Jalan Gombak 53100, Kuala Lumpur, Malaysia

[3]Université du 20 Août-1955- Skikda-Algérie, BP 26 Route El Hadaek Skikda (21000) Laboratoire de Mathématiques Appliquées et d'Histoire et Didactique des Mathématiques (LAMAHIS)

E-mail: sazzadbd@iium.edu.my(Corresponding author: M. S. H. Chowdhury)

Abstract. In this paper, we are interested in the numerical simulation of the mathematical model of Keller-Segel Elliptic-Parabolic problem using finite volume scheme. The finite volume scheme is applied to the elliptic-parabolic model's problem and we have shown under certain assumptions, the existence of a unique and positive approximate solution. Moreover, under adequate regularity assumption of the exact solution, the finite volume scheme is the first order accurate. A good agreement between our numerical simulation and the theoretical results has been obtained.

1. Introduction

Chemotaxis is an important means for cellular communication. It is the influence of chemical substances in the environment on the movement of mobile species. This can lead to strictly oriented movement or to partially oriented and partially tumbling movement. The movement towards a higher concentration of the chemical substance is called positive chemotaxis whereas the movement towards regions of lower chemical concentration is called negative chemotactical movement.

The classical chemotaxis model of the so-called Keller–Segel model defined in system (1) was first introduced by Paltak [9], E. Keller and L. Segel [7]

$$\begin{aligned} &u_t - \nabla(a\nabla u) + \nabla(\chi u \nabla v) = 0 \qquad && (t,x) \in \mathbb{R}^+ \times \mathbb{R}^d \\ &\alpha v_t - \Delta v + \tau v + \beta u = 0 && x \in \mathbb{R}^d \end{aligned} \tag{1}$$

where $u(t,x)$ denotes the density of bacteria in the position $x \in \mathbb{R}^d$ at time t, v the concentration of chemical signal substance, $\alpha \geq 0$ the relaxation time, the parameter χ the sensitivity of cells to the chemoattractant and a, τ, β are given smooth functions. As it can be seen, when $\alpha \neq 0$ the model is called Parabolic-Parabolic while it is an Elliptic-Parabolic model when $\alpha = 0$. This model is very simple. It exhibits a profound mathematical structure and only dimension 2 is well understood, especially chemotactic collapse. It has been extensively studied in the last few years, see [5, 6, 11, 10] for a recent status of the problem.

Finite volume method is a class of discretization schemes that has been proven highly successful in approximating the solution of a wide variety of conservation law systems. It is extensively used in fluid mechanics, meteorology, electromagnetic, semi-conductor device simulation, models of biological processes and many other engineering areas governed by conservative systems that can be written in an integral control volume form. Numerical simulations for biological Keller-Segel model's problems were investigated in the following references [1, 2, 4].

The aim of this paper is to study the finite volume scheme applied to the elliptic-parabolic model's problem defined as

$$(\mathrm{P})\left\{\begin{array}{l}(\mathrm{P}_1)\left\{\begin{array}{ll}u_t-\Delta u+\operatorname{div}(u\nabla v)=0 & (t,x)\in\mathbb{R}^+\times\Omega\\ u=0 & \partial\Omega\\ u(0,x)=u_0 & x\in\Omega\end{array}\right.\\ \\ (\mathrm{P}_2)\left\{\begin{array}{ll}-\Delta v+\tau v=0 & x\in\Omega\\ v=g & \partial\Omega.\end{array}\right.\end{array}\right. \tag{2}$$

where

Assumption 1

(i) Ω is an open bounded polygonal subset of $\mathbb{R}^2$ or $\mathbb{R}^3$

(ii) $\tau\geq 0$,

(iii) $g\in C(\partial\Omega,\mathbb{R})$

(iv) $u_0\in C^2(\overline{\Omega},\mathbb{R})$

First we state the existence and uniqueness of the solution.

Theorem 1.1. *Based on [8]. If $g\in L^{\frac{3}{2}}(\Omega)$, $u_0\in L^2(\Omega)$. Then the problem (P) has a unique solution.*

Remark 1.2. *If the function g is positive on $\partial\Omega$, then the third condition of the assumption assure existence and uniqueness as well as positive of variational solution to problem (P_2) via Lax- Milgran's Theorem and Maximum principle*

To obtain a finite volume discretization of problem (P), we introduce some important notations and definitions:

(i) size(T) =sup{diam(K), $K\in T$}.

(ii) For any $K\in T$ and $\sigma\in\Xi$, $m(K)$ is the d-dimensional Lebesgue measure of K and $m(\sigma)$ is the (d-1)-dimensional measure of σ.

(iii) Denoting the set of interior (resp. boundary) edges by Ξ_{int} (resp. byΞ_{ext})).

(iv) Let $N(K)$ be the set of neighbors of K. If $\sigma=K|L$, we denote d_σ or $d_{K,L}$ the Euclidean distance between x_K and x_L and $d_{K,\sigma}$ is the distance from x_K to σ. If $\sigma\in\Xi_K\cap\Xi_{ext}$, d_σ or $d_{K,\sigma}$ denote the Euclidean distance between x_K and y_σ, where y_σ is the point of intersection of the straight line σ and the straight line which is orthogonal to σ and contains the point x_K.

(v) The transmissibility through σ is defined by $\tau_\sigma=\frac{m(\sigma)}{d_\sigma}$ and $n_{K,\sigma}$ resp (n_K) denotes the outward normal unit vector to σ resp (∂K) .
We note that throughout this paper we consider T is an admissible mesh in the sense of Definition 3.1 in [3].

For the time discretization, we define a temporal partition $t^n=nk$, for $n\in\{0,...,N_k+1\}$ where $k\in Ţ$ is the time step and $N_k=max\{n\in\mathcal{N},nk<Ţ\}$. The value (u_K^n,v_K) is an approximation of $(u(nk,x_K),v(x_K))$ at $(t^n,x_K)\in(0\,Ţ)\times K$.

Definition 1.3. *Let Ω be an open bounded polygonal subset of $\mathbb{R}^d$, $d = 2$ or 3, and T is an admissible mesh. Define $X(T)$ as the set of functions from Ω to $\mathbb{R}$ which are constant over each control volume of the mesh.*

Definition 1.4. *(Discrete H_0^1 norm) Let Ω be an open bounded polygonal subset of $\mathbb{R}^d$, $d = 2$ or 3, and T an admissible finite volume mesh. For $v \in X(T)$, define the discrete H_0^1 norm by*

$$\|v\|_{1,T} = \left(\sum_{\sigma\in\Xi} \tau_\sigma \left(D_\sigma v\right)^2\right)^{\frac{1}{2}} \tag{3}$$

where

$$\begin{aligned} D_\sigma v &= |v_L - v_K| \ \forall\sigma \in \Xi_{int}, \quad \sigma = K\,|L\,, \\ D_\sigma v &= |v_K| \quad \textit{if } \sigma \in \Xi_K \cap \Xi_{ext}. \end{aligned}$$

The paper is organized as follows. In section 2, we give some results related to the existence and the uniqueness of the approximate solution of problem (P_2). Under adequate regularity condition on the exact solution of this problem, the finite volume scheme is the first order accurate. In section 3, we prove the same results as in section 2 (existence, the uniqueness, and C^2 error estimate) for problem (P_1) which is based on the proof of Theorem 2.3 and Theorem 4.1 in Ref. [3]. We conclude in section 4 the same results obtained in the previous sections for problem (p). Finally, in Section 5 we illustrate the methods performance against test problems from the literature, and verify that the results are in agreement with our numerical analysis.

2. Finite volume method for a elliptic problem (P_2)

A finite volume scheme for the problem (P_2) is given by

$$\sum_{\sigma\in\Xi_K} F_{K,\sigma} + \tau m\left(K\right) v_K = 0 \quad \forall K \in T \tag{4}$$

$$F_{K,\sigma} = -\tau_{K|L}\left(v_L - v_K\right) \quad \forall\sigma \in \Xi_{int} \quad \text{if } \sigma = K|L \tag{5}$$

$$F_{K,\sigma} = -\tau_\sigma\left(v_\sigma - v_K\right) \quad \forall\sigma \in \Xi_{ext} \cap \partial\Omega \ \text{ where } \ v_\sigma = g\left(y_\sigma\right) \tag{6}$$

and y_σ is the point of intersection of the straight line σ and the straight line which is orthogonal to σ and contains the point x_K.
We recall some important results of the approximate solution of elliptic problem (P2) (for proof we refer to [3]).

Proposition 2.1. *Based on [3]. Under assumption 1. Let Tbe an admissible mesh . If $g\left(y_\sigma\right) \geq 0$ for all $\sigma \in \Xi_{ext}$, then the solution $(v_K)_{K\in T}$ of (4)-(6) is positive for any $K \in T$.*

Lemma 2.2. *(Existence and uniqueness [3]). Under assumptions 1, let T be an admissible mesh ; there exists a unique solution $(v_K)_{K\in T}$ to the equations (4)-(6).*

Theorem 2.3. *(C^2 error estimate) [3]. Under assumption 1, let T be an admissible mesh, $(v_K)_{K\in T}$ is the solution to (4)-(6). Assume that the unique variational solution v of problem (P_2) satisfied $v \in C^2(\overline{\Omega})$. Let, $e_T \in X(T)$ defined by $e_T = e_K = (v(x_K) - v_K)$ for a.e. $x \in K$ and for all $K \in T$. Then, there exists $C > 0$ depends only on v, τ and Ω such that*

$$\|e_T\|_{1,T} \leq Csize(T), \tag{7}$$

$$\|e_T\|_{L^2(\Omega)} \leq Csize(T), \tag{8}$$

and

$$\sum_{\substack{\sigma\in\Xi_{int}\\ \sigma=K|L}} m(\sigma)\, d_\sigma \left(\frac{(v_L - v_K)}{d_\sigma} - \frac{1}{m(\sigma)}\int_\sigma \nabla v(x).n_{K,\sigma} d\delta(x)\right)^2$$
$$+ \sum_{\substack{\sigma\in\Xi_{ext}\\ \sigma=\overline{K}\cap\partial\Omega}} m(\sigma)\, d_\sigma \left(\frac{(g(y_\sigma) - v_K)}{d_\sigma} - \frac{1}{m(\sigma)}\int_\sigma \nabla v(x).n_{K,\sigma} d\delta(x)\right)^2$$
$$\leq C\,(size(T))^2.$$

3. Finite volume method for a parabolic problem ($\mathbf{P}_1$)

In this section, we applied the following implicit finite volume scheme for the discretization of parabolic problem (P_1) which can be written as follows.

$$m(K)\frac{u_K^{n+1} - u_K^n}{k} + \sum_{\sigma\in\Xi_K} F_{K,\sigma}^{n+1} + \sum_{\sigma\in\Xi_K} V_{K,\sigma} u_{\sigma,+}^{n+1} = 0, \qquad \forall K\in T,\ \forall n\in\{0,\ldots,N_k\}, \tag{9}$$

$$u_K^0 = u_0(x_K) \qquad \forall K\in T, \tag{10}$$

where

$$F_{K,\sigma}^n = -\tau_{K|L}\left(u_L^n - u_K^n\right) \text{ for all } \sigma\in\Xi_{int} \text{ such that } \sigma = K\,|L\,,\ \text{for } n\in\{1,...,N_k+1\}, \tag{11}$$

$$F_{K,\sigma}^n = \tau_\sigma\left(u_K^n\right) \qquad \forall\sigma\in\Xi_{ext}\cap\partial\Omega,\ \text{for } n\in\{0,...,N_k+1\}, \tag{12}$$

$$\begin{cases} u_{\sigma,+}^n = u_K^n, & \text{if } V_{K,\sigma}\geq 0,\\ u_{\sigma,+}^n = u_L^n, & \text{if } V_{K,\sigma} < 0,\end{cases} \quad \text{for all } \sigma\in\Xi_{int} \quad \text{such that } \sigma = K\,|L\,, \tag{13}$$

$$\begin{cases} u_{\sigma,+}^n = u_K^n & \text{if } V_{K,\sigma}\geq 0,\\ u_{\sigma,+}^n = 0 & \text{if } V_{K,\sigma} < 0,\end{cases} \quad \text{for all } \sigma\in\Xi_K \quad \text{such that } \sigma\subset\partial\Omega, \tag{14}$$

$$V_{K,\sigma} = \tau_\sigma\left(v_L - v_K\right), \quad \text{for all } \sigma\in\Xi_{int} \text{ such that } \sigma = K\,|L\,, \tag{15}$$

and

$$V_{K,\sigma} = \tau_\sigma\left(g(y_\sigma) - v_K\right), \quad \text{for all} \sigma\in\Xi_{ext} \text{ such that } \sigma\subset\partial\Omega, \tag{16}$$

with v_K satisfies the equations (4)-(6) for all $K\in T$.

3.1. Error estimation

In the following Theorem we give a L^∞ estimate as well as the estimate error.

Theorem 3.1. *Let Ω be an open bounded polygonal subset of $\mathbb{R}^d$, $\underset{\sim}{T}>0$. Let $u\in C^2(\mathbb{R}_+\times\overline{\Omega},\mathbb{R})$ be defined in (P_1) and $v\in C^2(\overline{\Omega},\mathbb{R})$ defined in (P_2) . Let $u_0\in C^2(\overline{\Omega},\mathbb{R})$ and $g\in C^0(\partial\Omega,\mathbb{R}_+)$. Let T be an admissible mesh and $k\in(0,\underset{\sim}{T})$. Then there exists a unique vector $(u_K)_{K\in T}$ satisfying (9-16). In addition to that, there exists C depends only on u_0, such that*

$$\sup\left\{\left|u_K^n\right|,\ K\in T,\ n\in\{1,...,N_k+1\}\right\}\leq C. \tag{17}$$

Furthermore, let $e_K^n = u(t^n, x_K) - u_K^n$, for $K\in T$, $n\in\{1,...,N_k+1\}$ and $h = size(T)$. Then there exists $C>0$ depends only on u, v, Ω and $\underset{\sim}{T}$ such that

$$\left(\sum_{K\in T} m(K)\left(e_K^n\right)^2\right)^{\frac{1}{2}} \leq C\,(h+k) \qquad \forall K\in T,\ \ \forall n\in\{1,...,N_k+1\}. \tag{18}$$

IOP Conf. Series: Journal of Physics: Conf. Series **949** (2017) 012015 doi:10.1088/1742-6596/949/1/012015

i) **Existence and uniqueness and L$^\infty$estimate**
For the existence and uniqueness of the solution of finite volume scheme , we can follow the proof of Lemma 3.2 page 42 in[3].
Let us now prove the estimate (17). Let $X_K \in K$, $n \in \{0, ..., N_k\}$ for all $K \in T$. Then, we claim that

$$\min\left\{u_K^{n+1},\ \ K \in T\right\} \geq \min\left\{\min\left\{u_K^n,\ \ K \in T\right\},\ \ 0\right\} \tag{19}$$

Indeed, if $\min\left\{u_K^{n+1},\ \ K \in T\right\} < 0$, let $K_0 \in T$ such that $u_{K_0}^{n+1} = \min\left\{u_K^{n+1},\ \ K \in T\right\}$. Since $u_{K_0}^{n+1} < 0$ and replacing in the relation (9) K by K_0 to get

$$m(K)\frac{u_{K_0}^{n+1} - u_{K_0}^n}{k} = -\sum_{\sigma\in\Xi_{K_0}} F_{K_0,\sigma}^{n+1} - \sum_{\sigma\in\Xi_{K_0}} V_{K_0,\sigma}\left(u_{\sigma,+}^{n+1} - u_{K_0}^{n+1}\right) - \sum_{\sigma\in\Xi_{K_0}} V_{K_0,\sigma}u_{K_0}^{n+1}$$

with $K_0 \in T$ and $\forall n \in \{0, ..., N_k\}$.
From the definition of $F_{K_0,\sigma}^{n+1}$ and $u_{\sigma,+}^{n+1}$, we deduce that

$$\sum_{\sigma\in\Xi_{K_0}} F_{K_0,\sigma}^{n+1} \leq 0 \ \text{ and } \sum_{\sigma\in\Xi_{K_0}} V_{K_0,\sigma}\left(u_{\sigma,+}^{n+1} - u_{K_0}^{n+1}\right) \leq 0.$$

Next, in view the Proposition 2.1, the function v is positive verifying $-\Delta v + \tau v = 0$ then we find

$$\sum_{\sigma\in\Xi_{K_0}} V_{K_0,\sigma}u_{K_0}^{n+1} = u_{K_0}^{n+1}\int_{\partial K_0}\nabla v.n_{K_0}(x)\gamma(x) = \tau u_{K_0}^{n+1}\int_{K_0} v\ dx \leq 0.$$

Thus, we have

$$u_{K_0}^{n+1} \geq u_{K_0}^n,$$

this proves (19) and by induction, we deduce that:

$$\min\left\{u_K^{n+1}, K \in T\right\} \geq \min\left\{\min\left\{u_K^0, K \in T\right\}, 0\right\}. \tag{20}$$

This implies that

$$\max\left\{u_K^{n+1}, K \in T\right\} \leq \max\left\{\max\left\{u_K^0, K \in T\right\}, 0\right\}.$$

So, the inequality (17) is satisfied with C depends only on u_0.
ii) **Error estimate**
One uses the regularity of the data and the solution to write an equation for the error $e_{K,}^n = u(t^n, x_K) - u_K^n$, defined for $K \in T$ and $n \in \{1, ..., N_k + 1\}$. Note that $e_K^0 = 0$ for $K \in T$. Let $n \in \{1, ..., N_k\}$. Integrating the first equation of problem (P$_1$) over each control volume K of T, at time $t = t^{n+1}$:

$$\int_K u_t\left(t^{n+1}, x\right)dx - \int_{\partial K}\left(\nabla u\left(t^{n+1}, x\right) - u\left(t^{n+1}, x\right)\nabla v(x)\right).n_K(x)\,d\gamma(x) = 0. \tag{21}$$

Subtracting (9) to (21) yields:

$$m(K)\frac{e_K^{n+1} - e_K^n}{k} + \sum_{\sigma\in\Xi_K}\left(G_{K,\sigma}^{n+1} + W_{K,\sigma}^{n+1}\right) = -\sum_{\sigma\in\Xi_K} mes(\sigma)\left(R_{K,\sigma}^n + r_{K,\sigma}^n\right) - S_K^n \ \ \forall K \in T, \tag{22}$$

where

$$\begin{aligned} G_{K,\sigma}^{n+1} &= -\tau_\sigma\left(e_L^{n+1}-e_K^{n+1}\right) \quad \forall\sigma\in\Xi_{int}\cap\Xi_K,\quad \sigma=K\,|L\,, \\ G_{K,\sigma}^{n+1} &= \tau_\sigma e_K^{n+1} \quad \forall\sigma\in\Xi_{ext}\cap\Xi_K, \\ W_{K,\sigma}^{n+1} &= W_{K,\sigma,1}^{n+1}+W_{K,\sigma,2}^{n+1}, \end{aligned}$$

with

$$\begin{cases} e_K=(v\left(x_K\right)-v_K) \\ W_{K,\sigma,1}^{n+1}=\tau_\sigma\left(e_L-e_K\right)u_{\sigma,+}^{n+1}=\tau_\sigma D_\sigma\left(e\right)u_{\sigma,+}^{n+1}, \\ W_{K,\sigma,2}^{n+1}=\tau_\sigma\left(v(x_L)-v(x_K)\right)\left(u\left(t^{n+1},x_{\sigma,+}\right)-u_{\sigma,+}^{n+1}\right)=\tau_\sigma D_\sigma\left(v\right)\left(e_{\sigma,+}^{n+1}\right), \end{cases}$$

where $x_{\sigma,+}=x_K$ (resp. x_L) $\sigma\in\Xi_{int}$, $\sigma=K\,|L$ and $\nabla v.n_{K,\sigma}\geq 0$ (resp. $\nabla v.n_{K,\sigma}<0$) and $x_{\sigma,+}=x_K$ (resp. y_σ) if $\sigma\in\Xi_{ext}\cap\Xi_K$, and $\nabla v.n_{K,\sigma}\geq 0$ (resp. $\nabla v.n_{K,\sigma}<0$).
Using a Taylor expansion and as the regularity of u is considered, then there exists C_1 and C_2 which C_1 depends only on u and Ţ and C_2 depends only on u, v and Ţ such that for all $x\in K$ and all $K\in T$ we have

$$S_K^n=\int_K u_t\left(t^{n+1},x\right)dx-m\left(K\right)\frac{\left(u\left(t^{n+1},x_K\right)-u(t^{n+1},x_K)\right)}{k},$$

$$\begin{aligned} -m\left(\sigma\right)R_{K,\sigma}^n &= \tau_\sigma\left(u\left(t^{n+1},x_K\right)-u(t^{n+1},x_L)\right)+\int_\sigma\nabla u\left(t^{n+1},x\right).n_{K,\sigma}d\delta(x) \text{ if } \sigma=K\,|L\in\Xi_{int}, \\ -m\left(\sigma\right)R_{K,\sigma}^n &= \tau_\sigma\left(u\left(t^{n+1},x_K\right)\right)+\int_\sigma\nabla u\left(t^{n+1},x\right).n_{K,\sigma}d\delta(x) \text{ if } \sigma\in\Xi_{ext}\cap\Xi_K, \end{aligned}$$

$$\begin{aligned} m\left(\sigma\right)r_{K,\sigma}^n &= \tau_\sigma\left(v\left(x_K\right)-v(x_L)\right)u\left(x_{\sigma,+}\right)+\int_\sigma\nabla v(x).n_{K,\sigma}u\left(t^{n+1},x\right)d\delta(x) \text{ if } \sigma=K|L\in\Xi_{int}, \\ m\left(\sigma\right)r_{K,\sigma}^n &= \tau_\sigma(v\left(x_K\right)-g\left(y_\sigma\right))u\left(x_{\sigma,+}\right)+\int_\sigma\nabla v(x).n_{K,\sigma}u\left(t^{n+1},x\right)d\delta(x) \text{ if } \sigma\in\Xi_{ext}\cap\Xi_K, \end{aligned}$$

where

$$S_K^n\leq m(K)C_1\left(h+k\right), \tag{23}$$

and

$$\left|R_{K,\sigma}^n\right|+\left|r_{K,\sigma}^n\right|\leq C_2h \qquad \forall K\in T \text{ and } \sigma\in\Xi_K. \tag{24}$$

Multiplying (22) by e_K^{n+1}, summing for $K\in T$, and noting that

$$\begin{aligned} \sum_{K\in T}\sum_{\sigma\in\Xi_K}G_{K,\sigma}^{n+1}e_K^{n+1} &= \sum_{\sigma\in\Xi}\tau_\sigma\left(e_K^{n+1}\right)^2=\left\|e_T^{n+1}\right\|_{1,\tau}^2, \\ \sum_{K\in T}\sum_{\sigma\in\Xi_k}W_{K,\sigma,2}^{n+1}e_K^{n+1} &\geq 0, \\ \sum_{K\in T}m\left(K\right)\left(e_K^{n+1}\right)^2 &= \left\|e_T^{n+1}\right\|_{L^2(\Omega)}^2, \end{aligned}$$

where $e_T^n\in X\left(T\right)$, $e_T^n\left(x\right)=e_K^n$ for a.e. $x\in K$, we obtain

$$\begin{aligned} \left\|e_T^{n+1}\right\|_{L^2(\Omega)}^2+k\left\|e_T^{n+1}\right\|_{1,T}^2 \leq -k\textstyle\sum_{K\in T}\sum_{\sigma\in\Xi_K}mes(\sigma)\left(R_{K,\sigma}^n+r_{K,\sigma}^n\right)e_K^{n+1}+\sum_{K\in T}k\left|S_K^n\right|e_K^{n+1} \\ -k\textstyle\sum_{K\in T}\sum_{\sigma\in\Xi_K}W_{K,\sigma,1}e_K^{n+1}+\sum_{K\in T}mes\left(K\right)e_K^{n+1}e_K^n. \end{aligned} \tag{25}$$

Using (23), (24) and Cauchy-Schwarz's inequality yields, with some C_3 only depending on u, v, Ω and Ţ

$$\begin{gathered}\left\|e_T^{n+1}\right\|_{L^2(\Omega)}^2+\tfrac{k}{2}\left\|e_T^{n+1}\right\|_{1,T}^2\\ \leq C_3kh^2+C_1k\left(h+k\right)\sum_{K\in T}m\left(K\right)\left|e_K^{n+1}\right|-k\sum_{K\in T}\sum_{\sigma\in\Xi_K}W_{K,\sigma,1}e_K^{n+1}+\sum_{K\in T}m\left(K\right)e_K^{n+1}e_K^n.\end{gathered}\tag{26}$$

Applying again the Cauchy-Schwarz's inequality and the estimate $\left(ab\leq\frac{1}{2}\left(a^2+b^2\right)\right)$ to estimate

$$\begin{aligned}(h+k)\,k\sum_{K\in T}mes(K)e_K^{n+1} &\leq (h+k)\,k\left(\sum_{K\in T}mes(K)\right)^{\frac{1}{2}}\left(\sum_{K\in T}mes(K)\left(e_K^{n+1}\right)^2\right)^{\frac{1}{2}}\\ &= (h+k)\,k\,(mes(\Omega))^{\frac{1}{2}}\left\|e_T^{n+1}\right\|_{L^2(\Omega)}.\end{aligned}$$

and

$$\begin{aligned}\left|\sum_{K\in T}m\left(K\right)e_K^{n+1}e_K^n\right| &\leq \frac{1}{2}\sum_{K\in T}m\left(K\right)\left(e_K^{n+1}\right)^2+\frac{1}{2}\sum_{K\in T}m\left(K\right)\left(e_K^n\right)^2\\ &= \frac{1}{2}\left\|e_T^{n+1}\right\|_{L^2(\Omega)}^2+\frac{1}{2}\left\|e_T^n\right\|_{L^2(\Omega)}^2.\end{aligned}$$

Using the estimation (7) of the Theorem 2.3 to find

$$\begin{aligned}\sum_{K\in T}\sum_{\sigma\in\Xi_K}W_{K,\sigma,1}e_{\sigma,+}^{n+1} &= \sum_{\sigma\in\Xi}\tau_\sigma D_\sigma\left(e\right)u_{\sigma,+}{}^{n+1}\left(e_{\sigma,+}^{n+1}-e_{\sigma,-}^{n+1}\right)\\ &\leq C\left|e_T\right\|_{1,T}\left\|e_T^{n+1}\right\|_{1,T}\\ &\leq C_4h\left\|e_T^{n+1}\right\|_{1,T},\end{aligned}$$

where

$$\begin{cases} e_{\sigma,-}^{n+1}=e_K^n & \text{if } V_{K,\sigma}\leq 0, \quad e_{\sigma,+}^{n+1}=e_K^n \text{ if } V_{K,\sigma}\geq 0\\ e_{\sigma,-}^{n+1}=e_L & \text{if } V_{K,\sigma}>0, \quad e_{\sigma,+}^{n+1}=e_L \text{ if } V_{K,\sigma}<0\end{cases}\quad\text{for all } \sigma\in\Xi_K$$

and $V_{K,\sigma}=\tau_\sigma D_\delta\left(v\right)$. We note that if $\sigma\in\Xi_{ext}$ then $e_{\sigma,-}^{n+1}=e_{\sigma,+}^{n+1}=0$.
Then, we obtain

$$\begin{gathered}\tfrac{1}{2}\left\|e_T^{n+1}\right\|_{L^2(\Omega)}^2+\tfrac{k}{2}\left\|e_T^{n+1}\right\|_{1,T}^2\\ \leq C_3kh^2+C_1\left(h+k\right)k\left(mes(\Omega)\right)^{\frac{1}{2}}\left\|e_T^{n+1}\right\|_{L^2(\Omega)}+C_4hk\left\|e_T^{n+1}\right\|_{1,T}+\tfrac{1}{2}\left\|e_T^n\right\|_{L^2(\Omega)}^2.\end{gathered}\tag{27}$$

So, applying again the estimate $\left(ab\leq\frac{1}{2}\left(a^2+b^2\right)\right)$ to $C_4kh\left\|e_T^{n+1}\right\|_{1,T}$ to deduce

$$\left\|e_T^{n+1}\right\|_{L^2(\Omega)}^2\leq C_5kh^2+2C_1\left(h+k\right)k\left(mes(\Omega)\right)^{\frac{1}{2}}\left\|e_T^{n+1}\right\|_{L^2(\Omega)}+\left\|e_T^n\right\|_{L^2(\Omega)}^2.\tag{28}$$

Since, we obtain

$$\left\|e_T^{n+1}\right\|_{L^2(\Omega)}^2\leq C_6\left(h^2k+\left(h+k\right)k\left\|e_K^{n+1}\right\|_{L^2(\Omega)}\right)+\left\|e_T^n\right\|_{L^2(\Omega)}^2.\tag{29}$$

Similarly as in proof of Theorem 4.1 in [3], we get the estimate (18).

Proposition 3.2. *Under assumption 1. Let T be an admissible mesh. If $u_0\left(x\right)\geq 0$ for all $K\in T$, then the solution $\left(u_K\right)_{K\in T}$ of (9)-(16) satisfied $u_K\geq 0$ for all $K\in T$.*

Proof. We have $u_0\left(x\right)\geq 0$ and from the relation (20), we deduce that

$$\min\left\{u_K^{n+1},K\in T\right\}\geq\min\left\{\min\left\{u_K^0,K\in T\right\},0\right\}=0,$$

therefore, $u_K^{n+1}\geq 0$, $\forall K\in T$. So this proves this Proposition. □

4. Finite volume method for a Elliptic-parabolic problem (P)

A finite volume scheme of the problem (P) can be defined by the following set of equations:

$$m(K)\frac{u_K^{n+1}-u_K^n}{k}+\sum_{\sigma\in\Xi_K}F_{K,\sigma}^{n+1}-\sum_{\sigma\in\Xi_K}F_{K,\sigma}u_{\sigma,+}^{n+1}=0 \quad \forall K\in T, \tag{30}$$

$$\sum_{\sigma\in\Xi_K}F_{K,\sigma}+\tau m(K)v_K=0 \quad \forall K\in T, \tag{31}$$

$$u_K^0=u_0(x_K) \quad \forall K\in T, \tag{32}$$

$$F_{K,\sigma}^n=-\tau_{K|L}\left(u_L^n-u_K^n\right)\ \forall\sigma\in\Xi_{int} \quad \text{if } \sigma=K|L\,, \tag{33}$$

$$F_{K,\sigma}^n=\tau_\sigma\left(u_K^n\right) \quad \forall\sigma\in\Xi_{ext}\cap\Xi_K, \tag{34}$$

$$F_{K,\sigma}=-\tau_{K|L}\left(v_L-v_K\right) \quad \forall\sigma\in\Xi_{int} \quad \text{if } \sigma=K|L\,, \tag{35}$$

$$F_{K,\sigma}=-\tau_\sigma\left(g\left(y_\sigma\right)-v_K\right) \quad \forall\sigma\in\Xi_{ext} \text{ such that } \sigma\in\Xi_K. \tag{36}$$

with $u_{\sigma,+}^{n+1}$ is defined by the relations (13)-(14),(we note that $F_{K,\sigma}=-V_{K,\sigma}$). In this work, we assume that the unknowns u and v are constants over each control volume K of the mesh T.

Now, from sections 2 and 3 we are able to conclude that the main result is represented by the following Theorem .

Theorem 4.1. *Let Ω be an open bounded polygonal subset of $\mathbb{R}^d$, $\mathcal{T}>0$. Let $(u,v)\in C^2(\mathbb{R}_+\times\overline{\Omega},\mathbb{R})\times C^2(\overline{\Omega},\mathbb{R})$ be defined in (P). Let $u_0\in C^2(\overline{\Omega},\mathbb{R})$ and $g\in C^0(\partial\Omega,\mathbb{R}_+)$. Let T be an admissible mesh and $k\in(0,\mathcal{T})$. Then there exists a unique vector $(u_K,v_K)_{K\in T}$ satisfying (30)-(36). Furthermore, let $e_K^n=u(t^n,x_K)-u_K^n$ and $e_K=v(x_K)-v_K$ for $K\in T$ and $n\in\{1,...,N_k+1\}$ and $h=size(T)$. Then there exists C only depending on u, v and $\mathcal{T}$ such that*

$$\left(\sum_{K\in T}m(K)\left(e_K^n\right)^2\right)^{\frac{1}{2}}+\left(\sum_{K\in T}m(K)\left(e_K\right)^2\right)^{\frac{1}{2}}\le C(h+k), \qquad \forall n\in\{1,...,N_k+1\}$$

The demonstration of this Theorem is a consequence of Lemma 2.2, Theorems 2.3 and 3.1.

Proposition 4.2. *Under assumption 1. Let T be an admissible mesh. If $u_0(x)\ge 0$ for all $K\in T$, then the solution $(u_K,v_K)_{K\in T}$ of (30)-(36) is positive for all $K\in T$, ($u_K\ge 0$ and $v_K\ge 0$ $\forall K\in T$)*

The proof of this Proposition follows from the Propositions 2.1 and 3.2.

5. Numerical simulations

In this section we present the results of numerical simulations. Two examples of problem (P) are considered.

- In the first example u_0 and g are positive such that

$$\begin{aligned} u_0 &= \sin(\pi x)\sin(\pi y)\,, \\ g &= \begin{cases} v(0,y)=0 \\ v(1,y)=1 \\ v(x,0)=v(x,1)=1-x^2 \end{cases} \end{aligned}$$

- In the second example we choose

$$u_0 = xy(1-x)(1-y)$$

$$g = \begin{cases} v(0,y) = 0 \\ v(1,y) = \sin\left(\frac{\pi}{2}y\right) \\ v(x,0) = 0 \\ v(x,1) = \sin\left(\frac{\pi}{2}x\right) \end{cases}$$

In our numerical scheme of the finite volume method, we consider $\Omega = [01]^2$ and the time (in second) $t \in [0, 0.2]$. The domain Ω is discretized on a regular square grid of 21×21 nodes so that the spatial step size is $h = 0.05$. For the time, we have used the time step size $k = 0.02s$.

Discussion

In both examples the function g is chosen to be positive.

(i) In figures 1(a) and 2(a), the function v is also positive. Therefore, the principal of maximum is verified.

(ii) In figures 1(b) and 2(b), the initial condition u_0 is positive and the point $(x, y) = (0.5, 0.5)$. The solution is given as a function of time t. It is clear in this case that the solution u of the problem (P) is positive. Therefore, the principal of maximum is verified.

(iii) In figures 1(c) and 2(c), the initial condition u_0 is positive. It is clear from these figures that the maximum of the solution u of the problem (P) remains positive during the time evolution $t \in [0\ 1]$. Moreover, the solution u is also positive. Therefore, the principal of maximum is verified.

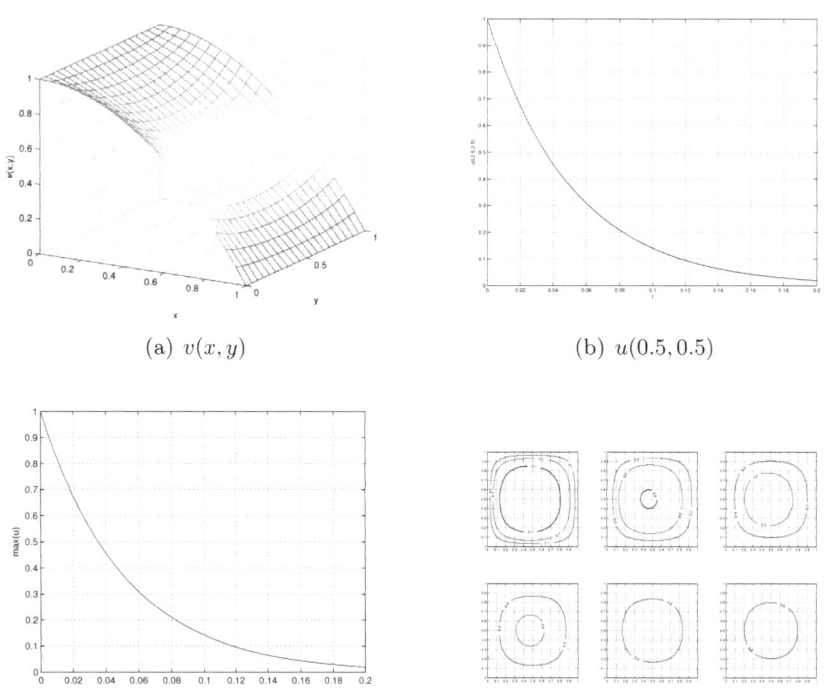

(a) $v(x, y)$

(b) $u(0.5, 0.5)$

(c) Max of u

(d) Countour plot

Figure 1. Numerical solution for the Example 1

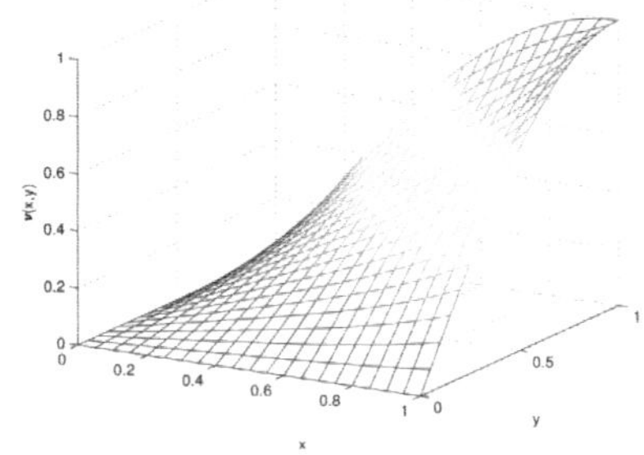

(a) $v(x, y)$

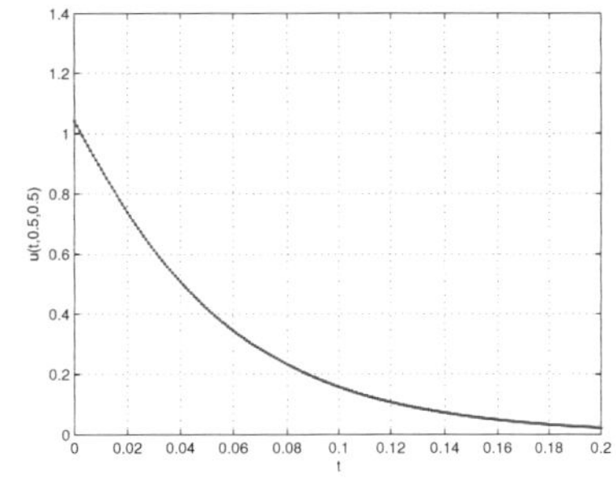

(b) $u(0.5, 0.5)$

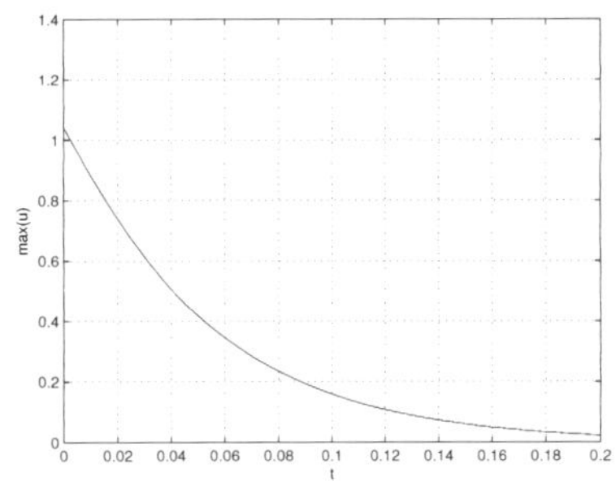

(c) Max of u

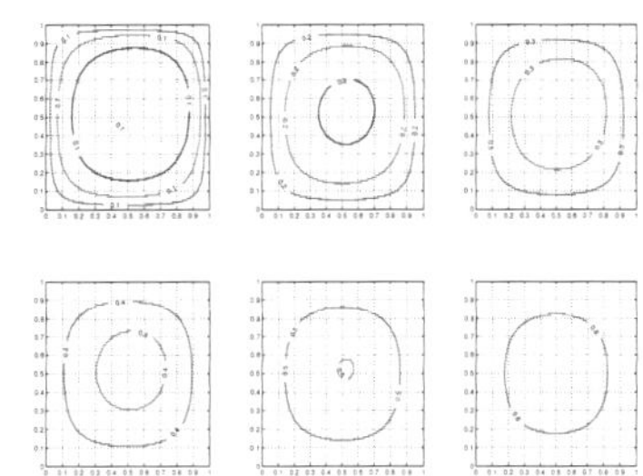

(d) Countour plot

Figure 2. Numerical solution for the Example 2

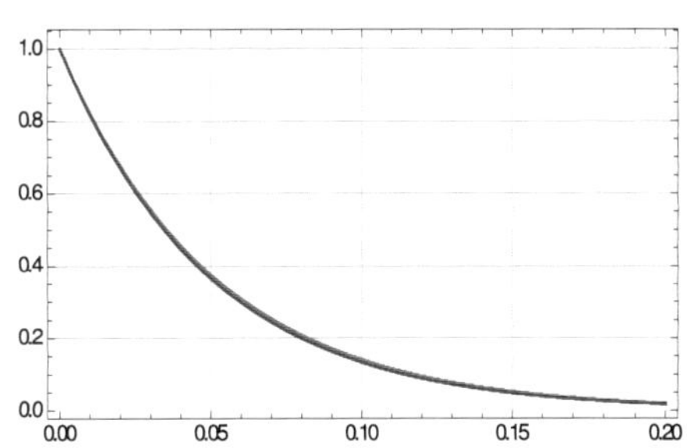

(a) Example 1: $u(t, 0.5, 0.5)$

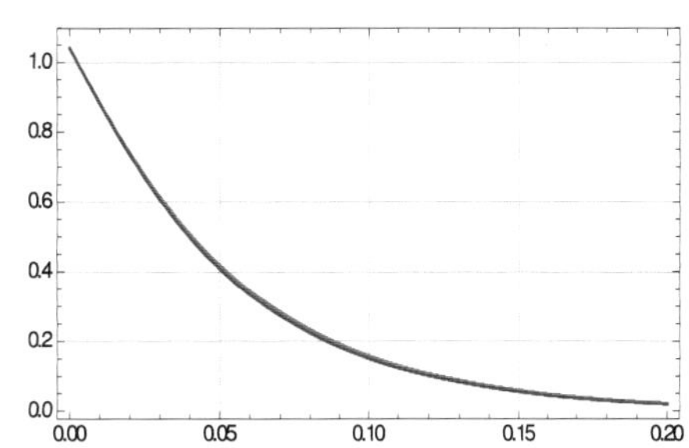

(b) Example 2: $100u(t, 0.5, 0.5)/6$

Figure 3. Comparison between the numerical solutions obtained by finite volume method and the finite element method.

(iv) The figures 1(d) and 2(d) represent the contour plot for the solution. One can see that the solution is always positive.

(v) We remark that for positive values of u_0 and g , the solution is also positive. This means that the principle of maximum remains numerically valid .
We compare the numerical solutions obtained by finite volume method and the finite element method (from Mathematica 10.2T) at the coordinates $(0.5, 0.5)$. There is a good agreement between these two solutions. (see Fig. 3a and 3b

(vi) From the figures. 1(d) and 2(d) we see that during the evolution the amplitude of u decreases.

ACKNOWLEDGMENT
The authors would like to acknowledge the financial supports received from the International Islamic University Malaysia, Ministry of higher education Malaysia through the research grant FRGS-14-143-0384.

References
[1] Marianne Bessemoulin-Chatard and Ansgar Jüngel. A finite volume scheme for a keller–segel model with additional cross-diffusion. *IMA Journal of Numerical Analysis*, page drs061, 2013.
[2] Adrien Blanchet, Vincent Calvez, and José A Carrillo. Convergence of the mass-transport steepest descent scheme for the subcritical patlak–keller–segel model. *SIAM Journal on Numerical Analysis*, 46(2):691–721, 2008.
[3] Robert Eymard, Thierry Gallouët, and Raphaèle Herbin. Finite volume methods. *Handbook of numerical analysis*, 7:713–1018, 2000.
[4] Ibrahim Fatkullin. A study of blow-ups in the keller–segel model of chemotaxis. *Nonlinearity*, 26(1):81, 2012.
[5] Miguel A Herrero and J J L Velázquez. A blow up mechanism for a chemotaxis model. *Annali della Scuola Normale Superiore di Pisa. Classe di Scienze. Serie IV*, 24(4):633–683, 1997.
[6] T Hillen and K Painter. Global existence for a parabolic chemotaxis model with prevention of overcrowding. *Advances in Applied Mathematics*, 26(4):280 – 301, 2001.
[7] Evelyn F Keller and Lee A Segel. Initiation of slime mold aggregation viewed as an instability. *J. Theor. Biol.*, 26:399–415, 1970.
[8] C Messikh, A Guesmia, and S Saadi. Global existence and uniqueness of the weak solution in keller segel model. *GJSFR*, 14:146–55, 2014.
[9] Clifford S Patlak. Random walk with persistence and external bias. *The bulletin of mathematical biophysics*, 15(3):311–338, 1953.
[10] Michael Winkler. Absence of collapse in a parabolic chemotaxis system with signal-dependent sensitivity. *Mathematische Nachrichten*, 283:1664–1673, 2010.
[11] Michael Winkler. Aggregation vs. global diffusive behavior in the higher-dimensional keller segel model. *Journal of Differential Equations*, 248(12):2889–2905, 2010.

IOP Conf. Series: Journal of Physics: Conf. Series **949** (2017) 012016 doi:10.1088/1742-6596/949/1/012016

The effect of convective boundary condition on MHD mixed convection boundary layer flow over an exponentially stretching vertical sheet

Siti Suzilliana Putri Mohamed Isa[1, 2*], Norihan Md. Arifin[1, 3], Roslinda Nazar[4], Norfifah Bachok[1, 3], and Fadzilah Md Ali[1, 3]

[1]Institute for Mathematical Research, Universiti Putra Malaysia, 43400 UPM Serdang, Selangor Darul Ehsan, MALAYSIA.
[2]Pusat Asasi Sains Pertanian, Universiti Putra Malaysia, 43400 UPM Serdang, Selangor Darul Ehsan, MALAYSIA.
[3]Department of Mathematics, Universiti Putra Malaysia, 43400 UPM Serdang, Selangor Darul Ehsan, MALAYSIA.
[4]School of Mathematical Sciences, Universiti Kebangsaan Malaysia, 43600 UKM, Bangi Selangor, MALAYSIA

*Corresponding author: ctsuzilliana@upm.edu.my

Abstract. A theoretical study that describes the magnetohydrodynamic mixed convection boundary layer flow with heat transfer over an exponentially stretching sheet with an exponential temperature distribution has been presented herein. This study is conducted in the presence of convective heat exchange at the surface and its surroundings. The system is controlled by viscous dissipation and internal heat generation effects. The governing nonlinear partial differential equations are converted into ordinary differential equations by a similarity transformation. The converted equations are then solved numerically using the shooting method. The results related to skin friction coefficient, local Nusselt number, velocity and temperature profiles are presented for several sets of values of the parameters. The effects of the governing parameters on the features of the flow and heat transfer are examined in detail in this study.

1. Introduction

Mixed convection flows, or combination of forced and natural convection, are often found in very high power output devices in which the forced convection does not dissipate the necessary heat. In such cases, combining natural convection with forced convection produces the desired results. Nuclear reactor technology and electronic cooling [1] are some of the examples of the engineering applications of mixed convection. In addition, the flow and thermal boundary layers of a continuous stretching surface have wide applications in metallurgical processes (drawing of continuous filaments using quiescent fluids, annealing and tinning of copper wires, glass blowing, manufacturing of plastic and rubber sheets, crystal growing, and continuous cooling and fibre spinning [2]). Many researchers

IOP Conf. Series: Journal of Physics: Conf. Series **949** (2017) 012016 doi:10.1088/1742-6596/949/1/012016

considered linear, nonlinear, quadratic, and exponential variations of stretching velocity and temperature distributions. However, Magyari and Keller [3] were the first who investigated the flow and thermal boundary layers on a stretching sheet, which has exponential variations of velocity and temperature. As a result, the findings due to the problems of flow induced by an exponential variation of stretching surface have been produced [4-10]. Furthermore, the cases of the mixed convection fluid flow over an exponentially stretching sheet have been reported [11-18]. The convective heat transfer involves in engineering procedures, namely, gas turbines, nuclear plants, storage of thermal energy etc [19]. These processes obtain high temperature, which the flow is subjected to the convective boundary condition. Therefore, a number of studies on boundary layer flows with a convective boundary condition, including mixed convection boundary layer flow [20–27] and a stretching sheet [28–35], have been reviewed since then.

Motivated by the above researchers, the purpose of this research paper is to develop the system of MHD mixed convection heat transfer of an electrically conducting fluid over an exponentially stretching continuous surface with an exponential temperature distribution. The system is maintained using applied convective boundary conditions, exponential magnetic field, viscous dissipation, buoyancy force, and internal heat generation. The mathematical formulation reported by Pal [13] is followed, which the impermeable surface is assumed to be stretched with an exponential velocity $U_w = U_0 e^{(x/L)}$ and the surface is maintained at a temperature $T_w(x) = T_\infty + (T_0 - T_\infty)e^{ax/2L}$. The nonlinear boundary value problems are solved numerically using the Runge–Kutta method with a shooting technique. Numerical solutions are obtained by the form of tables and graphs, to illustrate the characteristics of the boundary layer flow and heat transfer.

2. Mathematical formulation

Consider the two-dimensional flow of an incompressible, viscous, and electrically conducting fluid over an exponentially impermeable stretching surface (see Figure 1) with a magnetic field $B(x)$. The x-axis is parallel to the stretching surface in the direction of the movement of the sheet and the y-axis is perpendicular to it. A transverse magnetic field is applied in the y-axis. The induced magnetic field in equation (2) is assumed to be negligible since the magnetic Reynolds number for the flow is very small. The impermeable surface is assumed to be stretched with an exponential velocity $U_w = U_0 e^{(x/L)}$ in a quiescent fluid and the surface is maintained at a temperature $T_w(x) = T_\infty + (T_0 - T_\infty)e^{ax/2L}$, where T_∞ is the temperature of the ambient fluid and T_0 is the reference temperature. We have followed the formulation of U_w and $T_w(x)$ by Pal [13]. Considering the assumptions of Boussinesq and boundary layer approximations, the flow and heat transfer problems are governed by the following equations:

$$\frac{\partial u}{\partial x} + \frac{\partial v}{\partial y} = 0 \tag{1}$$

$$u\frac{\partial u}{\partial x} + v\frac{\partial u}{\partial y} = \upsilon\frac{\partial^2 u}{\partial y^2} + g\beta(T - T_\infty) - \frac{\sigma B^2(x)}{\rho}u \tag{2}$$

$$u\frac{\partial T}{\partial x} + v\frac{\partial T}{\partial y} = \alpha\frac{\partial^2 T}{\partial y^2} + \frac{\sigma}{\rho c_p}B^2(x)u^2 + \frac{\mu}{\rho c_p}\left(\frac{\partial T}{\partial y}\right)^2 + \frac{Q}{\rho c_p}(T - T_\infty) \tag{3}$$

where u and v are the components of velocity in the x and y directions, respectively, $\upsilon = \mu/\rho$ is the kinematic viscosity, μ is the viscosity, ρ is the fluid density, g is acceleration due to gravity, β is the thermal expansion coefficient, σ is the electrical conductivity, α is the thermal diffusivity, T is the temperature of the fluid, c_p is the specific heat of the fluid at a constant pressure, and $Q > 0$ is the internal heat generation coefficient. In Equation (2), we consider the special form of magnetic field, $B(x) = B_0 e^{(x/2L)}$.

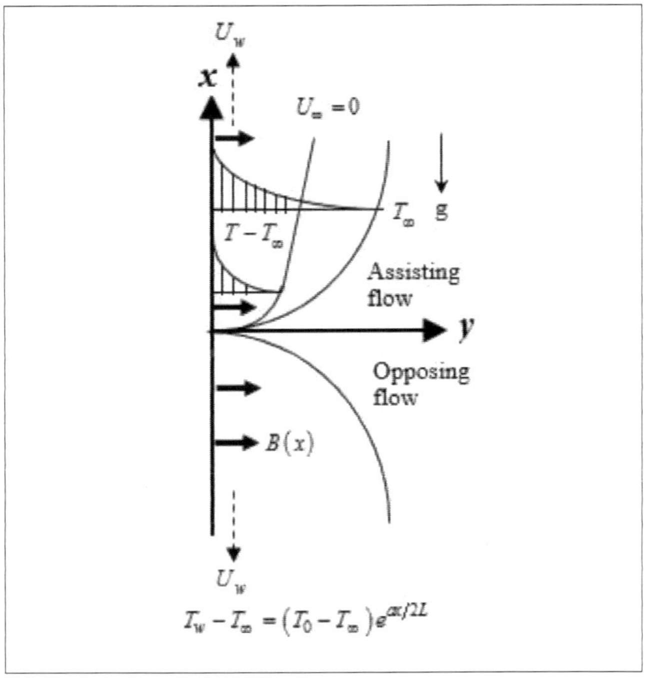

Figure 1. Physical model and coordinate system.

The appropriate boundary conditions are as follows:

$$u = U_w(x) = U_0 e^{(x/L)}, \qquad v = 0, \qquad -k\frac{\partial T}{\partial y} = h_0(T_0 - T) \qquad \text{at } y = 0$$
$$u \to 0 \qquad T \to T_\infty \qquad \text{as } y \to \infty \tag{4}$$

where k is the thermal conductivity of the fluid and h_0 is the convective heat transfer coefficient.

The new similarity variables that were introduced are as follows:

$$\eta = \frac{y}{L}\left(\frac{\text{Re}}{2}\right)^{\frac{1}{2}} e^{\left(\frac{x}{2L}\right)}, \quad \psi(x,y) = \sqrt{2\text{Re}}\,\upsilon e^{(x/2L)} f(\eta), \quad T(x,y) = T_\infty + (T_0 - T_\infty)e^{ax/2L}\theta(\eta) \tag{5}$$

where ψ is the stream function which is defined as $u = \partial\psi/\partial y$ and $v = -\partial\psi/\partial x$, and a as the similarity variable for temperature $T(x,y)$ is a parameter of the temperature distribution in the stretching surface.

Substituting similarity variables (5) into equations (1)–(3), we obtain

$$f''' + ff'' - 2(f')^2 - 2\frac{Ha^2}{\text{Re}} f' + 2\lambda\left(e^{aX/2}\right)\left(e^{-2X}\right)\theta = 0 \tag{6}$$

$$\frac{1}{\Pr}\theta'' + f\theta' - af'\theta + Ec\left(e^{2X-(aX/2)}\right)\left[2\frac{Ha^2}{\mathrm{Re}}\left(f'(\eta)\right)^2 + \left(f''(\eta)\right)^2\right] + 2A^*e^{-X}\theta = 0 \tag{7}$$

where the prime denotes differentiation with respect to η, and the boundary conditions reduce to

$$\begin{aligned} &f'(\eta)=1, \quad f(\eta)=0, \quad \theta(\eta)=(1/Bi)\theta'(\eta)+\left(1/e^{aX/2}\right) && \text{at } \eta=0 \\ &f'(\eta)\to 0 \quad \theta(\eta)\to 0 && \text{as } \eta\to\infty \end{aligned} \tag{8}$$

where $X = x/L$ is the dimensionless coordinate along the plate; $Ha = \sqrt{\sigma B_0{}^2 L^2/\rho\upsilon}$ is the Hartmann number; $\mathrm{Re} = U_0 L/\upsilon$ is the Reynolds number; $\lambda = Gr/\mathrm{Re}^2$ is the constant mixed convection parameter, where $Gr = \left[g\beta(T_0 - T_\infty)L\right]/U_0{}^2$ is the Grashof number; $\Pr = \upsilon/\alpha$ is the Prandtl number; $Bi = h_0 y/\eta k$ is the Biot number; and $A^* = QL^2/\left(\mu c_p \mathrm{Re}\right)$ is the dimensionless heat generation parameter.

In the above similarity equations, the effect of the magnetic field is considered as a ratio of the Hartmann number to the Reynolds number. Note that $\lambda > 0$, $\lambda < 0$, and $\lambda = 0$ correspond to assisting flow, opposing flow, and forced convection flow, respectively. The value $\lambda > 0$ refers to the assisting flow parameter, whereas the opposing flow parameter, denoted as $\lambda < 0$, refers to the value of mixed convection parameter.

The skin friction coefficient C_f and the Nusselt number Nu_x, which are the physical parameters of interest in the present problem, are given by

$$C_f = \frac{2\rho\upsilon}{\rho U_w{}^2(x)}\left(\frac{\partial u}{\partial y}\right)_{y=0}, \quad Nu_x = -\frac{x}{T_w - T_\infty}\left(\frac{\partial T}{\partial y}\right)_{y=0} \tag{9}$$

Substituting Equation (5) into Equation (9), we obtain

$$\frac{C_f\sqrt{\mathrm{Re}_x}}{\sqrt{2X}} = f''(0), \quad \frac{Nu_x}{\sqrt{\mathrm{Re}_x}\left(\sqrt{X/2}\right)} = -\theta'(0), \tag{10}$$

where $\mathrm{Re}_x = xU_w(x)/\upsilon$ is the Reynolds number.

2.1. Numerical method for solution

The boundary value problems in equations (6)–(8) are solved by converting them into an initial value problem using the shooting method. The fourth-order Runge–Kutta integration scheme is applied to solve the initial value problem. We set

$$\begin{aligned} f' &= fp, \\ fp' &= fpp, \\ fpp' &= -f(fpp) + 2(fp)^2 + 2\left(Ha^2/\mathrm{Re}\right)fp - 2\lambda\left(e^{aX/2}\right)\left(e^{-2X}\right)\theta. \end{aligned} \tag{11}$$

$$\begin{aligned} \theta' &= \theta p, \\ \theta p' &= -\Pr(f\theta p) + a\Pr(fp\theta) - \Pr Ec\left(e^{2X-(aX/2)}\right)\left[2\left(Ha^2/\mathrm{Re}\right)(fp)^2 + (fpp)^2\right] \end{aligned} \tag{12}$$

with the boundary conditions

$$f=0,\quad fp=1,\quad \theta=\left(1/Bi\right)\theta p+\left(1/e^{aX/2}\right)\qquad \text{at } \eta=0$$
$$fp\to 0,\quad \theta\to 0\qquad \text{as } \eta\to\infty \tag{13}$$

In order to carry out the integration in equations (11) and (12) as an initial value problem, we assume initial values for $fpp(0)$, i.e. $f''(0)$ and $\theta p(0)$, i.e. $\theta'(0)$. We also assume a suitable finite value for $\eta\to\infty$, say η_∞. Then, we calculate the values of $f'(\eta_\infty)$ and $\theta(\eta_\infty)$, by adjusting the assumed values of $f''(0)$ and $\theta'(0)$ to give a better approximation of the solution. We consider that the assumed values of η_∞, $f''(0)$ and $\theta'(0)$ satisfy the boundary conditions in Equation (13) $\left(f'(\eta_\infty)=0 \text{ and } \theta(\eta_\infty)=0\right)$. This steps is repeated with higher value of η_∞ until two consecutive values of $f''(0)$ and $\theta'(0)$ vary significantly by a specified value. Finally, the most appropriate value of η_∞ is selected, which is the largest η_∞ that satisfies the solution.

3. Results and discussion

Numerical computations have been performed for the velocity profile $f'(\eta)$, the temperature profile $\theta(\eta)$, the skin friction coefficient $f''(0)$ and the local Nusselt number $-\theta'(0)$ for various values of physical parameters, such as mixed convection parameter λ, parameter of the temperature distribution in the stretching surface a, Hartmann number Ha^2/Re, Biot number Bi, Eckert number Ec, dimensionless heat generation parameter A^* and dimensionless coordinate along the plate X. To examine the accuracy of the numerical method, we compare our results with those obtained by Magyari and Keller [3], Al-Odat et al. [5], and Pal [13] in Table 1. In this table, the comparison is made for the nonmagnetic case $Ha^2/\mathrm{Re}=0$, non-buoyant flow $\lambda=0$, and when $Bi=Ec=X=A^*=0$. This table also shows the comparison of the values of $-\theta'(0)$ for different values of Pr and a. From this table, it can be observed that the present results are in very good agreement with the previous results, which proves our theoretical study and numerical computation.

On the other hand, multiple solutions are obtained from the numerical shooting technique. Stability analysis determines which solutions correspond to physically meaningful solution. Merkin [36] have performed a stability analysis of dual solutions and stated that upper branch (or namely as the first solution) is stable and owns physical meaning in real situation. Otherwise, lower branch (second and the subsequent solutions) is unstable.

3.1. Mixed convection

The illustrations of $f''(0)$ and $-\theta'(0)$ versus mixed convection parameter λ are shown in Figures 2 and 3, respectively. For the case of the assisting flow $(\lambda>0)$, the occurrence of three solutions can be seen when the value of mixed convection parameter $\lambda\geq 0.01234$. The point at $\lambda=0.01234$ is the minimum point of $f''(0)$ and $-\theta'(0)$ for the third solution. Thus, it has only two solutions for the values of λ between 0 and 0.01234 for the assisting flow. For the case of the opposing flow $(\lambda<0)$, two solutions are obtained in the range $-0.24616<\lambda<0$. The first and second solutions are combined at the critical point $\lambda_c=-0.24616$. A unique solution is obtained at the critical point $\lambda_c=-0.24616$, whereas no solutions exist for a small value of mixed convection parameter $\lambda<-0.24616$. Finally, three solutions are obtained for a large value of assisting flow parameter. However, two solutions are obtained for a small value of assisting and opposing flow parameters. When a very small value of mixed convection parameter is considered (large opposing parameter), no solutions are found. In Figure 2, it is shown that the values of $f''(0)$ increase with an increase in λ for all solution profiles.

The values of $f''(0)$ are close to each other for all solution profiles for a larger λ. The values of $-\theta'(0)$ decrease when the value of λ approaches 0 for the third solution profile.

Table 1. Comparison of local Nusselt number $-\theta'(0)$ calculated by (...) Magyari and Keller [3], <...> Al-odat et al. [5], {...} Pal [13] and * * present method.

a	Pr					
	0.5	1	3	5	8	10
-1.5	(−0.20405)	(−0.37741)	(−0.92386)	(−1.35324)	(−1.88850)	(−2.20000)
	<−0.19191>	<−0.36152>	<−0.90309>	<−1.34143>	<−1.82858>	<−2.13693>
	{−0.20405}	{−0.37741}	{−0.92386}	{−1.35324}	{−1.88849}	{−2.20003}
	−0.20405	*−0.37741*	*−0.92386*	*−1.35324*	*−1.88849*	*−2.20003*
−0.5	(0.17582)	(0.29988)	(0.63411)	(0.87043)	(1.15032)	(1.30861)
	<0.18187>	<0.32697>	<0.67215>	<0.84156>	<1.08391>	<1.25074>
	{0.17582}	{0.29988}	{0.63411}	{0.87043}	{1.15032}	{1.30861}
	0.17582	*0.29988*	*0.63411*	*0.87043*	*1.15032*	*1.30861*
0.0	(0.33049)	(0.54964)	(1.12219)	(1.52124)	(1.99185)	(2.25743)
	<0.31006>	<0.53104>	<1.08522>	<1.47558>	<1.92633>	<2.18847>
	{0.33049}	{0.54964}	{1.12209}	{1.52124}	{1.99184}	{2.25742}
	0.33049	*0.54964*	*1.12209*	*1.52124*	*1.99184*	*2.25742*
1.0	(0.59434)	(0.95478)	(1.86908)	(2.50014)	(3.24213)	(3.66038)
	<0.57771>	<0.91903>	<1.81039>	<2.28864>	<3.00587>	<3.18620>
	{0.59434}	{0.95478}	{1.86907}	{2.50013}	{3.24212}	{3.66037}
	0.59434	*0.95478*	*1.86907*	*2.50013*	*3.24212*	*3.66037*
3.0	(1.00841)	(1.56029)	(2.93854)	(3.88656)	(5.00047)	(5.62820)
	<0.97665>	<1.46569>	<2.89007>	<3.78072>	<4.86245>	<5.58576>
	{1.00841}	{1.56030}	{2.93853}	{3.88656}	{5.00046}	{5.62820}
	1.00841	*1.56030*	*2.93853*	*3.88656*	*5.00046*	*5.62820*

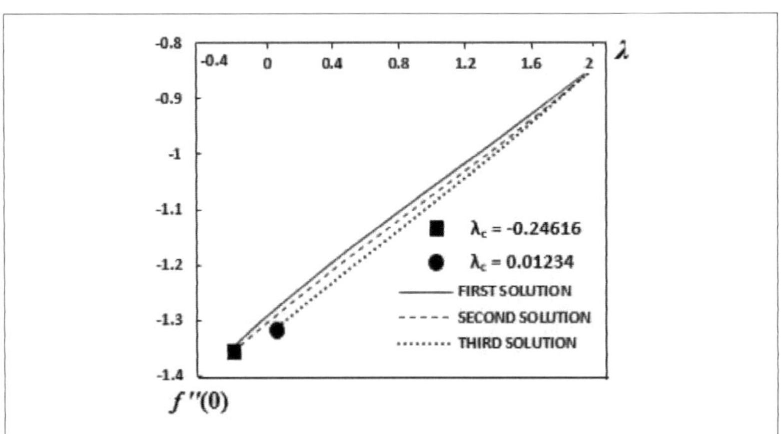

Figure 2. Variation of $f''(0)$ with mixed convection parameter λ.

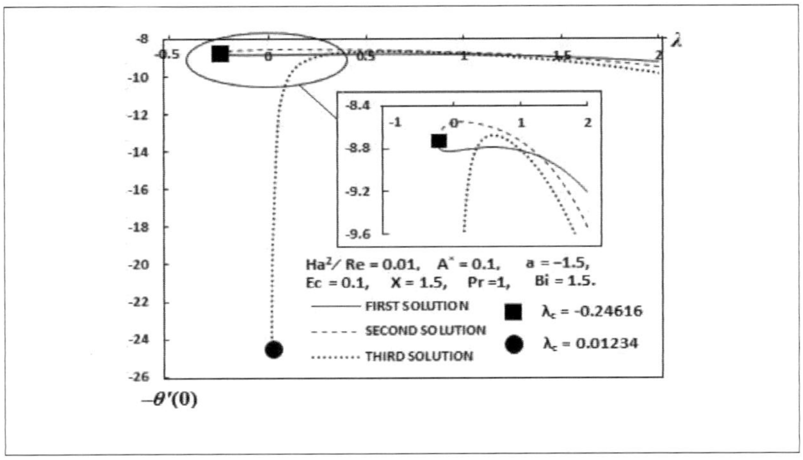

Figure 3. Variation of $-\theta'(0)$ with mixed convection parameter λ.

3.2 The assisting flow case, $\lambda > 0$

To understand the characteristics of the flow and heat transfer in the assisting flow case, the variations in skin friction coefficient $f''(0)$, local Nusselt number $-\theta'(0)$, velocity profile $f'(\eta)$, and temperature profile $\theta(\eta)$ are presented in figures and tables. The velocity profiles $f'(\eta)$ for $\lambda = 2$ are shown in Figure 4. The velocity of the first solution declines continuously with an extension in the distance of the point from the stretching sheet. In the second and third solution profiles, the velocity $f'(\eta)$ initially decreases at a point and reaches to a certain negative value, then for higher values of η it starts to increase. The velocity of the second and third solution profiles reaches a very high value in the boundary layer with a negative fluid velocity. In other words, reversal flow arises at the fluid near the edge of the stretching sheet. Figure 5 shows the variations in temperature profiles when $\lambda = 2$. In this figure, the temperature is found to increase initially at a point for all profiles, but it drops for a large value of η. The occurrence of a peak for a small value of η indicates that there is the highest temperature value at the point of the fluid near the stretched wall. The increment in the temperature difference between the shrieked wall and the adjacent fluid is indicated by the increase in the temperature peak value. As a result, the heat transfer process from the stretching sheet to the ambient fluid is improved.

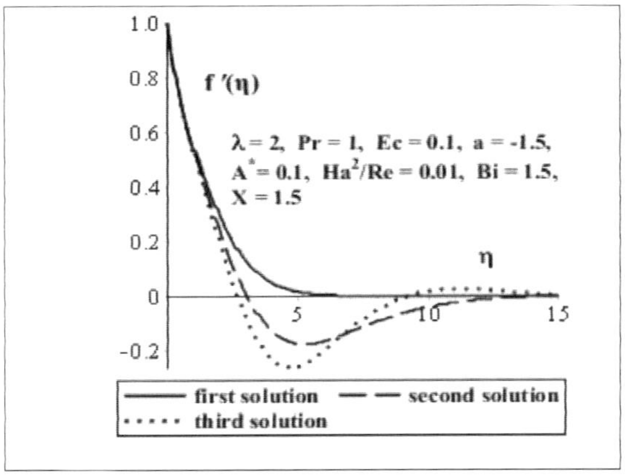

Figure 4. Velocity profiles when $\lambda = 2$.

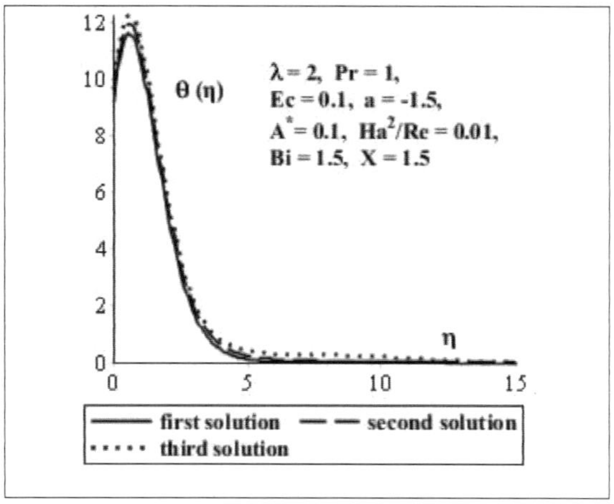

Figure 5. Temperature profiles when $\lambda = 2$.

Tables 2 and 3 show the values of skin friction coefficient $f''(0)$ and local Nusselt number $-\theta'(0)$ for different values of X, a, Ha^2/Re, Bi, Ec, and A^* for the case of the assisting flow. From these tables, it can be seen that for the increasing values of X, the values of $f''(0)$ and $-\theta'(0)$ decrease for all solution profiles. Table 3 shows that increasing parameter a increases the rate of local Nusselt number, whereas the effect of a decreases the rate of skin friction coefficient. It is found that for the first solution profile the value of $f''(0)$ reduces with an increase in magnetic field, whereas for other solution profiles it shows a reverse trend. This table also shows that the effect of magnetic field reduces the rate of local Nusselt number. The results from Table 2 indicate that the skin friction coefficient decreases with an increment in Biot number, but an opposite effect has been observed in the rate of local Nusselt number. However, the effect of Ec and A^* on the values of $f''(0)$ and $-\theta'(0)$ shows reverse characteristics when compared with the effect of the increase in Biot number.

Table 2. Values of the skin friction coefficient $f''(0)$ for several sets of values of the parameters when $\lambda = 2$ and $\text{Pr} = 1$.

Parameters	Values	$f''(0)$		
		First solution	Second solution	Third solution
X	1.47	−0.8172	−0.8199	−0.8183
	1.49	−0.8306	−0.8332	-0.8317
a	-1.49	−0.8486	−0.8502	−0.8481
	-1.47	−0.8690	−0.8691	−0.8662
Ha^2/Re	0.02	−0.8375	−0.8393	−0.8379
	0.04	−0.8385	−0.8376	−0.8367
Bi	1.6	−0.8670	−0.8701	−0.8695
	1.8	−0.9092	−0.9132	−0.9139
Ec	0.11	−0.8166	−0.8128	−0.8071
	0.13	−0.7752	−0.7545	−0.7416

A^*	0.094	−0.8439	−0.8467	−0.8456
	0.098	−0.8393	−0.8420	−0.8406

[a] While studying the effect of individual parameters the following values are used $a=-1.5$, $X=1.5$, $Bi=1.5$, $Ec=0.1$, $Ha^2/\mathrm{Re}=0.01$, and $A^*=0.1$.

Table 3. Values of the local Nusselt number $-\theta'(0)$ for several sets of values of the parameters when $\lambda=2$ and $\mathrm{Pr}=1$.

Parameters	Values	$-\theta'(0)$		
		First solution	Second solution	Third solution
X	1.47	−8.8429	−9.1517	−9.4545
	1.49	−9.0823	−9.4063	−9.7290
a	-1.49	−8.8106	−9.1613	−9.5034
	-1.47	−8.1078	−8.4835	−8.8366
Ha^2/Re	0.02	−9.4528	−9.8197	−10.1080
	0.04	−9.9440	−10.4318	−10.5848
Bi	1.6	−8.6862	−9.0038	−9.3208
	1.8	−7.9822	−8.2798	−8.5763
Ec	0.11	−9.8097	−10.3287	−10.7515
	0.13	−11.0357	−12.0666	−12.6371
A^*	0.094	−9.0167	−9.3370	−9.6766
	0.098	−9.1438	−9.4717	−9.8070

[a] While studying the effect of individual parameters the following values are used $a=-1.5$, $X=1.5$, $Bi=1.5$, $Ec=0.1$, $Ha^2/\mathrm{Re}=0.01$, and $A^*=0.1$.

3.3 The opposing flow case, $\lambda < 0$

To compare the findings of the previous section, the numerical calculations of $f''(0)$, $-\theta'(0)$, $f'(\eta)$ and $\theta(\eta)$ for $\lambda<0$ are shown in this section. Figure 6 and 7 show the velocity $f'(\eta)$ and temperature $\theta(\eta)$ profiles when $\lambda=-0.1$. Velocity profiles for $\lambda<0$ show the same characteristics as for $\lambda>0$, except the nonexistence of the third solution profile. In Figure 7, the temperature is found to enhance initially at a point, but it reduces for a large value of η. The temperature of the first solution profile near the stretching sheet is higher than that of the second solution profile; however, it shows a reverse behaviour for the distance at an adjacent fluid.

4.
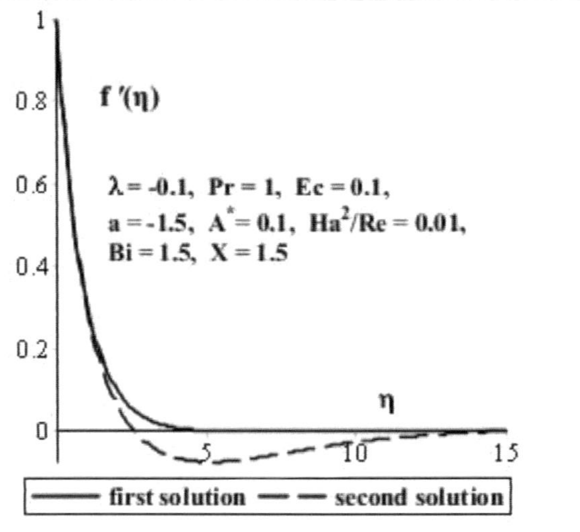

Figure 6. Velocity profiles when $\lambda = -0.1$

5.
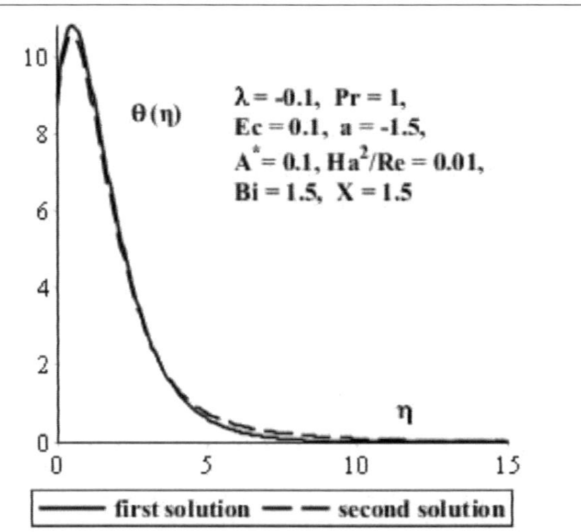

Figure 7. Temperature profiles when $\lambda = -0.1$

To study the effect of several physical parameters on the rate of skin friction coefficient and local Nusselt number, their values are presented in Table 4 and 5. In Table 5, the rate of local Nusselt number reduces due to the effect of parameter X, whereas this parameter increases the values of skin friction coefficient. From this, it is inferred that the effects of a and Bi increase the skin friction coefficient and local Nusselt number. It is evident from these two figures that the values of both $f''(0)$ and $-\theta'(0)$ reduce by enhancing the values of Ha^2/Re, Ec, and A^*.

IOP Conf. Series: Journal of Physics: Conf. Series **949** (2017) 012016 doi:10.1088/1742-6596/949/1/012016

Table 4. Values of the skin friction coefficient $f''(0)$ for several sets of values of the parameters when $\lambda = -0.2$ and $\Pr = 1$.

Parameters	Values	$f''(0)$	
		First solution	Second solution
X	1.47	−1.3459	−1.3527
	1.49	−1.3444	−1.3517
a	-1.49	−1.3431	−1.3507
	-1.47	−1.3418	−1.3499
Ha^2/Re	0.02	−1.3516	−1.3598
	0.04	−1.3673	−1.3768
Bi	1.6	−1.3407	−1.3489
	1.8	−1.3361	−1.3452
Ec	0.11	−1.3471	−1.3537
	0.13	−1.3544	−1.3583
A^*	0.04	−1.3362	−1.3463
	0.08	−1.3408	−1.3494

[a] While studying the effect of individual parameters the following values are used $a = -1.5$, $X = 1.5$, $Bi = 1.5$, $Ec = 0.1$, $Ha^2/\mathrm{Re} = 0.01$, and $A^* = 0.1$.

4. Conclusion

The problems of mixed convection heat transfer in a steady flow with the effect of convective surface boundary condition, magnetic field, viscous dissipation, and internal heat generation are theoretically discussed in this study. We focus on the flow of a viscous fluid, which passes an exponentially stretching continuous surface with an exponential temperature variation. It is observed that the three solution profiles exist for both skin friction coefficient and local Nusselt number when the flow is assisting, i.e. with the large assisting flow parameter, and only two solution profiles exist for the small values of opposing and assisting flow parameters. In order to show the solution profiles for mixed convection, curves are drawn for velocity profile, temperature profile, skin friction coefficient and local Nusselt number in the figures. As discussed in the previous section, it is found that the characteristics of the boundary layer flow and heat transfer are significantly affected by the physical parameters.

Table 5. Values of the local Nusselt number $-\theta'(0)$ for several sets of values of the parameters when $\lambda = -0.2$ and $\Pr = 1$.

Parameters	Values	$-\theta'(0)$	
		First solution	Second solution
X	1.47	−8.2950	−8.1372
	1.49	−8.6336	−8.4657
a	-1.49	−8.6229	−8.4664
	-1.47	−8.2623	−8.1372
Ha^2/Re	0.02	−8.9018	−8.6995
	0.04	−9.0826	−8.8138
Bi	1.6	−8.5655	−8.4059
	1.8	−8.1798	−8.0481
Ec	0.11	−9.4742	−9.3235
	0.13	−10.8043	−10.7178
A^*	0.04	−7.6429	−7.5003
	0.08	−8.3795	−8.2068

[a] While studying the effect of individual parameters the following values are used $a = -1.5$, $X = 1.5$, $Bi = 1.5$, $Ec = 0.1$, $Ha^2/\mathrm{Re} = 0.01$, and $A^* = 0.1$.

References

[1] Joye D D, Bushinsky J P, Saylor, P E 1989 Mixed convection heat transfer at high Grashof number in a vertical tube. *Ind. Eng. Chem. Res.* **28** 1899–1903.

[2] Hady F M, Ibrahim F S, Abdel-Gaied S M, Eid M R 2012 Radiation effect on viscous flow of a nanofluid and heat transfer over a nonlinearly stretching sheet. *Nanoscale Res. Lett.* **7** 229.

[3] Magyari E, Keller B 1999 Heat and mass transfer in the boundary layers on an exponentially stretching continuous surface. *J. Phys. D: Appl. Phys.* **32** 577–585.

[4] Elbashbeshy E M A 2001 Heat transfer over an exponentially stretching continuous surface with suction. *Arch. Mech.* **53** 643–651.

[5] Al-Odat M Q, Damseh R A, Al-Azab T A 2006 Thermal boundary layer on an exponentially stretching continuous surface in the presence of magnetic field effect. *Int. J. of Applied Mechanics and Engineering* **11** 289–299.

[6] Sajid M, Hayat T 2008 Influence of thermal radiation on the boundary layer flow due to an exponentially stretching sheet. *Int. Commun. Heat Mass* **35** 347–356.

[7] Ishak A 2011 MHD boundary layer flow due to an exponentially stretching sheet with radiation effect. *Sains Malays.* **40** 391–395.

[8] Fang T 2004 Influences of fluid property variation on the boundary layers of a stretching surface. *Acta Mech.* **171** 105–118.

[9] Mabood F, Khan W A, Ismail A I Md 2014 MHD flow over exponential radiating stretching sheet using homotopy analysis method. *JKSUES* **29** 68-74.

[10] Weidman P 2016 Flows induced by exponential stretching and shearing plate motions. *Phys. Fluids* 28 113602

[11] Partha M K, Murthy P V S N, Rajasekhar G P 2005 Effect of viscous dissipation on the mixed convection heat transfer from an exponentially stretching surface. *Heat Mass Transf.* **41** 360–366.
[12] El-Aziz M A 2009 Viscous dissipation effect on mixed convection flow of a micropolar fluid over an exponentially stretching sheet. *Can. J. Phys.* **87** 359–368.
[13] Pal D 2010 Mixed convection heat transfer in the boundary layers on an exponentially stretching surface with magnetic field. *Appl. Math. Comput.* **217** 2356–2369.
[14] Kumari M, Nath G 2014 Steady mixed convection flow of Maxwell fluid over an exponentially stretching vertical surface with magnetic field and viscous dissipation. *Meccanica* **49** 1263–1274.
[15] Hayat T, Ashraf B, Shehzad S A, Alsaedi A, Bayomi N 2015 Three-dimensional mixed convection flow of viscoelastic nanofluid over an exponentially stretching surface. *Int. J. Numer. Method H.* **25** 333-357.
[16] Sharada K, Shankar B 2015 MHD mixed convection flow of a Casson fluid over an exponentially stretching surface with the effects of Soret, Dufour, thermal radiation and chemical reaction. *WJM 5* 165–177.
[17] Ahmad K, Hanouf Z, Ishak A 2016 Mixed convection Jeffrey fluid flow over an exponentially stretching sheet with magnetohydrodynamic effect. *AIP Adv.* **6** 035024.
[18] Besthapu P, Haq R U, Bandaria S, Al-Mdallal Q M 2017 Mixed convection flow of thermally stratified MHD nanofluid over an exponentially stretching surface with viscous dissipation effect. *J. Taiwan Inst. Chem. Eng.* **71** 307–314.
[19] Malik R, Khan M, Munir A, Khan W A 2014 Flow and heat transfer in Sisko fluid with convective boundary condition. *PLoS One* **9** e107989.
[20] Makinde O D, Aziz A 2010 MHD mixed convection from a vertical plate embedded in a porous medium with a convective boundary condition. *Int. J. Therm. Sci.* ***49*** 1813–1820.
[21] Subhashini S V, Samuel N, Pop I 2011 Effects of buoyancy assisting and opposing flows on mixed convection boundary layer flow over a permeable vertical surface. *Int. Commun. Heat Mass* **38** (2011) 499–503.
[22] Subhashini S V, Samuel N, Pop I 2011 Double-diffusive convection from a permeable vertical surface under convective boundary condition, *Int. Commun. Heat Mass* **38** 1183–1188.
[23] Hamad M A A, Uddin M J, Ismail A I M 2012 Radiation effects on heat and mass transfer in MHD stagnation-point flow over a permeable flat plate with thermal convective surface boundary condition, temperature dependent viscosity and thermal conductivity. *Nucl. Eng. Des.* **242** 194– 200.
[24] Grosan T, Merkin J H, Pop I 2013 Mixed convection boundary-layer flow on a horizontal flat surface with a convective boundary condition. *Meccanica* **48** 2149–2158.
[25] Patil P M, Momoniat E, Roy S 2014 Influence of convective boundary condition on double diffusive mixed convection from a permeable vertical surface. *Int. J. Heat Mass Tran.* **70** 313–321.
[26] Hayat T, Ashraf M B, Alsulami H H, Alhuthali M S 2014 Three-dimensional mixed convection flow of viscoelastic fluid with thermal radiation and convective conditions. *PLoS One* **9** e0090038.
[27] Rahman M M, Merkin J H, Pop I 2015 Mixed convection boundary-layer flow past a vertical flat plate with a convective boundary condition. *Acta Mech.* **226** 2441–2460.
[28] Yao S, Fang T, Zhong Y 2011 Heat transfer of a generalized stretching/shrinking wall problem with convective boundary conditions. *Commun. Nonlinear. Sci. Numer. Simul.* **16** 752–760.
[29] Makinde O D, Aziz A 2011 Boundary layer flow of a nanofluid past a stretching sheet with convective boundary condition. *Int. J. Therm. Sci.* **50** 1326–1332.
[30] Ferdows M, Khan M S, Alam M M, Sun S 2012 MHD mixed convective boundary layer flow of a nanofluid through a porous medium due to an exponentially stretching sheet. *Math. Probl. Eng.* **2012** 408528.

[31] Bachok N, Ishak A, Pop I 2013 Stagnation point flow toward a stretching/shrinking sheet with a convective surface boundary condition. *J. Franklin Inst.* **350** 2736–2744.
[32] Hayat T, Imtiaz M, Alsaedi A, Mansoor R 2014 MHD flow of nanofluids over an exponentially stretching sheet in a porous medium with convective boundary conditions. *Chin. Phys. B* **23** 054701
[33] Ramesh G K, Gireesha B J, Gorla R S R 2015 Boundary layer flow past a stretching sheet with fluid‐particle suspension and convective boundary condition. Heat Mass Transf. 51(8) 1061–1066.
[34] Hayat T, Saeed Y, Alsaedi A, Asad S 2015 Effects of convective heat and mass transfer in flow of Powell-Eyring fluid past an exponentially stretching sheet. *PLoS One* **10** e0133831
[35] Khan M, Malik R, Munir A 2015 Mixed convection heat transfer to Sisko fluid over a radially stretching sheet in the presence of convective boundary conditions. *AIP Adv.* **5** 087178.
[36] Merkin J H 1986 On dual solutions occurring in mixed convection in a porous medium. *J Eng Math* **20** 171–9.

IOP Conf. Series: Journal of Physics: Conf. Series **949** (2017) 012017 doi:10.1088/1742-6596/949/1/012017

Application of one-way ANOVA in completely randomized experiments

Zaharah Wahid[1], Ahmad Izwan Latiff[2] and Kartini Ahmad[3]

[1, 3] Department of Science in Engineering, Faculty of Engineering, International Islamic University Malaysia, P.O. Box 10, Jalan Gombak, 53100, Kuala Lumpur, Malaysia.

[2] Department of Biotechnology in Engineering Faculty of Engineering, International Islamic University Malaysia. , P.O. Box 10, Jalan Gombak, 53100, Kuala Lumpur, Malaysia.

E-mail : [1] zaharahwahid@iium.edu.my

Abstract. This paper describes an application of a statistical technique one-way ANOVA in completely randomized experiments with three replicates. This technique was employed to a single factor with four levels and multiple observations at each level. The aim of this study is to investigate the relationship between chemical oxygen demand index and location on-sites. Two different approaches are employed for the analyses; critical value and p-value. It also presents key assumptions of the technique to be satisfied by the data in order to obtain valid results. Pairwise comparisons by Turkey method are also considered and discussed to determine where the significant differences among the means is after the ANOVA has been performed. The results revealed that there are statistically significant relationship exist between the chemical oxygen demand index and the location on-sites.

1. Introduction

The analysis of variance, often abbreviated as ANOVA is a statistical technique to compare three or more means. It is referred to a one-way ANOVA when the number of independent variable in the analysis of variance test is only one. The null hypothesis the means are all alike using the F distribution. The alternative hypothesis is that at least one of the means is different from the others. With the F-test, all the means are compared simultaneously. Even though one is comparing three or more means in the use of the F-test, variances are used in the test instead of means. The t and z tests are not used to compare the means of three or more samples since they tend to increase the estimated errors [1, 2].

The primary assumptions in applying ANOVA are that the response must follow a normal distribution. The variance of the response is constant and they are independent and randomly distributed.

The analysis of variance was originally invented by Ronald Fisher in the 1920's and has its root in agriculture sector. It has been further developed and has been applied in various fields such as engineering, chemical, business, management and human resource [3, 4, 5]. The basic idea remains the same.

IOP Conf. Series: Journal of Physics: Conf. Series **949** (2017) 012017 doi:10.1088/1742-6596/949/1/012017

2. Application of one-way ANOVA

2.1. Completely randomized experiments

The order of conducting the experimental trials can affect the result. The order is selected randomly or randomly assign by allocating each trial a number and using random numbers to choose the order. If there is only one factor, the experimental design is referred to as a completely randomized design. In this study the factor being investigated is the location where the leachates are collected. The location is subdivided into four groups or levels. The response variable of interest is the chemical oxygen demand index (COD). In the case of one way classification analysis of variance, data obtained from a designed experiment as described for the completely randomized experiments, CRD are usually considered to represent by a fixed effect model, since treatment of interests are included [1, 2]. The experiments were replicated three times.

2.2. Completely randomized experiments

The ANOVA test is computed based on decomposition of the total variances, σ2 into two distinct components. The first component is known as the between group variance, and it involves finding the variance of the means. The second component, the within group or (levels) variance, by computing the variance using all data and is not affected by differences in the means, [6]. If there is no difference in the COD means, the between group variance estimate will be approximately equal to the within group variance estimate, and the F statistics test will be equal to 1. The null hypothesis will not be rejected.

To detect differences in the four means of the chemical oxygen demand indexes, we test the null hypothesis, H_0 against the alternative hypothesis, H_1 at a significant level, $\alpha = 0.05$.The sum square between group SSB (S_B^2) and within group, SSW (S_w^2) can be computed as follow:-

The grand mean, GM

$$\bar{X}_{GM} = \frac{\Sigma x}{N} \tag{1}$$

Computational of the between –group variance
Sum square between groups, SSB

$$\mathrm{SSB} = \sum n_i(\bar{X}_i - \bar{X}_{GM})^2 \tag{2}$$

The mean square between, MSB is an estimate of the between groups mean square

$$MSB = \frac{\sum n_i\,(\bar{X}_i - \bar{X}_{GM})^2}{N-1} \tag{3}$$

The within-group variance, SSW

$$SSW = \sum (n_i - 1){S_i}^2 \tag{4}$$

Mean Square within, MSW is an estimate of the expected with within groups mean square

$$MSW = \frac{\Sigma(n_i-1){S_i}^2}{\Sigma(n_i-1)} \tag{5}$$

Compute the value of the test statistic, F

$$F_{cal} = \frac{MSB}{MSW}$$

When the F_{cal} calculated which is the ratio of equation (3) and (5) is larger than the F value obtained from the table, there is enough evidence to reject the null hypothesis and conclude that at least one of the means is different from the others. The p-value approached could be another alternative to analyze the ANOVA result. Both techniques should provide the same conclusion.

3. Results and discussion

Table 3.1 shows the design matrix which is the combinations of factor and levels that make up the experiment. The test results of the experimental run according to a completely randomized design are summarized in table 1. The ANOVA analysis showing the value of the F-statistic and p-value are depicted in table 2 and a boxplot of four different locations is illustrated in figure 1. The plot indicates that location 3 has the highest COD estimated mean value of 611.333. Mathematically,

Table 1. Mean Chemical Oxygen Demand in Four Locations

Location	*Sample*	R_1 *(mg/L)*	R_2 *(mg/L)*	R_3 *(mg/L)*	*Average(mg/L)*
1	1	256	239	230	241.667
1	2	135	158	124	139.000
1	3	135	141	129	135.000
2	1	202	224	218	214.667
2	2	233	199	141	191.000
2	3	186	181	200	189.000
3	1	591	589	581	587.000
3	2	519	504	486	503.000
3	3	628	574	632	611.333
4	1	260	285	222	255.667
4	2	399	386	375	386.667
4	3	235	231	185	217.000

the hypotheses are stated as follows:

$$H_0 = \mu_1 = \mu_2 = \mu_3 = \mu_4$$

$$H_1 = Not\ all\ \mu s\ are\ zero$$

The ANOVA table 2 showed that there is very strong evidence that the location influences the response (COD). Using the critical value approach with $\alpha = 0.05$, the F-test statistic shown in Table 3.2 is 92.09 with, $df_1 = 3$ and $df_2 = 32$ degrees of freedom, the S_B^2 is many times larger than the S_W^2 or error mean square (3198) .When F-statistic exceeds the critical value of 3.32, the null hypothesis is rejected and the result is statistically significant at the 5% level. This suggests that there is a difference between the mean chemical oxygen demand at different locations. Similarly using p-value approach, the ($p \leq 0.000$) is less than $\alpha = 0.05$. We have a sufficient evidence to suggest that at least one of the four means is different from one of the others.

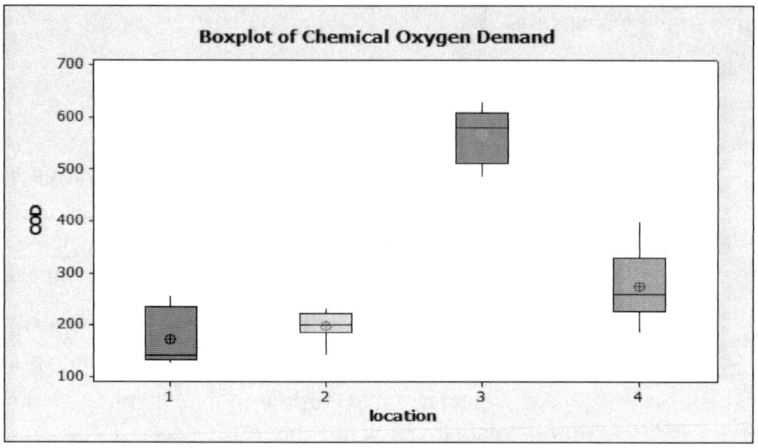

Figure 1. Boxplot of chemical oxygen demand index.

Table 2. One-way ANOVA of COD

Source	*DF*	*SS*	*MS*	*F*	*P*
Location	*3*	*883469*	*294490*	*92.09*	*0.000*
Error	*32*	*102334*	*3198*		
Total	***35***	***985803***			

S = 56.55 R-Square = 89.62% R-Square (adjusted) = 88.65%

The R^2-square (adjusted) value of 88.65% is the coefficient of determination of COD and fit. It indicates that the model fitted to the data can explain 89.5% of the variation occurs in the COD value. In addition, the estimated standard deviation (s =56.55) is used to obtain confidence intervals for the four levels as displayed in Table 3. Next, further investigation was conducted to determine which means are different from the others. In this study, pairwise comparisons were selected by the Turkey method with terms location. Confidence interval with confidence level 95% were performed to decide which levels have significantly different. As displayed in table 3, the first two confidence intervals do not include zero, however, the third interval does. This suggests that there is no evidence of any difference between location 1, 2 and 4. There is a significant difference between location 3 and 4. Based on these findings the locations do influence the oxygen demand index. The adequacy of the model for COD was assessed by graphing the residuals, as shown in figure 1. The structure-less residuals figure 2 and figure 3 verified that the normality assumptions, constant variance and randomly scattered points are validated and therefore the model is adequate.

IOP Conf. Series: Journal of Physics: Conf. Series **949** (2017) 012017 doi:10.1088/1742-6596/949/1/012017

Table 3. Individual 95% Confidence Interval for Mean (COD)

```
Tukey 95% Simultaneous Confidence Intervals
All Pairwise Comparisons among Levels of Location

Individual confidence level = 98.87%


Location = 1 subtracted from:

LOC    Lower   Center    Upper  ---------+---------+---------+---------+
2     -4.873   -0.400    4.073                  (-*)
3     -2.873    1.600    6.073                   (-*)
4     48.727   53.200   57.673                                  (-*)
                                ---------+---------+---------+---------+
                                       -30         0        30        60


Location = 2 subtracted from:

LOC    Lower   Center    Upper  ---------+---------+---------+---------+
3     -2.473    2.000    6.473                   (-*)
4     49.127   53.600   58.073                                  (-*)
                                ---------+---------+---------+---------+
                                       -30         0        30        60


Location = 3 subtracted from:

LOC    Lower   Center    Upper  ---------+---------+---------+---------+
4     47.127   51.600   56.073                                  (*-)
                                ---------+---------+---------+---------+
                                       -30         0        30        60
```

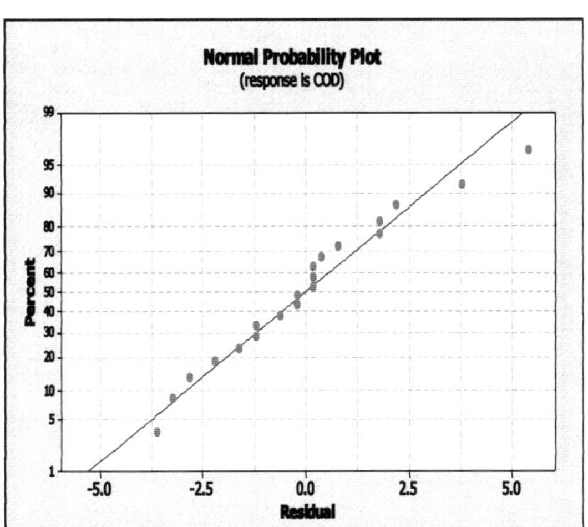

Figure 2. Normal probability plot.

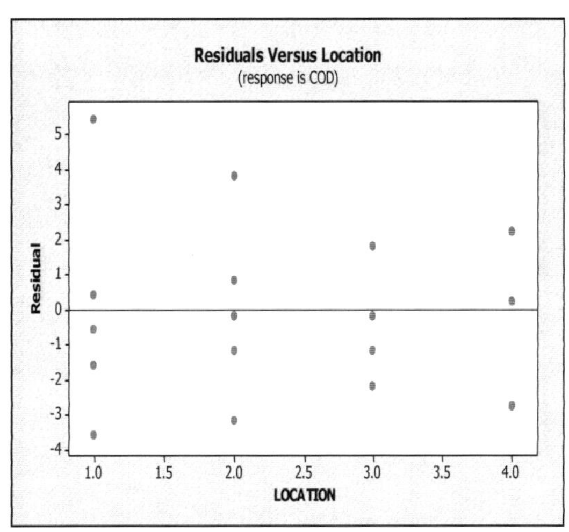

Figure 3. Residuals plot.

IOP Conf. Series: Journal of Physics: Conf. Series **949** (2017) 012017 doi:10.1088/1742-6596/949/1/012017

4. Results and discussion

This paper illustrates the application of one way ANOVA in completely randomized design. The experimental results have provided evidence that the location of the leachate has statistically significant effect on the index of COD. Thus, using the above procedure one can discover which means have significantly different, a common problem facing researchers could be answered.

References

[1] Montgomery, D. C. (2017). *Design and analysis of experiments*. John Wiley & Sons.
[2] Mead, R., Gilmour, S. G., & Mead, A. (2012). *Statistical principles for the design of experiments: applications to real experiments* (Vol. 36). Cambridge University Press.
[3] Ellendersen, L. D. S. N., Granato, D., Bigetti Guergoletto, K., & Wosiacki, G. (2012). Development and sensory profile of a probiotic beverage from apple fermented with Lactobacillus casei. *Engineering in Life Sciences*, *12*(4), 475-485.
[4] Wahid, Z., & Tham, M. T. (2012). Single Array and Response Modeling of Robust Design Experiments. *International Journal of Applied Physics and Mathematics*, *2*(5), 359.
[5] Nadir, N., Wahid, Z., Shafie, A. A., Muthalif, A. A., Malek, M. A., Aziz, N. N. A., Zainuddin, M.T. & Islam, N. M. (2013). Development of real time experimental system for investigating photochromic response to UV irradiation. In *IOP Conference Series: Materials Science and Engineering* (Vol. 53, No. 1, p. 012083). IOP Publishing.
[6] Faiz Elfaki and Zaharah Wahid (2015). Computational Methods and Statistics. IIUM Press. Kuala Lumpur, Malaysia.

IOP Conf. Series: Journal of Physics: Conf. Series **949** (2017) 012018 doi:10.1088/1742-6596/949/1/012018

The effect of dephasing on superadiabatic single-qubit rotation gates

Gharib Subhi Mahmoud

Dept. of Science in Engineering, Kulliyah of Engineering, IIUM, Kuala Lumpur, Malaysia

E-mail: gharib@iium.edu.my

Azeddine Messikh

Dept. of Computer Science, Kulliyah of Information and communication technology, IIUM, Kuala Lumpur, Malaysia

E-mail: messikh@iium.edu.my

Abstract. To implement quantum gates stimulated Raman adiabatic passage (STIRAP) can be used. This STIRAP requires high Rabi frequencies and to overcome this problem we use superadiabatic approach. Our model is a tripod consisting of four-level system driven by three resonant fields. These fields are modulated by Gaussian pulses with different amplitudes, phases and time delays. We investigate the robustness of our model against dephasing which are caused by collisions or phase fluctuations of the fields.

1. Introduction

The process of stimulated Raman adiabatic passage (STIRAP) is well known technique in quantum information. It is used to transfer adiabaticaly the population from one state to another state and it is based on the adiabatic theorem [1]. This theorem states that if the system is initially prepared in an instantaneous eigenstate of a time-depend Hamiltonian which varies slowly, the state of the system remain all the time in this instantaneous eigenstate.

Recently, STIRAP has been used to generate quantum gates [2, 3]. Single-qubit gates are quantum rotation gates which are given in matrix form by

$$R(a,\phi) = \begin{bmatrix} \cos a & e^{i\phi}\sin a \\ -e^{-i\phi}\sin a & \cos a \end{bmatrix}, \tag{1}$$

where a is the angle of rotation and ϕ is the phase of the gate. These rotation gates can be implemented efficiently in a three-level lambda system using a reversed STIRAP followed by a standard STIRAP [2]. In both STIRAPs, strong fields are required which is a disadvantage in many practical realizations. This technical problem can be reduced by superadiabatic or transitionless approach which has been recently proposed [4–7].

We have recently discussed the generation of single-qubit rotation gates using superadiabatic [3]. Our model is robust against decay rate of the excited state, but not against the dephasing of the ground state which caused by collisions or phase fluctuations of the driving fields. We will extend our previous paper to include other dephasing rate.

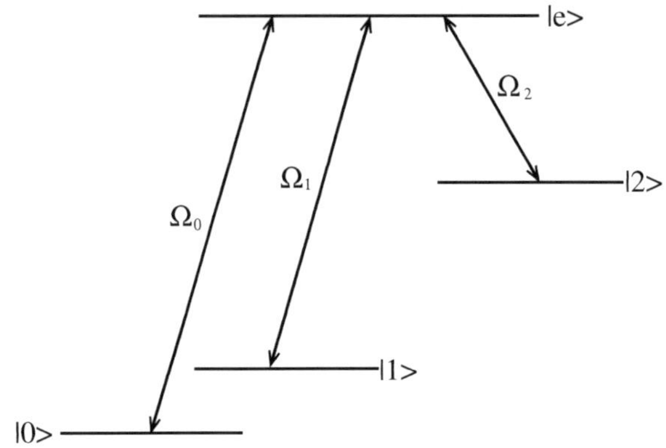

Figure 1. The three ground states $|0\rangle$, $|1\rangle$ and $|2\rangle$ of the tripod are coupled to the excited state $|e\rangle$ by three different coherent fields Ω_i, $i = 1, 2, 3$, respectively.

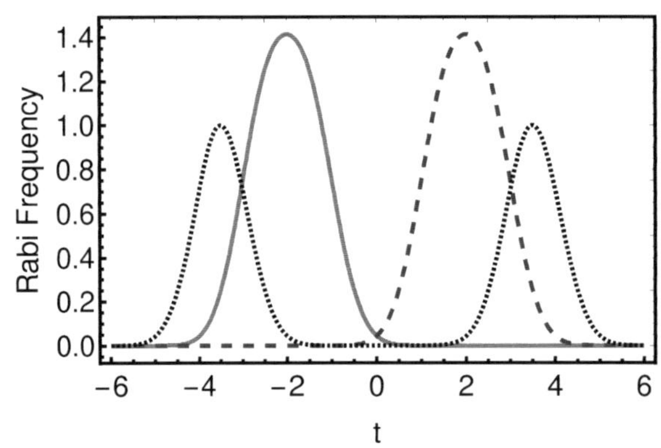

Figure 2. The Rabi frequencies as function of time. Ω_0 (solid line), Ω_1 (dashed line), and Ω_2 (dotted line).

2. STIRAP model

The model is a tripod consisting of four-level system driven by three resonant coherent fields with Rabi frequencies Ω_0, Ω_1, and Ω_2. These coherent fields couple the excited level $|e\rangle$ to the three lower levels $|0\rangle$, $|1\rangle$, and $|2\rangle$ as in Figure 1. The fields are modulated by Gaussian pulses with width δ, amplitudes A_j, phase ϕ_j, and time delay τ_j

$$\Omega_j(t - \tau_j) = A_j \mathrm{e}^{i\phi_j}\, \mathrm{e}^{-\dfrac{(t-\tau_j)^2}{\delta^2}}. \tag{2}$$

The three Rabi frequencies are given by

$$\begin{aligned}
\Omega_0 &= \Omega(t + T - \tau) + \Omega(t + T + \tau)\cos\theta + \Omega(t - T - \tau)\sin\theta, \\
\Omega_1 &= \Omega(t + T + \tau)\sin\theta + \Omega(t - T + \tau) + \Omega(t - T - \tau)\cos\theta \\
\Omega_2 &= \Omega(t + T + 3\tau) + \Omega(t - T - 3\tau).
\end{aligned} \tag{3}$$

They represent two STIRAPs separated by T and each STIRAP has two pulses separated by τ (see Figure. 2). The Hamiltonian of the system is given in the basis $\{|0\rangle, |1\rangle, |e\rangle, |2\rangle\}$ as follows.

IOP Conf. Series: Journal of Physics: Conf. Series **949** (2017) 012018 doi:10.1088/1742-6596/949/1/012018

$$H_0 = \frac{1}{2}\begin{bmatrix} 0 & 0 & \Omega_0 & 0 \\ 0 & 0 & -\Omega_1 & 0 \\ \Omega_0 & -\Omega_1 & 0 & \Omega_2 \\ 0 & 0 & \Omega_2 & 0 \end{bmatrix}, \tag{4}$$

where we have assumed that the Rabi frequencies Ω_i are all real numbers. The Hamiltonian H_0 has four instantaneous adiabatic eigenvalues

$$\lambda_\pm = \pm\frac{1}{2}\sqrt{\Omega_0^2 + \Omega_1^2 + \Omega_2^2} \quad \text{and} \quad \lambda_1 = \lambda_2 = 0. \tag{5}$$

The eigenstates that correspond to zero eigenvalue are called dark states. They are given by

$$\begin{aligned} |D_1\rangle &= -\cos\theta_1 \sin\theta_0 \,|0\rangle + \cos\theta_1 \cos\theta_0 \,|1\rangle + \sin\theta_1 \,|2\rangle, \\ |D_2\rangle &= \cos\theta_0 \,|0\rangle + \sin\theta_0 \,|1\rangle, \end{aligned} \tag{6}$$

where

$$\tan\theta_0 = \frac{\Omega_0}{\Omega_1}, \quad \text{and} \quad \tan\theta_1 = \frac{\sqrt{\Omega_0^2 + \Omega_1^2}}{\Omega_2}. \tag{7}$$

So, according to the adiabatic theorem [1], if the system starts in the superposition of the two dark states it evolution will remain in this superposition.

If we use the following conditions

$$\begin{aligned} \lim_{t\to-\infty} \tan\theta_0 &= \cot\theta & \lim_{t\to-\infty} \tan\theta_1 &= 0, \\ \lim_{t\to\infty} \tan\theta_0 &= \tan\theta & \lim_{t\to\infty} \tan\theta_1 &= 0, \end{aligned}$$

and the initial state is in a superposition of the two dark states, then we obtain the rotation gate

$$R(a,0) = \begin{bmatrix} \cos a & \sin a \\ -\sin a & \cos a \end{bmatrix} = \begin{bmatrix} \sin 2\theta & \cos 2\theta \\ -\cos 2\theta & \sin 2\theta \end{bmatrix} \tag{8}$$

So, $R(\pi/2, 0)$ corresponds to $\theta = 0$.

3. Superadiabatic approach

The process of superadiabatic or transitionless adiabatic passage was proposed to overcome the technical problem of high Rabi frequencies required by STIRAP. Additional pulse is required to couple between the two lower states. This coupling can be implemented using magnetic field with π-area or near π-area [7]. Thus, the total Hamiltonian becomes

$$H = H_0 + H_1, \tag{9}$$

where H_0 is given by eq. (4) and H_1 is the correction due to the superadiabatic process [4, 6, 7]

$$H_1 = i\sum_n \left[|\partial_t n\rangle\langle n| - \langle n|\partial_t n\rangle|n\rangle\langle n|\right]. \tag{10}$$

The sum in eq. (10) is taken over all the instantaneous eigenstates. The Hamiltonian H_1, in the basis $\{|0\rangle, |1\rangle, |e\rangle, |2\rangle\}$, takes the form [3].

$$H_1 = \begin{bmatrix} 0 & h_{0,1} & 0 & h_{0,2} \\ h_{0,1}^* & 0 & 0 & h_{1,2} \\ 0 & 0 & 0 & 0 \\ h_{0,2}^* & h_{1,2}^* & 0 & 0 \end{bmatrix}, \tag{11}$$

where

$$\begin{aligned}
h_{0,1} &= i\frac{\Omega_0\dot{\Omega}_1 - \Omega_1\dot{\Omega}_0}{\Omega_0^2+\Omega_1^2}, \\
h_{0,2} &= i\Omega_0\frac{(\Omega_0\dot{\Omega}_0 + \Omega_1\dot{\Omega}_1)\Omega_2 - (\Omega_0^2+\Omega_1^2)\dot{\Omega}_2}{(\Omega_0^2+\Omega_1^2)(\Omega_0^2+\Omega_1^2+\Omega_2^2)}, \\
h_{1,2} &= i\Omega_1\frac{(\Omega_0\dot{\Omega}_0 + \Omega_1\dot{\Omega}_1)\Omega_2 - (\Omega_0^2+\Omega_1^2)\dot{\Omega}_2}{(\Omega_0^2+\Omega_1^2)(\Omega_0^2+\Omega_1^2+\Omega_2^2)}.
\end{aligned}$$

With the Gaussian pulses given in eq. (2), the term $h_{0,1} = 0$. Thus, the Hamiltonian H_1 becomes

$$H_1 = \begin{bmatrix} 0 & 0 & 0 & h_{0,2} \\ 0 & 0 & 0 & h_{1,2} \\ 0 & 0 & 0 & 0 \\ h_{0,2}^* & h_{1,2}^* & 0 & 0 \end{bmatrix}. \tag{12}$$

which is equivalent to an additional driving field which couples the two lower states $|0\rangle$ and $|1\rangle$ to the level $|2\rangle$. So, this Hamiltonian does not induce any effects on the excited state $|e\rangle$.

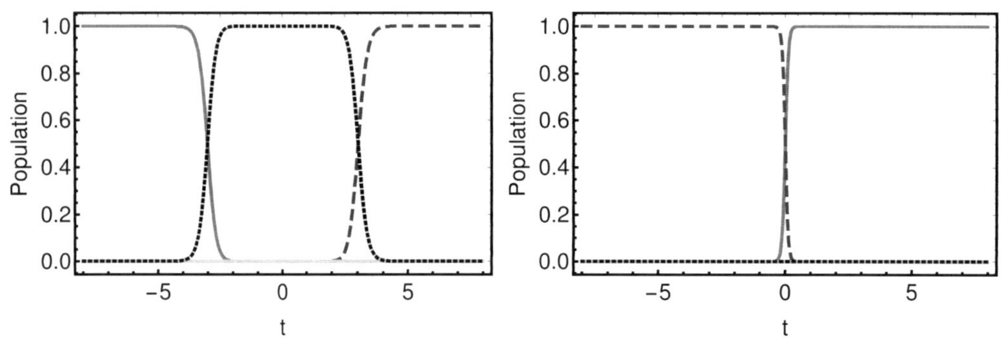

Figure 3. The population of the tripod in the absence of dephasing for the rotation gate $R(\pi/2, 0)$.

In Figure. 3 we plot the population of the four states of the tripod for the rotation gate $R(\pi/2, 0)$. The left figure, the initial state is at $|0\rangle$ and the right figure, the initial state is $|1\rangle$. It is clear from the figure that the state $|2\rangle$ and the excited state $|e\rangle$ are not populated during the evolution of the system.

4. The effect of dephasing

In the absence of any sort of decoherence the evolution is given by the shrödinger equation

$$i\frac{d}{dt}|\psi(t)\rangle = H|\psi(t)\rangle. \tag{13}$$

However, in the presence of decoherence, this equation (13) is replaced by the the Lindblad master equation

$$\dot{\rho} = -i\,[H,\rho] + \frac{1}{2}\sum_i \left(2\,C_i\rho C_i^\dagger - C_i^\dagger C_i\rho - \rho C_i^\dagger C_i\right), \tag{14}$$

IOP Conf. Series: Journal of Physics: Conf. Series **949** (2017) 012018 doi:10.1088/1742-6596/949/1/012018

where ρ is the atomic reduced density operator, H is the total Hamiltonian operator $H = H_0 + H_1$, and C_i are the Lindblad operators associated with the decoherence.

The effect of dephasing on quantum rotation gates has been investigated by various authors [3, 8, 9]. In this paper we restrict ourselves to the dephasing of the form $C_i = \sqrt{2\Gamma}\sigma_{zi}$, where Γ is the dephasing rate and $\sigma_{zi} = |i\rangle\langle i| - |e\rangle\langle e|$ is the Pauli operator. The measure of the performance of the rotation gate is given by the fidelity

$$F = |\langle\psi(t_f)|R(a,\phi)|\psi(t_i)\rangle|, \tag{15}$$

where the initial state is $|\psi(t_i)\rangle$ and the final state is $|\psi(t_f)\rangle$. The final state is obtained by solving the master equation (14). In our numerical computations of fidelity, we focus on the generation of the rotation gate $R(\pi/2, 0)$.

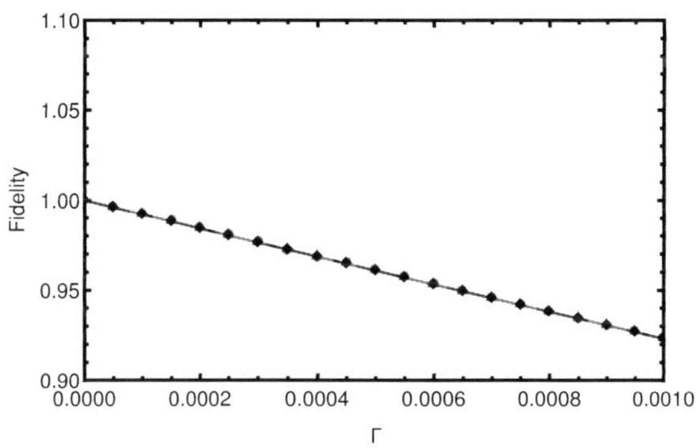

Figure 4. The Maximum, minimum and average fidelities for the rotation gate $R(\pi/2, 0)$ as a function of the dephasing rate Γ.

If the dephasing rate is small enough we can use the quantum trajectory approach [10–14]. This approach unravels the Lindblad master equation (14) as trajectories and each trajectory consists of the non jump evolution processes, and the quantum jump processes. The non jump evolution is a continuous coherent evolution and takes the form

$$i\frac{d|\psi(t)\rangle}{dt} = H_{\text{eff}}(t)|\psi(t)\rangle, \tag{16}$$

where the operator H_{eff} is non Hermitian operator

$$H_{\text{eff}}(t) = H(t) - \frac{i}{2}\sum_i C_i^\dagger C_i. \tag{17}$$

The quantum jump is a discrete random jump resulted from emission or absorption of energy quanta [15]. These jumps are described by the operators C_i which act on the state according to the following rule [15]

$$|\psi\rangle \longleftarrow C_i|\psi\rangle. \tag{18}$$

The maximum, minimum, and average fidelities are shown in Figure 4 as a function of Γ for the rotation gate corresponds to $a = \pi/2$ and phase $\phi = 0$. These fidelities are computed numerically for 1000 initial random states uniformly distributed on the Bloch Sphere as follows.

$$|\psi(t_i)\rangle = \sqrt{u}|0\rangle + e^{i2\pi v}\sqrt{1-u}|1\rangle,$$

where the two random variable u and v are uniformly distributed on the unit interval $[0, 1]$. We have used quantum jump approach which is valid for very small dephasing rates. As the figure shows, the fidelity does not depend on the initial state. We should note that the quantum jump has two processes, the coherent evolution and the quantum jump. We have neglected the quantum jump due to the fact that the dephasing is very small. By increasing the dephasing rate it is better to use the master equation which is the average over many trajectories, coherent evolution and random quantum jumps.

5. Conclusion

Quantum rotation gates are single-qubit gates. To implement these gates we can use STIRAP. However, this scheme require high Rabi frequencies which is a disadvantage in many experiments. In this study we have shown that using superadiabatic approach can overcome this technical problem. Furthermore, we have investigated the effect of a realistic dephasing on the performance of the rotation gate $R(\pi/2, 0)$. We have shown that it is robust to the dephasing rate of the excited state due to the fact that the excited state is not populated during the evolution of the system. However, the dephasing which cause by collisions or phase fluctuations of the fields can leads to imperfect gate. In our study we have focused on the dephasing proportional to the σ_z and we have used the quantum trajectory approach. We have found that for small values of the dephasing rate, the fidelity decreases linearly. So, to obtain a practical useful gate one must keep the dephasing as small as possible.

References

[1] Messiah A 1962 *Quantum Mechanics* (Amsterdam: North-Holland Publishing Company)
[2] Lacour X, Guerin S, Vitanov N V, Yatsenko L P and Jauslin H R 2006 *Optics Communications* **264** 362–367
[3] Issoufa Y H, Mahmoud G S and Messikh A 2015 *Quant. Inf. Rev.* **3** 17–21
[4] Lim R and Berry M V 1991 *J. Phys. A: Math. Gen.* **24** 3255
[5] Berry M 2009 *J. Phys. A: Math. Theor.* **42** 365303
[6] Demirplak M and Rice S A 2008 *J. Phys. Chem. A* **129** 154111
[7] Giannelli L and Arimondo E 2014 *Phys. Rev. A* **89** 033419
[8] Ivanov P A, Vitanov N V and Bergmann K 2004 *Phys. Rev. A* **70** 063409
[9] Issoufa Y H and Messikh A 2010 *J. Phys. B: At. Mol. Opt. Phys.* **43** 215506
[10] Carmichael H 2009 *Statistical Methods in Quantum Optics 2: Non-Classical Fields* (*Theoretical and Mathematical Physics* no v. 2) (Springer) ISBN 9783540713203
[11] Dalibard J, Castin Y and Mølmer K 1992 *Phys. Rev. Lett.* **68**(5) 580–583 URL `http://link.aps.org/doi/10.1103/PhysRevLett.68.580`
[12] Mølmer K, Castin Y and Dalibard J 1993 *J. Opt. Soc. Am. B* **10** 524
[13] Mølmer K and Castin Y 1996 *Quantum Semiclassic. Opt* **8** 49
[14] Plenio M B and Knight P L 1998 *Rev. Mod. Phys.* **70**(1) 101–144 URL `http://link.aps.org/doi/10.1103/RevModPhys.70.101`
[15] Ficek Z and Wahiddin M R 2014 *Quantum Optics for Beginners* (Pan Stanford Publishing) ISBN 978-981-4411-75-2

Electric Quadrupole $E2$- Transitions of $^{170-174}Yb$ Isotopes

Mohd Kh. M. Abu El Sheikh[1], Abdurahim A Okhunov[2], Ph.N. Usmanov[3], and Torla HJ Hassan[2]

[1] Department of Physics, Faculty of Science, University of Malaya, 50603, Kuala Lumpur, Malaysia

[2] Department of Science in Engineering, International Islamic University Malaysia, 57528, P.O. 10, Kuala Lumpur, Malaysia

[3] Institute of Nuclear Physics, Academy of Science of Uzbekistan, 100132, Tashkent, Uzbekistan

E-mail: abdurahimokhun@iium.edu.my , aaokhunov@gmail.com

Abstract. The non-adiabatic effects which is manifested in the electric properties of low-lying states of even-even deformed nuclei are studied. A simple phenomenological model which takes into account the Coriolis mixing of $K^{\pi} = 0^+_n$, 2^+_n and $K^{\pi} = 1^+_{\nu}$ state bands.

The Calculations for isotopes $^{170-174}Yb$, are carried out. The reduced probability of electric quadrupole transitions from the states 0^+_{ν} and 2^+_{ν} - bands to the ground (gr) state band is calculated and non adiabatic effect is discussed. The ratio of $E2-$ transitions R_{IK} from 0^+_2, 0^+_3, 2^+_1, and 2^+_2 bands are calculated and compared with the experimental data.

1. Introduction

Despite the fact that the structure of deformed nuclei and nature of low excited levels have been substantially studies over more than four decades, this still occupies a central part of today's research Refs. [1]-[3].

The isotopes of Yb is the deformed nuclei in rare-earth region and the spherical nuclei near $Z = 70$. For well deformed nuclei the states associated with ground state rotation produce regular sequences which dominate the yrast line. The nuclei $^{170,172,174}Yb$ have been well studied. In highspin spectroscopy studies of rare-earth nuclei, the isotopes of ytterbium have been a central focus for many experiments, and a large body of data has been established [4].

It is important to note that these are investigated in a number of ways such as radioactive decay of $^{170,172,174}Lu$, and different nuclear reactions. In these isotopes, many 1^+ states and $K^{\pi} = 0^+, 2^+$ bands have been observed.

Numerous conducted experiments on defining spectroscopic characteristics of low-lying exciting states, particular $K^{\pi} = 1^+$ in deformed nuclei [5], have motivated the further theoretical investigations. In this case, investigations influence of $K^{\pi} = 1^+$ states to the properties of low-lying levels is actual.

In the present paper analyzing properties of positive parity low-lying state in the transuranium nuclei, within the phenomenological model [6]-[8], which into account Coriolis mixture states of ground, 0^+_m- $(m = 2, 3, ..)$, $2^+_{\nu}(\gamma_{\nu})-$ and $K^{\pi} = 1^+_{\nu}$ rotational bands. The calculation are performed for the $^{170,172,174}Yb$ isotopes.

2. Electric Quadrupole $E2-$ Transitions

We shall calculate the reduced probability for $E2-$ transitions, using the wave function obtained by describing the energy of states. The expression for the reduced probability of $E2-$ transitions between states I_iK_i and ground state band and also intraband transitions of ground band within the framework our model [9] as follows:

$$\begin{aligned} B(E2; I_iK_i \to I_f0_1) &= \Bigg\{ \sqrt{\frac{5}{16\pi}} eQ_0 \left[\Psi^{I_f}_{0_1,0_1} \Psi^{I_i}_{0_1,K_i} C^{I_f0}_{I_i0;20} + \sum_n \Psi^{I_f}_{K_n,0_1} \Psi^{I_i}_{K_n,K_i} C^{I_fK_n}_{I_iK_n;20} \right] \\ &+ \sqrt{2} \left[\Psi^{I_f}_{0_1,0_1} \sum_n \frac{(-1)^{K_n} m_{K_n} \Psi^{I_i}_{K_n,K_i}}{\sqrt{1+\delta_{K_n,0}}} C^{I_f0}_{I_iK_n;2-K_n} \right. \\ &+ \left. \Psi^{I_i}_{0_1,K_i} \sum_n \frac{m_{K_n} \Psi^{I_f}_{K_n,0_1}}{\sqrt{1+\delta_{K_n,0}}} C^{I_fK_n}_{I_i0;2K_n} \right] \Bigg\}^2 \end{aligned} \tag{1}$$

here $m_{K_n} =< 0_1^+|\hat{m}(E2)|K_n^+ >-$ is matrix elements between intrinsic wave functions of ground (0_1^+) and $K_n^+ = 0_m^+, 2_\ell^+, 1_\nu^+$ bands, whose values are defined from from experimental data, Q_0- is intrinsic quadrupole moment of nucleus; and $C^{I_fK_f}_{I_iK_i;2(K_i+K_1)}-$ Clebsch-Gordan coefficients.

In the adiabatic approximation, the following equations are valid for $B(E2)$ factors from the $I = 2$ states to the K_n^+ rotational band:

$$B^{adiabatic}(E2; 2K_n \to 00_1) = (2 - \delta_{K_n,0})\, |m_{K_n} C^{00}_{2K_n;2-K_n}|^2 \tag{2}$$

which allows us to calculate the empirical values of the parameters m_{K_n} from the experimental data.

The $K^\pi = 0_2^+$, 0_3^+ and 2_1^+ bands are located close to each other in isotopes $^{170,172}Yb$, which leads to a strong mixing of states even $I = 2$. In this case, the adiabatic approximation (2) becomes inapplicable to determine m_{K_i}.

Therefore, to describing the experimental data for $B(E2)$ in $^{170,172}Yb$ isotopes, the value of m_0 and m_2 parameters are varied slightly. The values of the parameters m_K which are used in calculating probability of $E2-$ transitions are given in Table 1.

Table 1. The values of the parameters m_K and the intrinsic quadrupole moment Q_0, which are used in calculations (in efm^2).

A	m_{0_2}	m_{0_3}	m_{0_4}	m_{0_5}	m_{1_ν}	m_{2_1}	m_{2_2}	Q_0[10]
170	2	24	3	8	-5	19	780(4)	
172	10	1	-6.9	-8	-5	15	-8	791(4)
174	8	1	-6.9	8	-10 (m_{1_1}=-1.7)	15	8	782(4)

The empirical values of parameters m_K have been defined by formula (2), using the experimental data of the reduced probabilities of $E2$- transitions $B(E2; 2K_n \to 00_1)$. In the case where the bandhead energy is close to each other the Coriolis mixing of state is manifested significantly even at low spins $I = 2$. For example, $K^\pi = 0_2^+$ and $K^\pi = 0_3^+$ bands in nucleus

^{170}Yb and also the $K^\pi = 0_3^+$ and $K^\pi = 2_1^+$ bands in nucleus ^{172}Yb are close to each other. Therefore, for these bands the value of parameter m_K is different from other bands (see Table 1). This mixture effect is more manifested in the electromagnetic than energy characteristics.

The magnitude and sign of the parameters $m_{1_1} = m_{1_\nu}$ were determined from the best agrement of the ratios $R_{IK} = B(E2; IK \to I+10_1)/B(E2; IK \to I-10_1)$ from odd states of the $K^\pi = 2_\ell^+$ and 1_ν^+ (states with the negative signature $\sigma = -1$).

The parameters m_{0_m} and m_{2_ℓ} are determined, requiring the best agreement between calculated and experimental values of the ratios $B(E2)$ transitions from the β_m- and γ_ℓ bands correspondingly, with positive signature $\sigma = +1$.

The comparison of the values of calculated reduced probabilities of $E2-$ transitions with experimental data [11]-[14] are shown in Table 2.

Table 2. Reduced probability of $E2-$ transitions in the isotopes Yb (e^2fm^4).

A	$I_iK_i \to I_fK_f$	Exp.	Thory	$I_iK_i \to I_fK_f$	Exp.	Thory
^{170}Yb	$22_1 \to 00_1$	151(35)[11]	90	$00_3 \to 20_1$	60(15)[11]	43
	$\to 20_1$	269(60)[11]	60	$00_4 \to 20_1$	567(118)[11]	567
	$\to 40_1$	27(6)[11]	10			
^{172}Yb	$22_1 \to 00_1$	75.6(63)[12]; 74.6(57)[14]	82	$00_2 \to 20_1$	205(60)[12]	100
	$\to 20_1$	121(12)[14]	130	$20_2 \to 00_1$	14(1)[12]; 14(1)[14]	13
	$\to 40_1$	7.3(6)[12]; 6.8(7)[14]	8.6	$\to 20_1$	45(7)[12]; 52(8)[14]	23
	$42_1 \to 20_1$	398(284)[12]	15	$\to 40_1$	142(20)[12]; 140(20)[14]	74
	$\to 40_1$	739(512)[12]	81	$00_3 \to 20_1$	0.14(3)[12]	1
	$32_1 \to 20_1$	152(11)[14]	154	$20_3 \to 00_1$	0.4(1)[12]; 3.4(2)[14]	3.6
	$\to 40_1$	79(6)[14]	73	$\to 20_1$	0.6(4)[12]; 11.9(8)[14]	3.0
	$22_2 \to 00_1$	20(4)[12]; 32(4)[14]	23	$\to 40_1$	1.0(1)[14]	1.2
	$\to 20_1$	31(2)[12]; 51(7)[14]	38	$00_4 \to 20_1$	>0.25[12]	48
	$\to 40_1$	3.3(4)[14]	2.2	$20_4 \to 00_1$	10(6)[12]	12
	$32_2 \to 20_1$	54(7)[14]	42	$00_5 \to 20_1$	> 0.27[12]	64
	$\to 40_1$	22(3)[14]	21	$20_5 \to 00_1$	19(9)[12]	21
				$\to 20_1$	> 0.18[12]	23
^{174}Yb	$00_2 \to 20_1$	81^{+64}_{-29}[13]	64	$22_1 \to 20_1$	144(30)[13]	133

In Table 3 provided theoretical values of the reduced matrix elements of $E2-$ transitions for ^{172}Yb, which are compared with experimental data and values taken by other models [15]-[17].

We note that our calculations were performed sequentially, i.e., first describes the energy state and then corresponding their wave functions are determined. Further, using these wave functions are calculated probabilities $E2-$ transitions. From Table 4, one can see that, the results of calculation in the framework of our model for the majority cases provide a good agreement with experimental data.

To evaluate the degree of significant nonadiabaticity manifests in the values of reduced probabilities of $E2-$ transition in Table 5 is given the comparison of theoretical ratios $R_{IK}^{theory} = \frac{I_\gamma(IK \to I_1 0_1)}{I_\gamma(IK \to I_2 0_1)}$ with experimental data [11]-[13], [10, 14, 18] and their adiabatic values $R_{IK}^{adiabtic}$ which defined by the formula:

$$R_{IK} = \frac{I_\gamma\,(IK \to I_1 0_1)}{I_\gamma\,(IK \to I_2 0_1)} \left(\frac{E_\gamma\,(IK \to I_2 0_1)}{E_\gamma\,(IK \to I_1 0_1)} \right)^5 \tag{3}$$

IOP Conf. Series: Journal of Physics: Conf. Series **949** (2017) 012019 doi:10.1088/1742-6596/949/1/012019

Table 3. Reduced Matrix Elements of $E2-$ transitions in ^{172}Yb, calculated within our model which comparison with experimental data [15] and are calculated using the rotational-vibrational model (RVM2) [16] and the IBA-1 model [17] (eb).

$I_iK_i \to I_fK_f$	Exp.	RVM2	IBA-1	Theory	$I_iK_i \to I_fK_f$	Exp.	RVM2	IBA-1	Theory
$20_1 \to 20_1$	$-2.63^{+0.28}_{-0.27}$	-2.92	-2.92	-2.93	$20_1 \to 00_1$	$2.45^{+0.12}_{-0.12}$	2.45^0	$2.45^{a)}$	2.45^0
$40_1 \to 40_1$	$-3.54^{+0.84}_{-0.18}$	-3.73	-3.69	-3.74	$40_1 \to 20_1$	$3.76^{+0.19}_{-0.19}$	3.93	3.91	3.93
$60_1 \to 60_1$	$-4.31^{+0.22}_{-0.62}$	-4.43	-4.33	-4.46	$60_1 \to 40_1$	$5.34^{+0.27}_{-0.27}$	4.97	4.90	4.96
$80_1 \to 80_1$	$-4.49^{+0.23}_{-0.77}$	-5.05	-4.83	-5.08	$80_1 \to 60_1$	$5.90^{+0.30}_{-0.30}$	5.80	5.60	5.80
$100_1 \to 100_1$	$-6.32^{+0.74}_{-0.32}$	-5.60	-5.22	-5.63	$100_1 \to 80_1$	$6.71^{+0.34}_{-0.34}$	6.54	6.29	6.54
$120_1 \to 120_1$	$-6.15^{+0.64}_{-0.74}$	-6.08	-5.53	-6.15	$120_1 \to 100_1$	$7.01^{+0.35}_{-0.35}$	7.19	6.79	7.20
$140_1 \to 140_1$	–	–	–	6.62	$140_1 \to 120_1$	$8.12^{+0.63}_{-0.43}$	7.80	7.18	7.81
$22_1 \to 00_1$	$0.208^{+0.010}_{-0.040}$	0.21	$0.20^{a)}$	0.203	$00_2 \to 20_1$	$0.166^{+0.018}_{-0.018}$	0.16	0.27	0.01
$\to 20_1$	$0.250^{+0.016}_{-0.018}$	0.25	0.31	0.255	$20_2 \to 00_1$	$0.090^{+0.010}_{-0.040}$	0.16	0.26	0.082
$\to 40_1$	$0.063^{+0.009}_{-0.004}$	0.062	0.10	0.066	$\to 20_1$	$-0.162^{+0.071}_{-0.008}$	0.19	-0.31	0.108
$42_1 \to 20_1$	$0.22^{+0.07}_{-0.05}$	0.20	0.13	0.11	$\to 40_1$	$0.27^{+0.02}_{-0.08}$	0.26	0.45	0.19
$\to 40_1$	$0.46^{+0.08}_{-0.13}$	0.38	0.45	0.27	$40_2 \to 40_1$	–	–	–	0.13
$32_1 \to 20_1$	0.32(11)	–	–	0.328					
$\to 40_1$	0.235(6)	–	–	0.226					

$^{a)}$ This matrix element was used to normalize the results of the model calculations.

where $I_\gamma\,(IK \to I_1 0_1)-$ is intensity and $E_\gamma\,(IK \to I_1 0_1)-$ is energy of $\gamma-$ transition.

Table 4. The ratios of reduced probabilities of $E2-$ transitions $R_{IK} = B(E2; I_iK_i \rightarrow I_f 0_1)/B(E2; I_iK_i \rightarrow I'_f 0_1)$ for ^{170}Yb isotope.

A	I_iK_i	I_fK_f	$I'_fK'_f$	Experiments		Theory	Alaga
^{170}Yb	22_1^+	20_1^+	00_1^+	1.77(8)[11]		1.86	1.43
		40_1^+	20_1^+	0.098(11)[11]		0.043	0.050
	32_1^+	40_1^+	20_1^+	0.78(4)[11]		0.75	0.40
	52_1^+	60_1^+	40_1^+	1.39(46)[11]		1.50	0.57
	72_1^+	80_1^+	60_1^+	1.27(24)[11]		2.42	0.67
	20_2^+	20_1^+	00_1^+	1.94(52)[11]		1.1	1.43
	40_2^+	40_1^+	20_1^+	–		1.82	0.91
	60_2^+	60_1^+	40_1^+	3.73(90)[11]		1.67	0.81
	80_2^+	80_1^+	60_1^+	10.7(19)[11]		2.45	0.77
	100_2^+	100_1^+	80_1^+	25.3(85)[11]		29.2	0.74
	120_2^+	120_1^+	100_1^+	29.9(71)[11]		2.0	0.73
	20_3^+	20_1^+	00_1^+	1.81(11)[11]		2.5	1.43
		40_1^+	20_1^+	3.00(15)[11]		2.3	1.80
		40_1^+	00_1^+	5.43(18)[11]		5.75	2.57
	20_4^+	40_1^+	00_1^+	4.0(2)[11]		3.9	2.57
^{172}Yb	22_1^+	20_1^+	00_1^+	1.62(12) [14]	1.71(84) [12]	1.59	1.43
		40_1^+	20_1^+	0.056(5) [14]	0.056(20) [12]	0.066	0.072
	32_1^+	40_1^+	20_1^+	0.52(4) [14]	0.56(3) [12]	0.48	0.40
	42_1^+	40_1^+	20_1^+		3.35(69) [12]	6.0	2.94
	22_2^+	20_1^+	00_1^+	1.69(21) [14]	1.55(30) [12]	1.61	1.43
		40_1^+	20_1^+	0.015(8) [14]	0.064(38) [12]	0.058	0.072
	32_2^+	40_1^+	20_1^+	0.163(56) [14]	0.409(77) [12]	0.504	0.40
	42_2^+	40_1^+	20_1^+		4.11(17) [12]	3.89	2.94
	52_2^+	60_1^+	40_1^+		$<$ 1.10 [12]	0.82	0.57
	20_2^+	20_1^+	00_1^+	3.71(24) [14]	2.88(36) [12]	1.80	1.43
		40_1^+	20_1^+	2.70(38) [14]	2.61(11) [12]	3.20	1.80
	40_2^+	40_1^+	20_1^+		6.78(1.36) [12]	1.56	0.91
^{174}Yb	22_1^+	20_1^+	00_1^+	$>$ 0.49[10, 18]	2.4(5)[13]	1.59	1.43
		40_1^+	20_1^+	0.167(75)[10, 18]	0.256(92)[13]	0.055	0.072
	32_1^+	40_1^+	20_1^+	$>$ 0.325[10, 18]	0.67(13)[13]	0.49	0.40
	42_1^+	40_1^+	20_1^+	4.83(5)[10, 18]	4.77(63)[13]	3.75	2.94
		60_1^+	40_1^+	$\leq$ 0.14[10, 18]		0.09	0.086
	20_2^+	20_1^+	00_1^+	$>$ 8.8[10, 18]	$>$ 9.7[13]	2.63	1.43
		40_1^+	20_1^+	3.45(5)[10, 18]	2.9(2)[13]	6.1	1.80
	40_2^+	40_1^+	20_1^+	7.82(26)[10, 18]	11.8(2.5)[13]	6.21	0.91
		60_1^+	40_1^+	1.50(22)[10, 18]	0.59(7)[13]	11.3	1.75
	20_4^+	20_1^+	00_1^+	$>$ 1.7[10, 18]	1.99(27)[13]	1.54	1.43
		40_1^+	20_1^+	1.24(16)[10, 18]	1.12(12)[13]	2.11	1.80

From Table 4 we can see in ^{170}Yb, the experimental values of the ratios $R_{I0_2^+}$ for $E2-$ transitions from the $K^{\pi} = 0_2^+$ states is a differ from the adiabatic theory of a several dozen times. This is associated with strong mixing of $K^\pi = 0_2^+$ and $K^\pi = 2_1^+$ bands. Then arises the question, why the value of rations R_{I2_1} for the transitions from the 2_1^+ bands is not so much different from the adiabatic theory compared R_{I0_2}?. This can be explained by the fact that the matrix element $m_{2_1} =< 0_1^+|\hat{m}(E2)|2_1^+ >$ is 10 times greater than $m_{0_2} =< 0_1^+|\hat{m}(E2)|0_2^+ >$ (cm. Table 2)

One may see from these comparison that the mixing effect of states low-lying bands plays a crucial role which considerably demonstrates that $E2-$ transitions even in low values of angular momentum I.

3. Conclusion

In the present work, non-adiabatic effects in energies and electric characteristics of excited states are studied within the phenomenological model which taking into account Coriolis mixing of all experimentally known rotational bands with $K^\pi < 3^+$.

The energy and structure of wave functions of excited states are calculated. And also the reduced probabilities of $E2-$ transitions is calculated. The ratio of $E2-$ transitions probability from $K^\pi = 0_m^+$ and 2_ℓ^+ bands are calculated and compared with experimental data which gives the satisfactory result.

If matrix elements of $E2-$ transitions m_K one of two strongly mixing bands K is less than matrix element of $m_{K'}$ of K' ($m_K < m_{K'}$), then the difference in the ratio R_{IK} for the first band K from Alaga rule is bigger than the difference in $R_{IK'}$ from Alaga rule. In other words, if $m_K < m_{K'}$, nonadiabaticity in the ratio R_{IK} is stronger than that of $R_{IK'}$.

This work has been financial supported by the MOHE, Fundamental Research Grant Scheme FRGS13-074-0315, and also by the Committee for the Coordination of the Development of Science and Technology under the Cabinet of Ministers of the Republic of Uzbekistan, Fundamental Grant "OT-F2-75" (2017-2020).

References

[1] Bohr A and Mottelson B R 1997 *Nuclear Structure* **1,2** (Benjamin, New York)
[2] Soloviev V G 1974 *Theory of difficult atomic nuclei* (M. Nauk)
[3] Burke D G, Soloviev V G, Sushkov A V and Shirikova N Yu 1999 *Nucle. Phys.* Vol. A **656** 287
[4] Firestone R B and Shirley V S 1996 *Table of Isotopes* (edited by Wiley, New York)
[5] Zilges A, Brentano P von, C.Wesselborg C and et al 1990 *Nuc.Phys.* A**507** 399
[6] Usmanov P N, Adam I, Salikhbaev U S and Solnyshkin A A 2010 *Physics of Atomic Nuclei* **73** (12) 1990-1996
[7] Okhunov A A, Usmanov Ph N and Hasan Abu Kassim 2014 *J. Phys. G: Nucl. Part. Phys.* **41** 075102
[8] Okhunov A A, Sharrad F I, A. Anwer Al-Sammarraie and Khandaker M U 2015 *Chinese Physics* C **39**(8) 084101
[9] Usmanov Ph N, I.N.Mikhailov IN 1997 *Phys. Part. Nucl. Lett.* **28** 348
[10] Begzhanov R B, Belinkiy V M et al. 1989 *The directory on the nuclear physics* Vol.**1,2** (Tashkent, FAN)
[11] Baglin M 2002 *Nucl.Data Sheets* **96**, 611 [1996 *Nucl.Data Sheets*, **77**(2) 125].
[12] Singh B 1995 *Nucl.Data Sheets* **75** 199
[13] Browne E, Junde H 1999 *Nucl.Data Sheets* **87** 15
[14] Reich C W, Greenwood R C, Lokken R A 1974 *Nucle. Phys.* **A228** 365
[15] Fahlander C, Varnestig B, Backlin F *et al.* 1992 *Nucl. Phys.* **A147** 157
[16] Faessler A, Greiner W, and R.K.Sheline R K 1965 *Nucl. Phys.* **70** 33
[17] Arima A and Iachello F 1984 *Adv. Nucl. Phys.* **13** 139
[18] Gasten R F, Von Brentano P, Kane W R 1973 *Phys. Rew.* **C8** 1035

IOP Conf. Series: Journal of Physics: Conf. Series **949** (2017) 012020 doi:10.1088/1742-6596/949/1/012020

A PWM Controller of a Full Bridge Single-Phase Synchronous Inverter for Micro-Grid System

Tawfikur Rahman[1], S. M. A. Motakabber[2], M. I. Ibrahimy[3], Aliza 'Aini Binti Md Ralib@ Md Raghib[4]

[1, 2, 3, 4] Department of Electrical and Computer Engineering, International Islamic University Malaysia, Kuala Lumpur, Malaysia.

E-mail : [1] tawfikurr@gmail.com

Abstract. Nowadays, microgrid system technology is becoming popular for small area power management systems. It is essential to be less harmonic-distortion and high efficiency of the inverter for microgrid applications. Pulse width modulation (PWM) controller is a conventional switching control technique which is suitable to use in the microgrid connected power inverter system. The control method and algorithm of this technique are challenging, and different approaches are required to avoid the complexity for a customized solution of the microgrid application. This paper proposes a comparative analysis of different controller and their operational methods. A PWM controller is used to reduce the ripple voltage noise while a continuous current mode provides a small output ripple which gives steady-state error as zero on fundamental and cutoff frequency. To reduce the ripple current, higher frequency harmonic distortion, switching loss and phase noise, LC low pass filter is used on either side of input and output terminals. The proposed inverter is designed by MATLAB 2016a simulation software. A balanced load resistance (RL = 20.5 Ω) of star configuration and a dual input DC voltage of ± 35V are considered. In this design, the circuit parameters, the fundamental frequency of 50 Hz, the PWM duty cycle of 95%, the cutoff frequency of the switching controller of 33 kHz are considered. The inverter in this paper exhibits THD of 0.44% and overall efficiency approximately of 98%. The proposed inverter is expected to be suitable for microgrid applications.

1. Introduction

The microgrid connected inverter can be found as both half-bridge and full-bridge configurations. While the voltage and current range are a small one cover, usually utilized in uninterruptible power supplies, high-power static power topologies, short-long distance transmission and distribution systems, islanding systems and power supply systems [1]. A half-bridge inverter (voltage source) generates an output of the square wave signal for a resistive load with single-phase. The resistive load is connected to the two-capacitor midpoint [2]. The inverters have simple switching technique to operate at much lower frequency compared to others inverter. On the other hand, a single phase full bridge inverter comprises of a couple of terminals which are connected with two IGBT switches. The parallel freewheeling diode is connected to each IGBT to pass the current in terms of different loads [3]. Additional diodes are employed to direct the load current when resistive-inductive loads are used. These diodes deliver an alternative way to deliver the inductive current is directed by these additional diodes when turn 'OFF' condition is occurred. A typical inverter control strategy is a sine-triangle, pulse width modulation (SPWM) control [4]. By SPWM control, the inverter of the semiconductor switch regulated in funding for an analysis of a sinusoidal wave signal and a triangular switching signal. The sinusoidal regulated waveform secures the required fundamental frequency through the output. Meanwhile, the triangular waveform builds the switching frequency of the circuit [5]. The proportion between the frequencies of the sinusoid and the triangle wave also is indicated because of the alteration percentage of the frequency. However, the reference waveform might be available in a completely different form to suit the inverter topology, for example, sine wave and twisted sine wave. A sinusoidal waveform may be utilized for PWM as a part of input DC to output AC inverter wherever it is utilized to form the AC voltage to be near to a sine wave [6].

The PWM could be a technique that is described by the generation of steady amplitude pulse by adjusting the duty cycle through balancing the pulse length of time. A PWM controller simply involves the generation of both carrier signals, and a reference signal which is sustained into the comparator devices and created by logical and absolute output is provided [7]. The referred signal (output frequency) may be a square wave. The carrier signal could also be a triangular wave into the frequency completely more observable rather than the reference signal. These signals are utilized as a part of the primary PWM control techniques such as the single, multiple and sinusoidal PWM system. The single PWM has generated a standard output gate pulse every one-half cycle. The output signal has modified by unsteady the size of the pulse [8]. The reference signal between a rectangular reference signal and a triangular reference is twisted by the gate signal. The two signs are approximately equivalent to the frequency. Multiple modulation systems around various quantities of pulses every one-half cycle and all pulses are of equal size. The output signals have created through matching a reference which is a rectangular and triangular references wave. The reference signal frequency generated the carrier frequency and output frequency [9]. Whereas, the sinusoidal modulation approach is the smaller size of pulse output for every half cycle and the individual pulses are measured. This size of every single pulse shifts by 180^0 out of phase with the measured sinusoidal wave.

The current controllers can reduce the harmonics on the output filter inductors while the voltage controller must decrease the harmonic voltage on the capacitors. However, the harmonics of the current and voltage that result from PWM switching can be lowered not by a controller but by a lowpass LC output filter [10]. The LC filters decrease outcome can improve by reducing the filter cutoff frequency compared to the inverter switching frequency. But, the cutoff frequency has confined the inverter control bandwidth. Improving the bandwidth is significant not only for inverter system operation, but also for specific voltage restitution without a phase delay at maximum harmonics. Therefore, there is a trade-off between the control bandwidth and the reduction effect in the design of the output LC filter [11].

This paper has proposed a PWM based, full bridge single phase PSI for transformer fewer microgrid systems to improve switching loss and minimize the attenuation effect. Although, considering the LC lowpass filter are used to decrease the ripple current exceed problem without reducing the damping performance of the inverter systems. Based on the study, a new controller design method that assurance steady state control methods and that reduces the switching loss and ripple current has proposed. This technique improves the switching loss and decreases the switching delay. In turn, the controller response in transient condition is further improved by this technique.

2. Design of single-phase H-bridge PSI systems

The planned single-phase PSI circuit is presented in figure 1. Here, the output terminal voltage of the bridge, V_a and V_b are proportionally related to the DC input voltage and the inverter duty cycle. The voltage can have any value of $+V_{DC}$, 0 or $-V_{DC}$ which is determined by the controlling signal of switches. It is noteworthy that the switches in a leg cannot be turned on simultaneously in order to prevent short circuit across the DC source. Precisely controlling the switch is the primary challenge; therefore, PWM technique is used to create the proper gate signal which has done by controlling the phase of an output signal with a specific frequency.

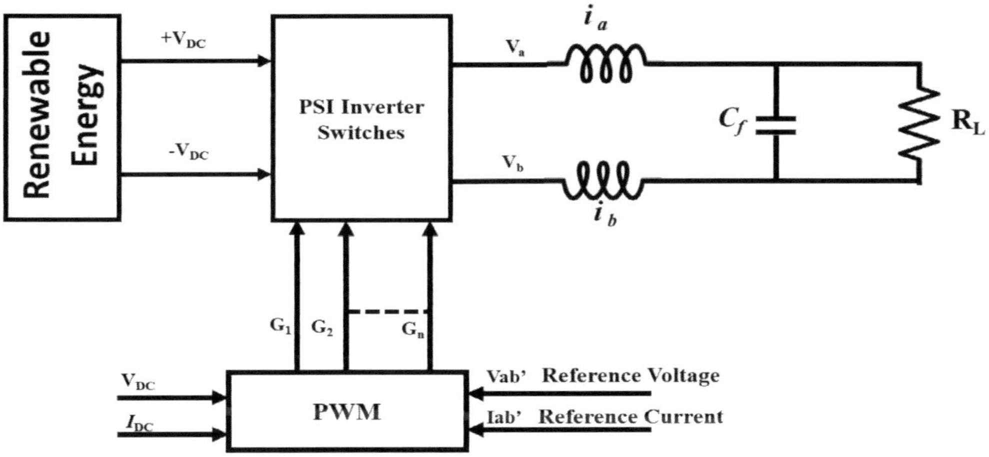

Figure 1. Block diagram of a PSI system

Also, a low pass filter is used to filter out the high-order harmonics and the switching harmonics of the output signal. It is observed that the filter size depends on cutoff switching frequency, modulating voltage and current, fundamental frequency and phase angle. The switching loss depends on the filter designed with different values of inductor and capacitor. The state space expressions of full-bridge PSI are presented in equations (1) and (2) [12].

$$\begin{bmatrix} i_L \\ v_c \end{bmatrix} = \begin{bmatrix} \frac{Z(R_L+R_C)-R_L R_C}{L(Z+R_C)} & \frac{-Z}{L(Z+R_C)} \\ \frac{1}{C(1+\frac{R_C}{Z})} & \frac{-1}{ZC(1+\frac{R_C}{Z})} \end{bmatrix} \begin{bmatrix} i_L(t) \\ v_c(t) \end{bmatrix} + \begin{bmatrix} \frac{1}{L} \\ 0 \end{bmatrix} V_{AB} U(t) \tag{1}$$

$$V_{OUT} = \begin{bmatrix} \frac{ZR_C}{Z+R_C} & \frac{Z}{Z+R_C} \end{bmatrix} \begin{bmatrix} i_L \\ v_c \end{bmatrix} \tag{2}$$

Here, a state variable i_L denotes the inductor current and another state variable v denotes the capacitor voltage. The load impedance Z and full-bridge voltage V_{ab} are defined by equation (3).

$$U(t) = \begin{cases} 1 & V_{AB}(t) = +\frac{V_{DC}}{2} \\ 0 & V_{AB}(t) = 0 \\ -1 & V_{AB}(t) = +\frac{V_{DC}}{2} \end{cases} \tag{3}$$

3. Pulse width modulation controller

The PWM controller is the most popular method of using carrier-based modulation technique. In PWM technique, the carrier signal with a high frequency and triangular shape are superpositioned on the reference sinusoidal signal. The carrier-based PWM technique is less complex and it provides a dynamic response which is suitable for PS inverter [13]. The switching ratio (SW_r) of the PWM technique has been given in equations (4) and (5).

$$SW_r = \frac{\text{Carrier Wave Frequency}}{\text{Output Voltage Frequency}} \tag{4}$$

SW_r is related to harmonic frequency, and the harmonics have standard located at:

$$f_c = k\, SW_R\, f_m \tag{5}$$

In equation (5), the frequency of the modulating signal is termed by f_m and k is an integer (1, 2, 3…). Furthermore, modulation index (M_i) is found by the ratio of the amplitudes of the modulating reference signal (A_r) and the carrier signal (A_c). This is expressed by,

$$M_i = \frac{A_r}{A_c} \tag{6}$$

The PWM controller modulation signal is applied to control the phase of the inverter. This is illustrated in figure 2.

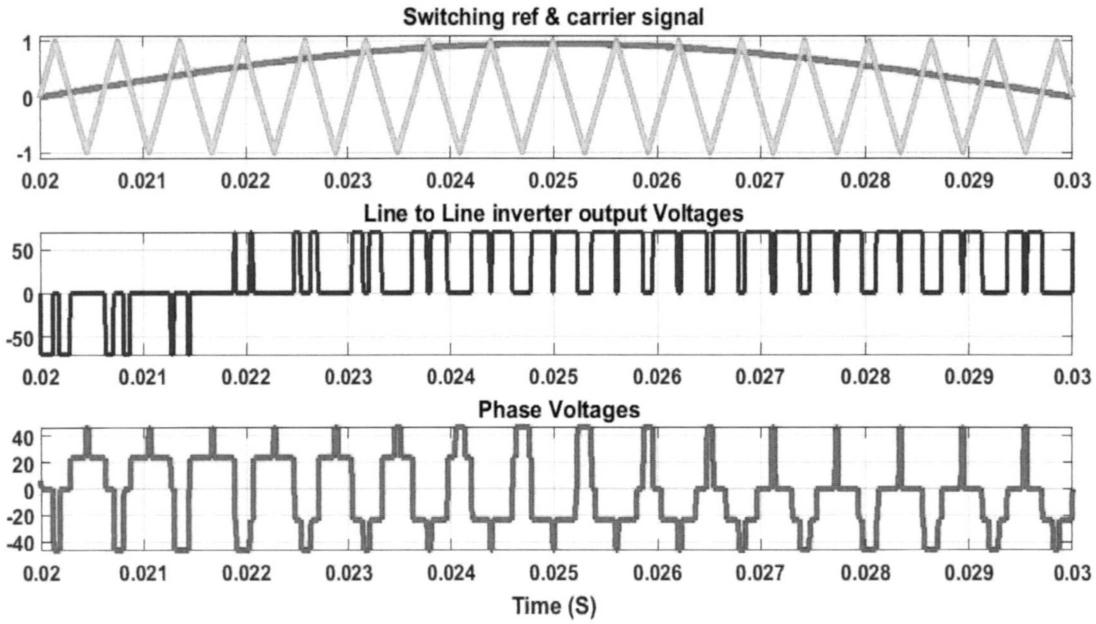

Figure 2. The modulation signal of a PWM controller

In figure 2, initial phase angles of the reference and the first carrier are considered as zero. On the basis of the relation expressed by equations (7) and (8), the generated voltage by every half-phase from the middle of the DC link is:

$$V_A(t) = \frac{V_{DC}}{2}\left[\frac{M}{2}\cos(\omega_1) + \sum_{m=1}^{+\infty}\frac{2}{m\pi}J_0\left(\frac{m\pi M}{2}\right) sin\frac{m\pi}{2}\cos(m\omega_c t)\right. \tag{7}$$

$$\left. + \sum_{m=1}^{+\infty}\sum_{n=\pm 1}^{\pm\infty}\frac{2}{m\pi}J_n\left(\frac{m\pi M}{2}\right) sin\frac{(m+n)\pi}{2}\cos(m\omega_c t + n\omega_1 t)\right]$$

$$V_B(t) = \frac{V_{DC}}{2}\left[\frac{M}{2}\cos(\omega_1) + \sum_{m=1}^{+\infty}\frac{2}{m\pi}J_0\left(\frac{m\pi M}{2}\right) sin\frac{m\pi}{2}\cos(m\omega_c t - \pi)\right. \quad (8)$$

$$\left. + \sum_{m=1}^{+\infty}\sum_{n=\pm 1}^{\pm\infty}\frac{2}{m\pi}J_n\left(\frac{m\pi M}{2}\right) sin\frac{(m+n)\pi}{2}\cos[m\,(\omega_c t - \pi) + n\omega_1 t)\right]$$

Like the case of full bridge phase synchronous inverters, harmonic cancellation between i_{L1} and i_{L2} can be considered the basis of the higher frequency harmonic cancellation voltage between v_1 and v_2. By observing the above comparison between a couple of expressions, it is found that the entire odd harmonics and their side-bands of the carrier are cancelled out through interleaving. Furthermore, it is observed that the amplitudes of each even-order harmonic become zero. Consequently, only the harmonic of $v_1(t) + v_2(t)$ remains as the sideband component with odd-order regarding the carrier harmonics with even-order. This relationship is presented in equation below,

$$\frac{4V_{DC}}{2}\sum_{m=2,4,..n}^{+\infty}\sum_{n=\pm 1,\pm 3,...}^{\pm\infty}\frac{1}{m}J_n\left(\frac{m\pi M}{2}\right) sin\frac{(m+n)\pi}{2}\cos(m\omega_c t + n\omega_1 t) \quad (9)$$

The relationship with the ripple cancelation principle expressed in equation (9) is extendable for parallel or series inverters of any number as N. This can be characterized by using the harmonic scaling factor [14]:

$$K_m = \frac{1}{N}\sum_{K=1}^{N} e^{jm\theta_{ck}} \quad (10)$$

Here, K_m is the scaling factor which permits the combined harmonic spectrum of N parallel or series inverter to select a single inverter through multiplying harmonics by N_{km}. Figure 3 shows the PWM MATLAB circuit diagram of full bridge PSI inverter system.

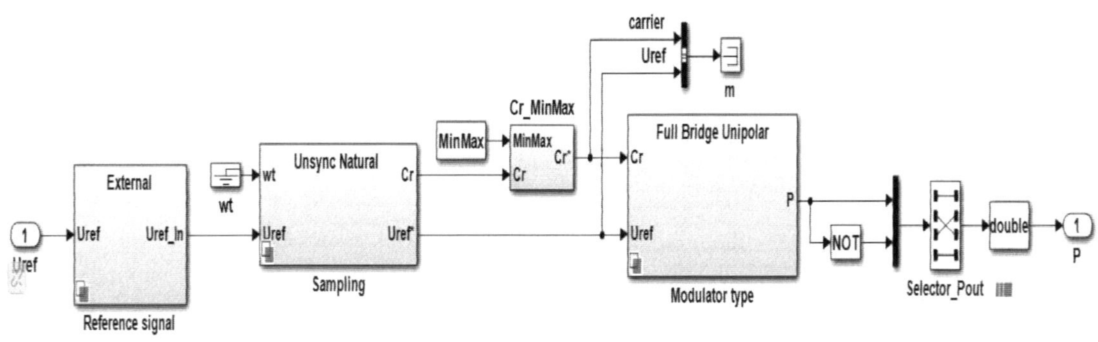

Figure 3. MATLAB circuit of a PWM controller

4. The PSI switching topology

Figure 4 shows that the full bridge PSI, which comprises of two input and output terminals, IGBT switches, lowpass *LC* filter and microgrid resistive load. IGBT switches are controlled by the PWM gate pulse signal to reduce switching loss and phase delay. The lowpass *LC* filter is used to convert the sinusoidal wave and to reduce THD and synchronize the phase between the inverter and microgrid. In this system, two half phases are connected in parallel, which operates at 1.65 kHz of the carrier frequency, 95% of the duty cycle and 50 Hz of the fundamental frequency. However, switching performance significantly depends on the intrinsic parameters of the switches. In this IGBT switches, $1.6e^5$ Ω of the snubber resistance (R_s), 0.001 Ω of the internal resistance (R_{on}) and $1e^{-6}$ μF of the snubber capacitance (C_s) are assumed.

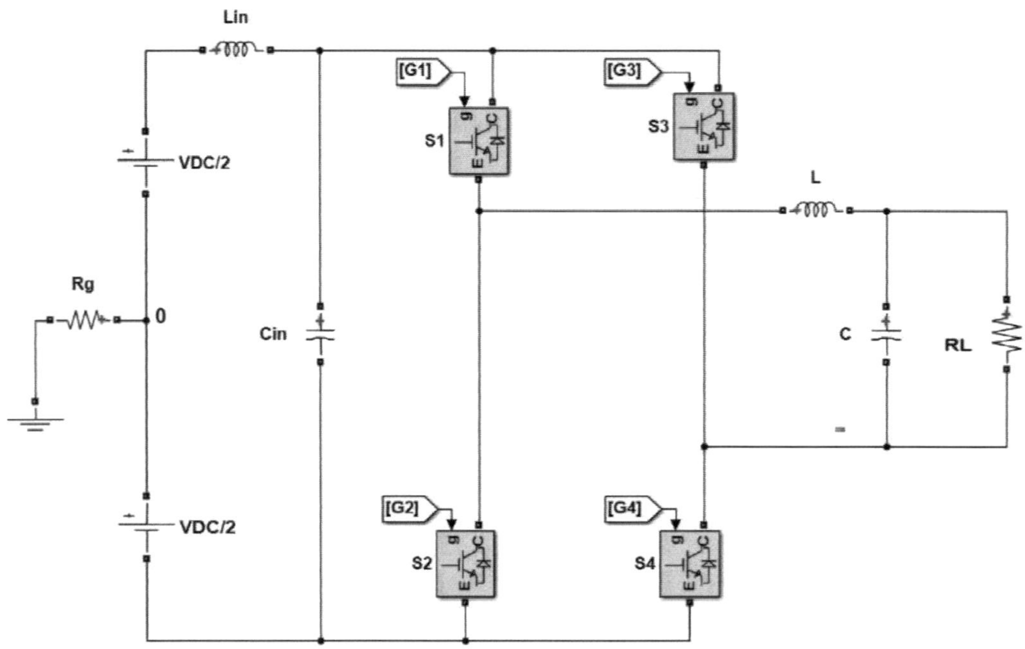

Figure 4. Full bridge PSI circuit

The inverter takes an input DC voltage, then transforms to sinusoidal AC voltage along with same microgrid frequency. There are four switching states S_1, S_2, S_3 and S_4 have presented in Table 1. The semiconductor switches in every branch should work on the opposite hand means they are not in the similar mode of operation (ON/OFF) particularly at the same time. Besides that, each switch remains OFF for a little period known as blanking time to avoid short-circuiting.

Table 1. Switching states of the PSI

S_1	S_2	S_3	S_4	V_A	V_B	V_{AB}
ON	OFF	OFF	ON	$\frac{V}{2}$	$\frac{V}{2}$	+V
OFF	ON	ON	OFF	$-\frac{V}{2}$	$-\frac{V}{2}$	-V
ON	OFF	ON	OFF	$\frac{V}{2}$	$-\frac{V}{2}$	0
OFF	ON	OFF	ON	$-\frac{V}{2}$	$+\frac{V}{2}$	0

IOP Conf. Series: Journal of Physics: Conf. Series **949** (2017) 012020 doi:10.1088/1742-6596/949/1/012020

However, S_1 and S_2 switches should work in a pair to get the inverter output and S_3, S_4 also made another pair. These legs have switched such an OFF and ON conditions that the output voltage and current have moved from one to another and therefore the shift in polarity follows in voltage waveform. If the movement angle is zero, the voltage output is additionally zero and maximal after change plot is 'π'.

5. Output filter design

Figure 5 shows an *LC* type of lowpass filter, a circuit that consists parameters of inductor and capacitor. The proper value of that parameter has needed in creating an *LC* filter.

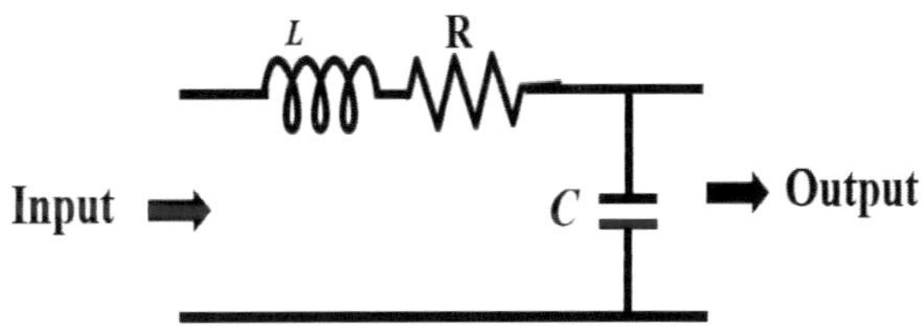

Figure 5. Output LC filter

Inductor Selection: The inductor is utilized to cut off current swell. The estimation of inductor depends on the passable swell current of the inductor. Then again, the swell current of the inductor is due to PWM switching (15). Determining the value of the inductor has done by the following equations:

$$L = \frac{V \times dt}{di} \tag{11}$$

$$L = \frac{2 \times V_{dc} \times D_{max}}{\Delta I_{max} \times f_c} \tag{12}$$

Where, L is the output filter inductor, V_{dc} is the input DC voltage, f_c is the switching frequency, ΔImax if the permissible ripple current and D_{max} is the duty cycle maximum.

Capacitor selection: The capacitor value depends on the peak values of the inverter; focuses the measurement of the ripple voltage of the output [7]. Essentially, the capacitor output retains the greater part of the ripple in the inductor current. The ripple voltage can be found out by

$$\Delta V_{out} = \Delta V_c \tag{13}$$

$$\cong \frac{\Delta I_L}{C \times f_s} + r_{ESR} \times \Delta I_L \tag{14}$$

Where, ΔV_{out} is the output voltage ripple, ΔI_L is the ripple current of the inductor and r_{ESR} is the capacitor equivalent series resistance. Figure 6 shows the LC lowpass filter phase response waveform.

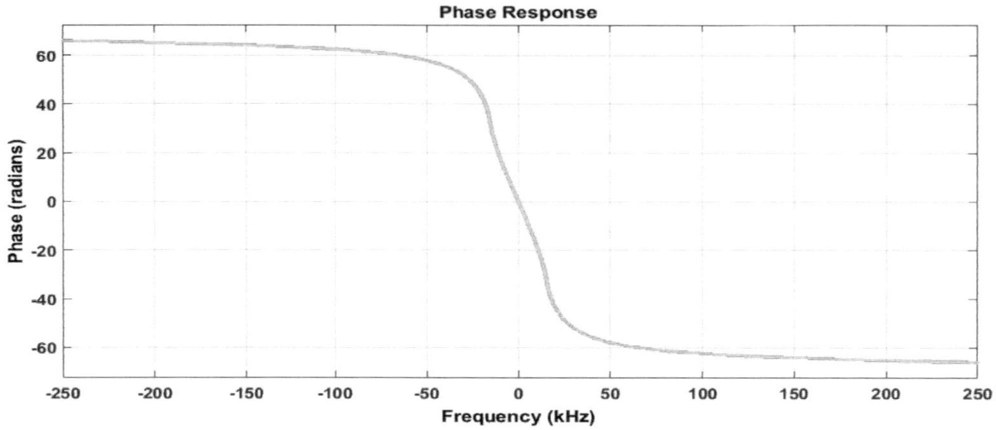

Figure 6. Output LC filter magnitude response

6. Simulated result and discussion

To solve this problem, of four IGBT/Diodes has used (Figure 4) that relates to output filter and microgrid. Besides that, proposed design can work as a single phase VSI design and make parallel and series resonant, based on the inverter. In this system have been used two level IGBT/Diode, carrier frequency 1.6625 kHz, micro-grid frequency 50 Hz, input voltage ± 35 V_{DC}, simulation time = 0.1s and modulation index 0.95, etc.

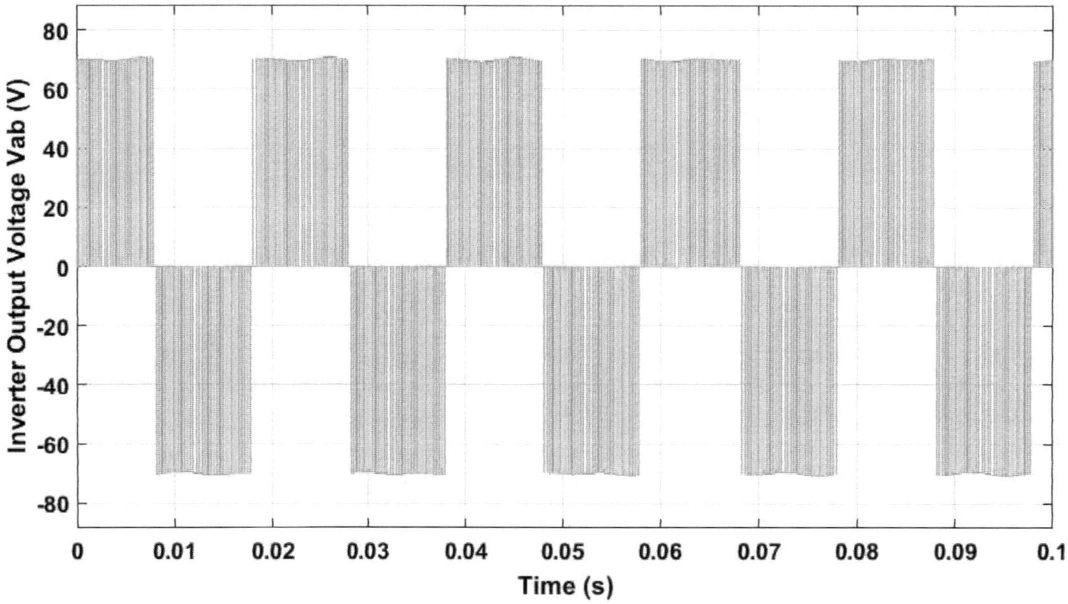

Figure 7. Inverter output voltage waveform without filtering

Simulated results of the proposed PSI circuit (without filtering condition), It is seen that the output voltage V_{ab}, is achieved about 138.94V (peak to peak), the output current I_{ab} is around 6A (peak to peak). The fundamental frequency is 50Hz square wave output voltage waveform and distorted current waveform, presenting in figure 7 and figure 8.

IOP Conf. Series: Journal of Physics: Conf. Series **949** (2017) 012020 doi:10.1088/1742-6596/949/1/012020

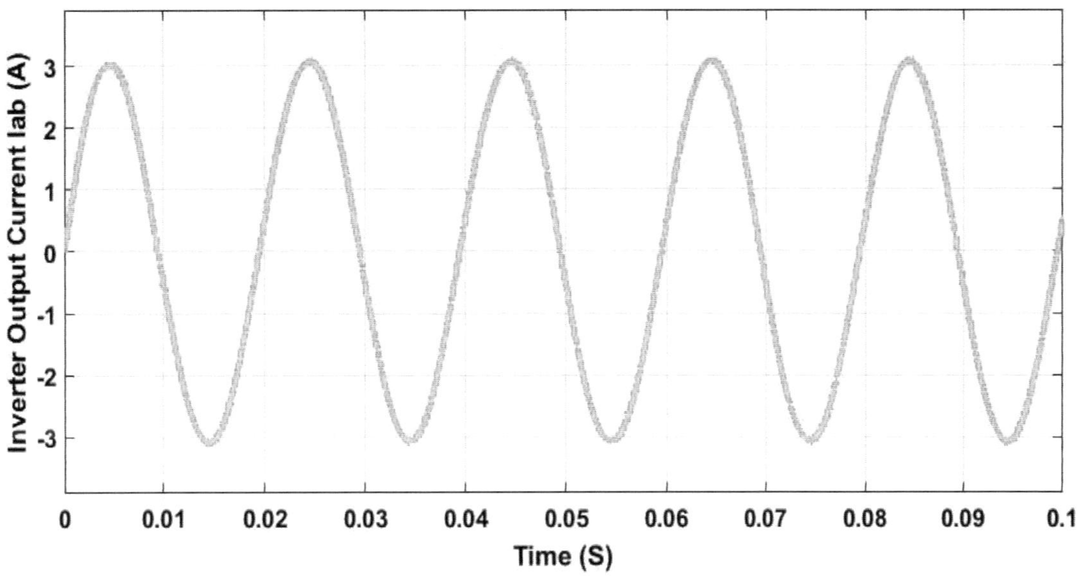

Figure 8. Inverter output current waveform

For filtering state, the output voltage of the microgrid has been attained almost 120 V_{pp} of the microgrid voltage, approximately 6 A_{pp} of the microgrid current and 50Hz of the microgrid frequency, showing in Figure 9 and Figure 10.

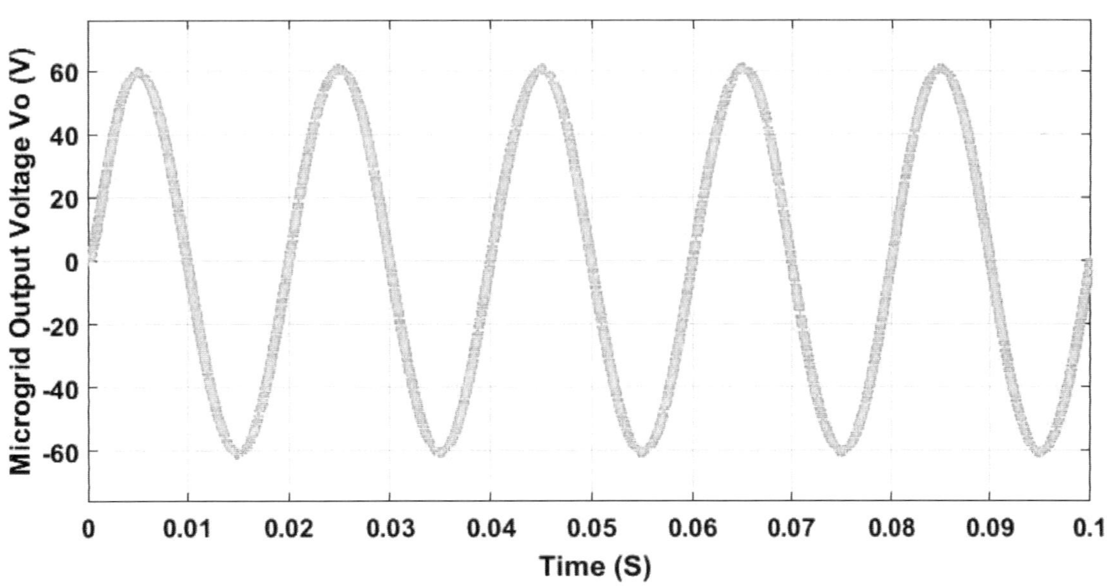

Figure 9. Inverter output voltages after filtering

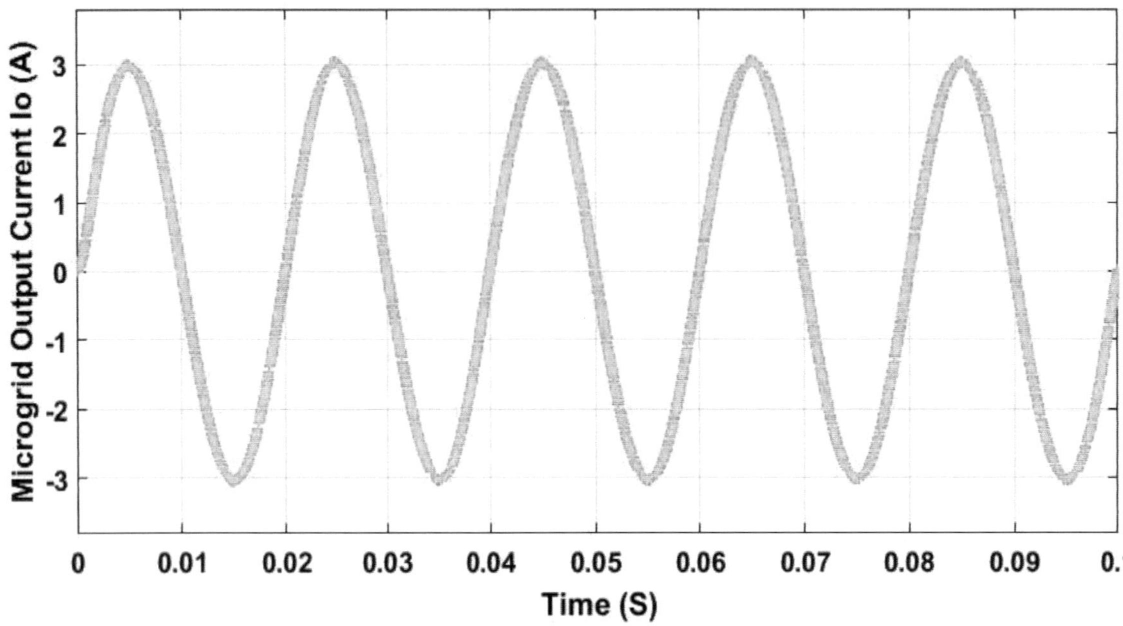

Figure10. Output current waveform

Also, FFT analysis has done on the output waveform of the voltage and current, without filtering effect that is illustrated in Figure 11 and Figure 12. This review makes certain that the inverter output voltage has harmonics without filter cases through significant, but with the filtered condition, the output voltage is as like as the micro-grid voltage which has only fundamental harmonics where 50Hz has deceived, and rest of harmonics are insignificant.

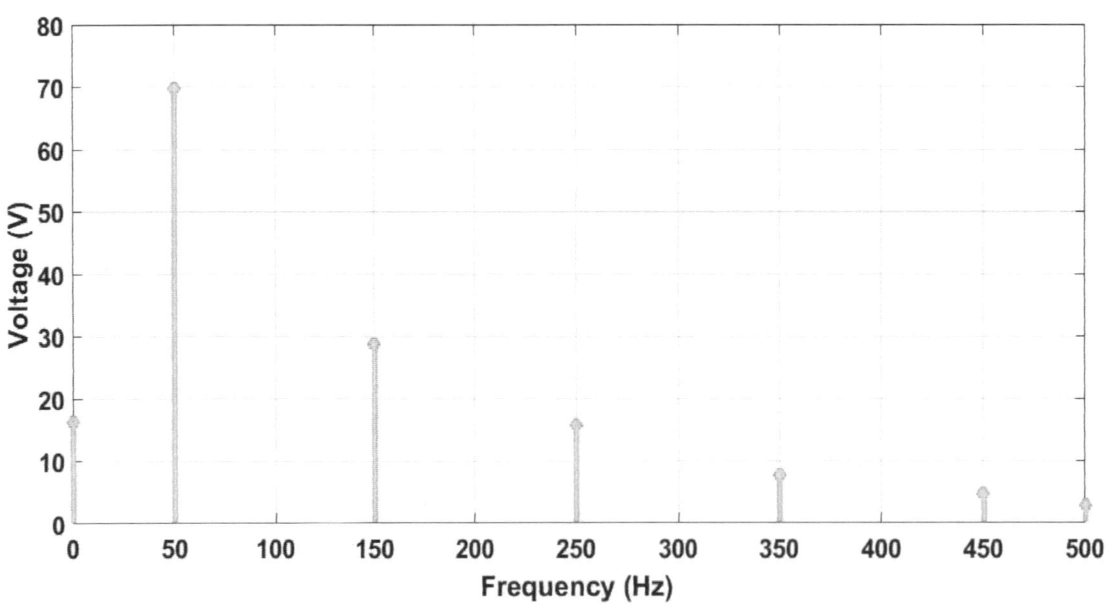

Figure 11. The FFT of the inverter output voltage without filtering

Figure 11 demonstrates that the output voltage of the inverter V_{ab} that attained about 138.94 V (peak to peak), corresponding THD is roughly 49.56%. Figure12 indicates that the inverter output current I_{ab} is around 6A (peak to peak), similar THD is almost 48.95%, whereas, the fundamental frequency is 50Hz square wave output voltage waveform and distorted current waveform (without filtering condition).

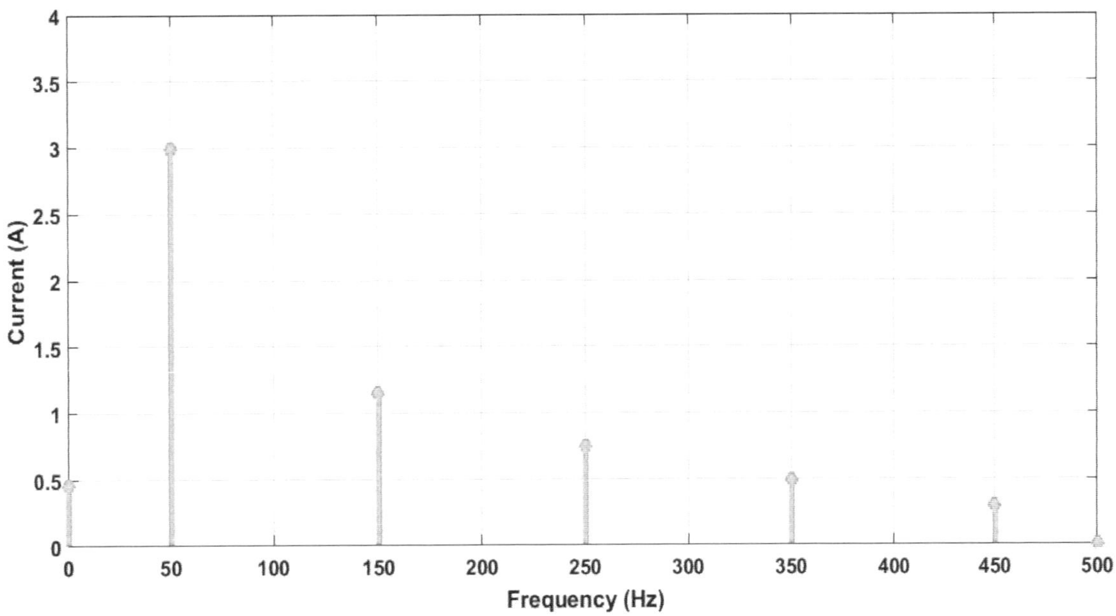

Figure 12. The FFT of the inverter output current without filtering

Figure 13 and Figure 14 have represented the output voltage and current along with FFT analysis. In the case of with filtering effect, the output voltage and current of the inverter has a low level of THD, which is 0.43% because this system is transformerless.

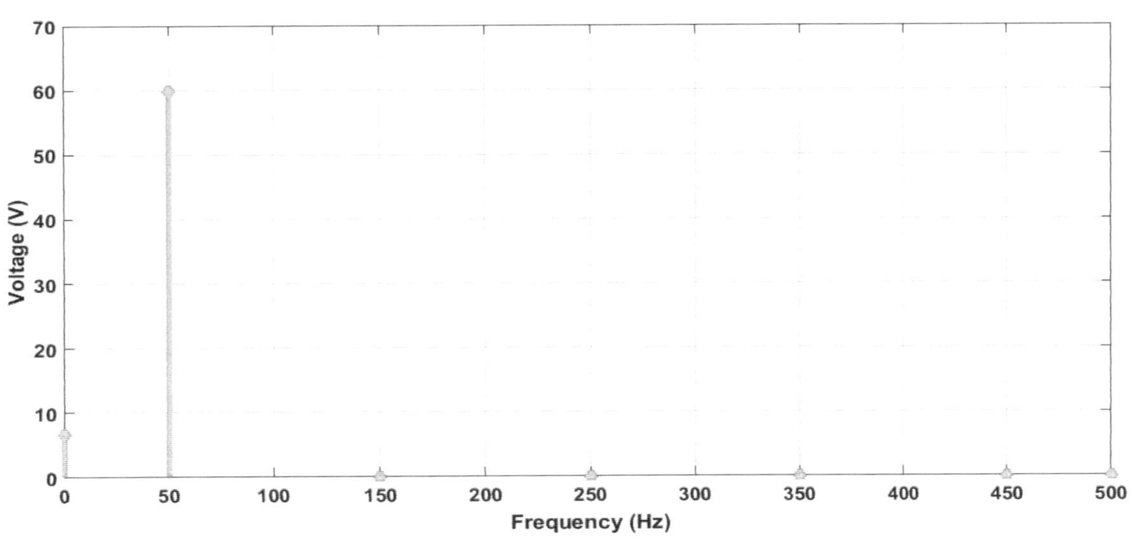

Figure 13. The FFT of the output voltage after filtering

It is found from the FFT analysis that THD of the output voltage is approximately 0.43%, THD of the output current is 0.46% while the fundamental frequency is 50 Hz.

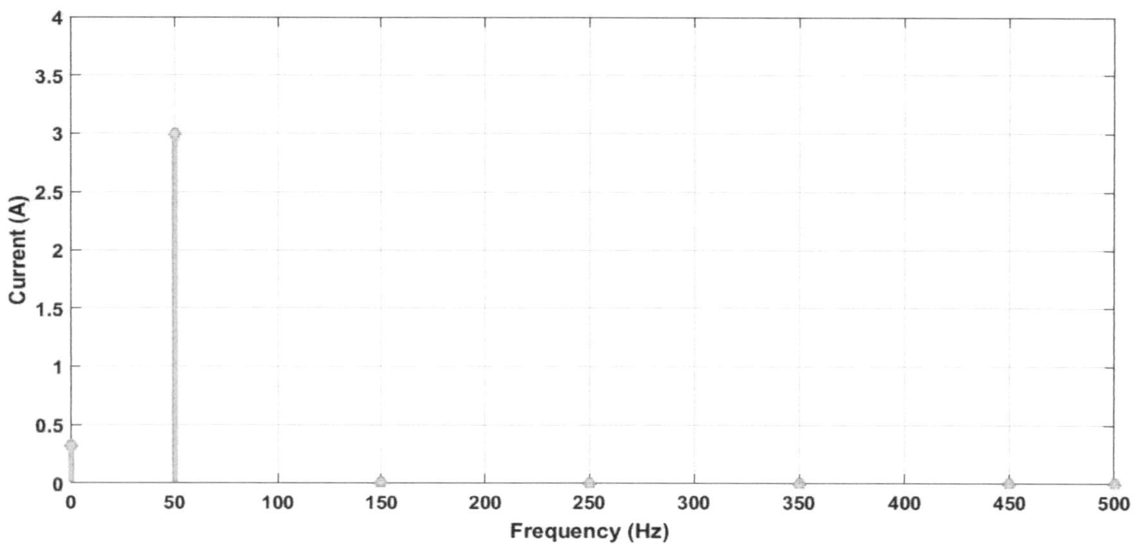

Figure 14. The FFT of the output current after filtering

7. Conclusions

The major concern of a microgrid power system is to deliver a maximum power from the inverter to the load. Simulated results of the proposed PSI indicate that the value of ripple current is at an acceptable limit due to phase synchronization of the micro-grid system. THD ratio is found as 0.44%, which is much less than the maximum permissible distortion limit of IEEE standard (THD <5%). It has concluded that proposed system needed fewer filter parameters. Therefore, more power can have transferred power from the inverter to the micro-grid load is gained. At that time, the inverter efficiency was about 98%, while the output voltage is 120V AC and load impedance is 20.5Ω, the inverter power is around 100w. Therefore, it is justified that the proposed technique is efficient and operative rather than the traditional reference signal generation by the switching logic controller and identical with the micro-grid voltage at a future stage.

Acknowledgments

This research has been supported by the Malaysian Ministry of Education through the Fundamental Research Grant Scheme under the project ID: FRGS15-190-0431.

8. References

[1] Rahman, T., Ibrahimy, M. I., Motakabber, S. M., & Mostafa, M. G. 2014 Three Phase Three Layer Phase Synchronous Inverter for Microgrid System *In Computer and Communication Engineering ICCCE*, pp. 44-47 IEEE.

[2] Soomro, J., Memon, T. D., & Shah, M. A. 2016 Design and analysis of single-phase voltage source inverter using Unipolar and Bipolar pulse width modulation techniques *In Advances in Electrical, Electronic and Systems Engineering ICAEES* 277-282 IEEE.

[3] Hasan, M. M., Abu-Siada, A., & Islam, M. R. 2016 Design and implementation of a novel three-phase cascaded half-bridge inverter *IET Power Electronics* ***9(8)*** 1741-1752.

[4] Panda, B., Panda, B., & Hota, P. K. 2017 A Comparative Analysis of PWM Methods of Z-Source Inverters Used for Photovoltaic System. *In Proceedings of the 5th International Conference on Frontiers in Intelligent Computing: Theory and Applications* pp. 281-290.

[5] Fang, J., Xiao, G., Yang, X., & Tang, Y. 2017 *Singapore* Parameter design of a novel series-parallel-resonant LCL filter for single-phase half-bridge active power filters. *IEEE Transactions on Power Electronics* **32(1)** 200-217.

[6] Saonerkar, A. K., Thakre, A., Podey, A., Chimote, A., Kadao, P., Kadu, P., & Satpute, R. (2016, August). Single phase residential multilevel inverter using supercapacitor *In Inventive Computation Technologies ICICT* **Vol. 3** pp. 1-4 IEEE.

[7] Rahman, T., Motakabber, S. M. A., & Ibrahimy, M. I. 2016 Phase Synchronous Inverter for Microgrid System *In Computer and Communication Engineering (ICCCE)* pp. 167-171 IEEE.

[8] Thenathayalan, D., Ahmed, A., Choi, B. M., Park, J. H., & Park, J. H. 2017 Independent MPP Tracking Method of Hybrid Solar-Wind Power Conditioning Systems Using Integrated Dual-Input Single-PWM-Cell Converter Topology *Journal of Electrical Engineering and Technology* Vol **12** 790.

[9] Jana, J., Saha, H., & Bhattacharya, K. D. 2017 A review of inverter topologies for single-phase grid-connected photovoltaic systems *Renewable and Sustainable Energy Reviews* Vol **72** 1256-1270.

[10] Madhusoodhanan, S., Mainali, K., Tripathi, A., Patel, D., Kadavelugu, A., Bhattacharya, S., & Hatua, K. 2017 Harmonic Analysis and Controller Design of 15 kV SiC IGBT-Based Medium-Voltage Grid-Connected Three-Phase Three-Level NPC Converter *IEEE Transactions on Power Electronics* **32(5)** 3355-3369.

[11] Lamo, P., López, F., Pigazo, A., & Azcondo, F. J. 2017 An Efficient FPGA Implementation of a Quadrature Signal-Generation Subsystem in SRF PLLs in Single-Phase PFCs *IEEE Transactions on Power Electronics* **32(5)** 3959-3969.

[12] Roshan, A., Burgos, R., Baisden, A. C., Wang, F., & Boroyevich, D. 2007 A DQ frame controller for a full-bridge single-phase inverter used in small distributed power generation systems. *In Applied Power Electronics Conference, APEC 2007-Twenty-Second Annual IEEE* pp. 641-647 IEEE.

[13] Vijayakumar, K., Raj, R. S., & Kannan, S. 2015 Application of Sinusoidal Pulse Width Modulation Based Matrix Converter as Revolutionized Power Electronic Converter *In Power Electronics and Renewable Energy Systems* pp. 9-17.

[14] Sun, J. 2008 Modeling and analysis of pulse-width modulation. In Professional Education Seminar Workbook *IEEE Applied Power Electronics Conference*.

[15] Rahman, T., Ibrahimy, M. I., Motakabber, S. M., & Mostafa, M. G. 2015 Simulation and evaluation of a phase synchronous inverter for the micro-grid system. *ARPN J. Eng. Sci.* **11(1)** 1-2.

IOP Conf. Series: Journal of Physics: Conf. Series **949** (2017) 012021 doi:10.1088/1742-6596/949/1/012021

Periodic and Weakly Periodic Ground States for the λ-Model on Cayley Tree

Farrukh Mukhamedov[1], Chin Hee Pah[2], Muzaffar Rahimatullaev[3], Hakim Jamil[4].

[1]Department of Mathematical Sciences, College of Science, The United Arab Emirates University, P.O. Box, 15551, Al Ain, Abu Dhabi, UAE
[2]Department of Computational & Theoretical Sciences Faculty of Science, International Islamic University Malaysia P.O. Box, 141, 25710, Kuantan
[3]Institute of Mathematics, National University of Uzbekistan, 29, Do'rmon Yo'li str., Tashkent, 100125, Uzbekistan.
[4]Department of Computational & Theoretical Sciences Faculty of Science, International Islamic University Malaysia P.O. Box, 141, 25710, Kuantan

E-mail: m.hakimjamil@yahoo.com.my

Abstract. In this paper we consider the λ-model on the Cayley tree of order two. We describe the periodic and weakly periodic ground states for the considered model.

1. Introduction

Each Gibbs measure is associated with a single phase of a physical system [1]. The existence of two or more Gibbs measures means that phase transitions exist. One of fundamental problems is to describe all the extreme Gibbs measures corresponding to a given Hamiltonian.

As is known, the phase diagram of Gibbs measures for a Hamiltonian is close to the phase diagram of isolated (stable) ground states of this Hamiltonian. At low temperatures, a periodic ground state corresponds to a periodic Gibbs measure, see [1], [2]. The problem naturally arises on description of periodic and weakly periodic ground states.

For the Potts model with competing interactions on the Cayley tree of order two, periodic ground states were studied in [3, 4, 5] (see also [6]). The notion of a weakly periodic ground state was introduced in [7]. In [11] it was proposed a new look at Gibbs measures corresponding to weakly periodic ground states. Unlike the classical case of the Z^d lattices, the free energy of the Ising model on the Cayley tree does depend on the boundary conditions. It was shown that the free energies of the Gibbs measures associated with weakly periodic ground states are larger than free energy of the translation invariant Gibbs measures. Moreover, it is conjectured that the disordered phase is a convex combination of Gibbs measures of weakly-periodic ground states. Therefore, it is necessary to describe weakly-periodic ground states of lattice models.

For the Ising model with competing interactions, weakly periodic ground states were described in [7], [8]. For the Potts model, such states for normal subgroups of index 2 were studied in [9], while such state for normal subgroups of index 4 were studied in [10].

On the other hand, it is natural to consider a model which is more general than Potts one, therefore, it was proposed in [12] to study so-called λ-model on the Cayley tree (see also [13, 6]).

In that model, many possible interactions (nearest-neighbor) could be taken into account. In the mentioned paper, for a special kind of λ-model, its disordered phase has been studied (see [14]) and some its algebraic properties were investigated. Furthermore, in [15] we have described ground states of the λ-model on the Cayley tree of order two. This model is much more general than Potts model, and exhibits interesting structure of ground states.

In the present paper, we continue an investigation of the λ-model on the Cayley tree of order two and study periodic and weakly periodic ground states for normal subgroups of index 2. The paper is organized as follows: In Section 2, we recall the main definitions and the known facts. In Section 3, we study periodic and weakly periodic ground states of the model. Note that our results extend the known ones [16, 7]. Hence, the proposed model opens new perspectives in the theory of statistical mechanics over trees.

2. Preliminaries

Let $\tau^k = (V, L)$ be a Cayley tree of order k, i.e, an infinite tree such that exactly $k+1$ edges are incident to each vertex. Here V is the set of vertices and L is the set of edges of τ^k.

Let G_k denote the free product of $k+1$ cyclic groups $\{e, a_i\}$ of order 2 with generators $a_1, a_2, \ldots, a_{k+1}$, i.e., let $a_i^2 = e$ (see[17]).

There exists a one-to-one correspondence between the set V of vertices of the Cayley tree of order k and the group G_k, see [6], [18].

We show how to construct this correspondence. We choose an arbitrary vertex $x_0 \in V$and associate it with the identity element e of the group G_k. Since we may assume that the graph under consideration is planar, we associate each neighbor of x_0 (i.e., e) with a single generator $a_i, i = 1, 2, \ldots, k+1$, where the order corresponds to the positive direction, see Figure 1.

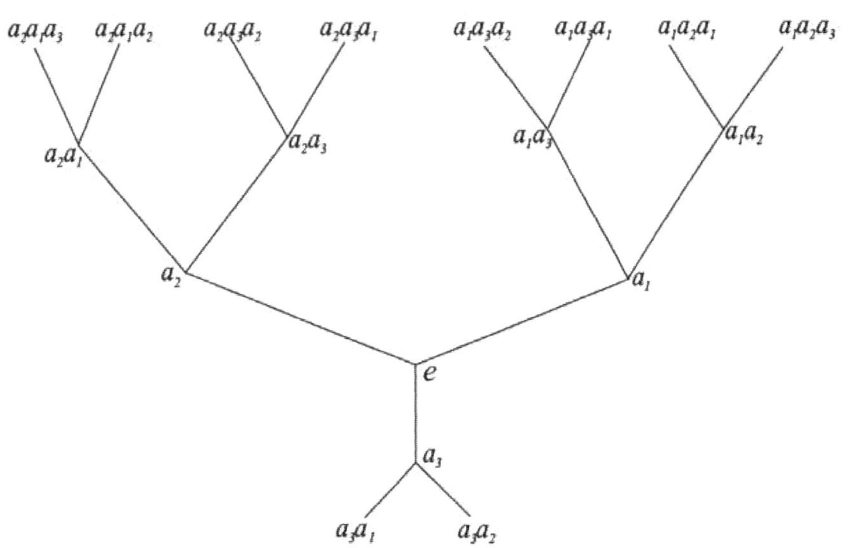

Figure 1. The Cayley tree τ^2 and elements of the group representation of vertices

For every neighbor of a_i, we introduce words of the form a_ia_j . Since one of the neighbors of a_i is e, we put $a_ia_i = e$. The remaining neighbors of a_i are labeled according to the above order. For every neighbor of a_ia_j , we introduce words of length 3 in a similar way. Since one of the neighbors of a_ia_j is a_i, we put $a_ia_ja_j = a_i$. The remaining neighbors of a_ia_j are labeled by words of the form $a_ia_ja_l$, where $i, j, l = 1, 2, ..., k+1$, according to the above procedure. This agrees with the previous stage because $a_ia_ja_j = a_ia_j^2 = a_i$. Continuing this process, we obtain a one-to-one correspondence between the vertex set of the Cayley tree τ^k and the group G_k.

The representation constructed above is said to be *right* because, for all adjacent vertices x and y and the corresponding elements $g, h \in G_k$, we have either $g = ha_i$ or $h = ga_j$ for suitable i and j. The definition of the *left* representation is similar.

For the group G_k (or the corresponding Cayley tree), we consider the left (right) shifts. For $g \in G_k$, we put

$$T_g(h) = gh\ (T_g(h) = hg)\ for\ all\ h \in G^k$$

The group of all left (right) shifts on G_k is isomorphic to the group G_k.

Each transformation S on the group G_k induces a transformation S on the vertex set V of the Cayley tree τ^k. In the sequel, we identify V with G_k.

The following assertion is quite obvious (see [5, 7]).

Theorem 2.1. *The group of left (right) shifts on the right (left) representation of the Cayley tree is the group of translations.*

By the group of translations we mean the automorphism group of the Cayley tree regarded as a graph. Recall (see, for example, [19]) that a mapping ψ on the vertex set of a graph G is called an automorphism of G if ψ preserves the adjacency relation, i.e., the images $\psi(u)$ and $\psi(v)$ of vertices u and v are adjacent if and only if u and v are adjacent.

For an arbitrary vertex $x_0 \in V$, we put

$$W_n = \{x \in V \mid d(x^0, x) = n\},\quad V_n = \bigcup_{m=0}^{n} W_m,\quad L_n = \{l =< x, y >\in L \mid x, y \in V_n\}.$$

where $d(x, y)$ is the distance between x and y in the Cayley tree, i.e., the number of edges of the path between x and y.

For each $x \in G_k$, let $S(x)$ denote the set of immediate successor of x, i.e., if $x \in W_n$ then

$$S(x) = \{y \in W_{n+1} : d(x, y) = 1\}.$$

For each $x \in G_k$, let $S_1(x)$ denote the set of all neighbors of x, i.e., $S_1(x) = \{y \in G_k :< x, y >\in L\}$. The set $S_1(x) \setminus S(x)$ is a singleton. Let $x_\downarrow$ denote the (unique) element of this set.

Assume that spin takes its values in the set $\Phi = \{1, 2, \ldots, q\}$. By a configuration σ on V we mean a function taking $\sigma : x \in V \to \sigma(x) \in \Phi$. The set of all configurations coincides with the set $\Omega = \Phi^V$.

Consider the quotient group $G_k/G_k^* = \{H1, \ldots, H_r\}$, where G_k^* is a normal subgroup of index r with $r \geq 1$.

Definition 2.2. *A configuration $\sigma(x)$ is said to be G_k^*-periodic if $\sigma(x) = \sigma_i$ for all $x \in G_k$ with $x \in H_i$. A G_k-periodic configuration is said to be translation invariant.*

By *period* of a periodic configuration we mean the index of the corresponding normal subgroup.

Definition 2.3. *A configuration $\sigma(x)$ is said to be G_k^*-weakly periodic if $\sigma(x) = \sigma_{ij}$ for all $x \in G_k$ with $x_\downarrow \in H_i$ and $x \in H_j$.*

The Hamiltonian of the λ-model [12, 15] has a form,

$$H(\sigma) = \sum_{<x,y>\in L} \lambda(\sigma(x), \sigma(y)) \tag{1}$$

IOP Conf. Series: Journal of Physics: Conf. Series **949** (2017) 012021 doi:10.1088/1742-6596/949/1/012021

In what follows, we restrict ourself to the case $k = 2$ and $\Phi = \{1, 2, 3\}$, and for the sake of simplicity, we consider the following function:

$$\lambda(i, j) = \begin{cases} \overline{a} & if \quad |i - j| = 2, \\ \overline{b} & if \quad |i - j| = 1, \\ \overline{c} & if \quad i = j, \end{cases} \tag{2}$$

where $\overline{a}, \overline{b}, \overline{c} \in \mathbb{R}$ for some given numbers.

3. Ground States

In this section, we describe ground state of the λ-model on a Cayley tree. For a pair of configurations σ and φ which coincide almost everywhere, i.e., everywhere except finitely many points, we consider a relative Hamiltonian $H(\sigma, \varphi)$ determining the energy differences of the two configurations σ and φ:

$$H(\sigma, \varphi) = \sum_{\substack{<x,y> \\ x,y \in V}} (\lambda(\sigma(x), \sigma(y)) - \lambda(\varphi(x), \varphi(y))) \tag{3}$$

Let M be the set of unit balls with vertices in V, i.e. $M = \{x, S_1(x), \forall x \in V\}$. We call the restriction of a configuration σ to the ball $b \in M$ a *bounded configuration* σ_b.

We define the energy of the configuration σ_b on b as follows:

$$U(\sigma_b) = \frac{1}{2} \sum_{\substack{<x,y> \\ x,y \in V}} (\lambda(\sigma(x), \sigma(y))).$$

From (3), we got the following lemma.

We shall say that two bounded configurations σ_b and $\sigma'_{b'}$ belong to the same class if $U(\sigma_b) = U(\sigma'_{b'})$ and we write $\sigma'_{b'} \sim \sigma_b$.

Lemma 3.1. *Relative Hamiltonian* (3) *has the form*

$$H(\sigma, \varphi) = \sum_{b \in M} (U(\sigma_b) - U(\varphi_b)).$$

Lemma 3.2. *The inclusion*

$$U(\varphi_b) \in \left\{ \frac{\alpha + \beta + \gamma}{2}, \forall\ \alpha, \beta, \gamma \in \{\overline{a}, \overline{b}, \overline{c}\} \right\} \tag{4}$$

holds for every configuration φ_b *on* $b (b \in M)$.

Definition 3.3. *A configuration* φ *is called a ground state of the relative Hamiltonian* H *if*

$$U(\varphi_b) = \min \left\{ \frac{\alpha + \beta + \gamma}{2}, \forall\ \alpha, \beta . \gamma \in \{\overline{a}, \overline{b}, \overline{c}\} \right\} \tag{5}$$

for any $b \in M$

If a ground state is a periodic (resp. weakly periodic, translation invariant) configuration then we call it a periodic (resp. weakly periodic, translation invariant) ground state.

For any configuration σ_b, we have

$$U(\sigma_b) \in \{U_1, U_2, U_3, U_4, U_5, U_6, U_7, U_8, U_9, U_{10}\},$$

where

$$\begin{aligned} U_1 &= 3\overline{a}/2, & U_2 &= 3\overline{b}/2, \\ U_3 &= 3\overline{c}/2, & U_4 &= \overline{a}/2+\overline{b}, \\ U_5 &= \overline{a}/2+\overline{c}, & U_6 &= \overline{b}/2+\overline{a}, \\ U_7 &= \overline{b}/2+\overline{c}, & U_8 &= \overline{c}/2+\overline{a}, \\ U_9 &= \overline{c}/2+\overline{b}, & U_{10} &= (\overline{a}+\overline{b}+\overline{c})/2. \end{aligned} \quad (6)$$

By denoting

$$A_m = \{(\overline{a},\overline{b},\overline{c}) \in \mathbb{R}^3 |\ U_m = \min_{1\le k\le 10}\{U_k\}\}, \quad (7)$$

we obtain each A_i as:

$$\begin{aligned} A_1 &= \left\{(\overline{a},\overline{b},\overline{c}) \in \mathbb{R}^3 |\ \overline{a} \le \overline{b} \le \overline{c}\right\} \cup \left\{(\overline{a},\overline{b},\overline{c}) \in \mathbb{R}^3 |\ \overline{a} \le \overline{c} \le \overline{b}\right\}, \\ A_2 &= \left\{(\overline{a},\overline{b},\overline{c}) \in \mathbb{R}^3 |\ \overline{b} \le \overline{c} \le \overline{a}\right\} \cup \left\{(\overline{a},\overline{b},\overline{c}) \in \mathbb{R}^3 |\ \overline{b} \le \overline{a} \le \overline{c}\right\}, \\ A_3 &= \left\{(\overline{a},\overline{b},\overline{c}) \in \mathbb{R}^3 |\ \overline{c} \le \overline{b} \le \overline{a}\right\} \cup \left\{(\overline{a},\overline{b},\overline{c}) \in \mathbb{R}^3 |\ \overline{c} \le \overline{a} \le \overline{b}\right\}, \\ A_4 &= \left\{(\overline{a},\overline{b},\overline{c}) \in \mathbb{R}^3 |\ \overline{a} = \overline{b} \le \overline{c}\right\}, \\ A_5 &= \left\{(\overline{a},\overline{b},\overline{c}) \in \mathbb{R}^3 |\ \overline{a} = \overline{c} \le \overline{b}\right\}, \\ A_6 &= \left\{(\overline{a},\overline{b},\overline{c}) \in \mathbb{R}^3 |\ \overline{a} = \overline{b} \le \overline{c}\right\}, \\ A_7 &= \left\{(\overline{a},\overline{b},\overline{c}) \in \mathbb{R}^3 |\ \overline{b} = \overline{c} \le \overline{a}\right\}, \\ A_8 &= \left\{(\overline{a},\overline{b},\overline{c}) \in \mathbb{R}^3 |\ \overline{a} = \overline{c} \le \overline{b}\right\}, \\ A_9 &= \left\{(\overline{a},\overline{b},\overline{c}) \in \mathbb{R}^3 |\ \overline{b} = \overline{c} \le \overline{a}\right\}, \\ A_{10} &= \left\{(\overline{a},\overline{b},\overline{c}) \in \mathbb{R}^3 |\ \overline{a} = \overline{b} = \overline{c}\right\}, \mathbb{R}^3 = \cup_{i=1}^{10} A_i. \end{aligned}$$

Let us introduce some notations. We put

$$C_i = \{\sigma_b \in \Omega_b :\ U(\sigma_b) = u_i\}, \quad i = \overline{1,10}$$

and $B^{(i)} = |\{x \in S_1(k) :\ \varphi_b(x) = i\}|$ for $i = \overline{1,3}$.

Let $A \subset \{1,2,...,k+1\}$, $H_A = \{x \in G_k : \sum_{j\in A} w_j(x) - \text{even}\}$, where $w_j(x)$-is the number of letters a_j in the word x.

It is obvious, that H_A is a normal subgroup of index two [6]. Let $G_k/H_A = \{H_A, G_k \setminus H_A\}$ be the quotient group. We set $H_0 = H_A, H_1 = G_k \setminus H_A$.

4. Periodic Ground States

In this section, we describe H_0-periodic ground states. We note that each H_0-periodic configurations has the following form:

$$\sigma(x) = \begin{cases} \sigma_1, & \text{if} \quad x \in H_0, \\ \sigma_2, & \text{if} \quad x \in H_1 \end{cases} \quad (8)$$

where $\sigma_i \in \Phi = \{1,2,3\}$ $(i = 1,2)$.

Theorem 4.1. *Let k=2 and $|A| = 1$. Then for the λ-model given by* (2) *the following statements hold:*

(i) let $|\sigma_1 - \sigma_2| = 0$, then H_0-periodic ground state exists if and only if the parameters $\overline{a}, \overline{b}, \overline{c}$ belong to A_3;

(ii) If $|\sigma_1 - \sigma_2| = 1$, then H_0-periodic ground state exists if and only if the parameters $\overline{a}, \overline{b}, \overline{c}$ belong to A_7;

(iii) If $|\sigma_1 - \sigma_2| = 2$, then H_0-periodic ground state exists if and only if the parameters $\overline{a}, \overline{b}, \overline{c}$ belong to A_5.

Proof. Let us consider the following configuration

$$\varphi(x) = \begin{cases} i, & \text{if} \quad x \in H_0 \\ i, & \text{if} \quad x \in H_1, \end{cases} \tag{9}$$

where $i = 1, 2, 3$.

By C_b we denote the center of $b \in M$. Let $C_b \in H_0$, then we have

$$\varphi_b(C_b) = i, \ B^{(i)} = 3;$$

Hence, $\varphi_b(x) \in C_3$, i.e. if $(\overline{a}, \overline{b}, \overline{c}) \in A_3$ then the corresponding configuration is a ground state.

Now consider the following configuration

$$\varphi(x) = \begin{cases} i, & \text{if} \quad x \in H_0, \\ j, & \text{if} \quad x \in H_1, \end{cases} \tag{10}$$

where $|i - j| = 1$

Assume that $C_b \in H_0$, the one finds that

$$\varphi_b(C_b) = i, \ B^{(i)} = 2, \ B^{(j)} = 1;$$

Hence, $\varphi_b(x) \in C_7$.

Let $C_b \in H_1$, one has

$$\varphi_b(C_b) = j, \ B^{(i)} = 1, \ B^{(j)} = 2;$$

Hence, $\varphi_b(x) \in C_7$.

We conclude that, if $(\overline{a}, \overline{b}, \overline{c}) \in A_7$ then the corresponding periodic configuration $\varphi(x)$ is a H_0-periodic ground state.

Let us consider the following configuration

$$\varphi(x) = \begin{cases} i, & \text{if} \quad x \in H_0, \\ j, & \text{if} \quad x \in H_1, \end{cases} \tag{11}$$

where $|i - j| = 2$

Let $C_b \in H_0$, we have

$$\varphi_b(C_b) = i, \ B^{(i)} = 2, \ B^{(j)} = 1;$$

Hence, $\varphi_b(x) \in C_5$.

Let $C_b \in H_1$, one gets

$$\varphi_b(C_b) = j, \ B^{(i)} = 1, \ B^{(j)} = 2;$$

So, $\varphi_b(x) \in C_5$. Therefore, if $(\overline{a}, \overline{b}, \overline{c}) \in A_5$ then the corresponding periodic configuration $\varphi(x)$ is H_0-periodic ground state. This completes the proof. □

Remark 4.1. *Notice that periodic configurations (ground states) described in (i) Theorem 4.1 are translation-invariant. Full description of translation-invariant ground states has been carried out in [15].*

IOP Conf. Series: Journal of Physics: Conf. Series **949** (2017) 012021 doi:10.1088/1742-6596/949/1/012021

5. Weakly Periodic Ground States

In this section, we describe H_A-weakly periodic ground states. In what follows, H_A stands for the normal subgroup of index two. Due to the definition of weakly periodic configuration, we infer that each H_A-weakly periodic configuration has the following form:

$$\varphi(x) = \begin{cases} \sigma_1 & if \quad x_\downarrow \in H_0, \quad x \in H_0, \\ \sigma_2 & if \quad x_\downarrow \in H_0, \quad x \in H_1, \\ \sigma_3 & if \quad x_\downarrow \in H_1, \quad x \in H_0, \\ \sigma_4 & if \quad x_\downarrow \in H_1, \quad x \in H_1, \end{cases} \tag{12}$$

where $\sigma_i \in \Phi = \{1, 2, 3\}$, $(i = 1, 2, 3, 4)$.

Hence, all possible weakly periodic (non periodic) ground states have the following forms:

$$\varphi_1(x) = \begin{cases} 1 & if \quad x_\downarrow \in H_0, \quad x \in H_0, \\ 1 & if \quad x_\downarrow \in H_0, \quad x \in H_1, \\ 2 & if \quad x_\downarrow \in H_1, \quad x \in H_0, \\ 2 & if \quad x_\downarrow \in H_1, \quad x \in H_1, \end{cases} \qquad \varphi_2(x) = \begin{cases} 1 & if \quad x_\downarrow \in H_0, \quad x \in H_0, \\ 2 & if \quad x_\downarrow \in H_0, \quad x \in H_1, \\ 2 & if \quad x_\downarrow \in H_1, \quad x \in H_0, \\ 1 & if \quad x_\downarrow \in H_1, \quad x \in H_1, \end{cases}$$

$$\varphi_3(x) = \begin{cases} 1 & if \quad x_\downarrow \in H_0, \quad x \in H_0, \\ 2 & if \quad x_\downarrow \in H_0, \quad x \in H_1, \\ 2 & if \quad x_\downarrow \in H_1, \quad x \in H_0, \\ 3 & if \quad x_\downarrow \in H_1, \quad x \in H_1, \end{cases} \qquad \varphi_4(x) = \begin{cases} 2 & if \quad x_\downarrow \in H_0, \quad x \in H_0, \\ 1 & if \quad x_\downarrow \in H_0, \quad x \in H_1, \\ 1 & if \quad x_\downarrow \in H_1, \quad x \in H_0, \\ 1 & if \quad x_\downarrow \in H_1, \quad x \in H_1, \end{cases}$$

$$\varphi_5(x) = \begin{cases} 2 & if \quad x_\downarrow \in H_0, \quad x \in H_0, \\ 1 & if \quad x_\downarrow \in H_0, \quad x \in H_1, \\ 2 & if \quad x_\downarrow \in H_1, \quad x \in H_0, \\ 2 & if \quad x_\downarrow \in H_1, \quad x \in H_1, \end{cases} \qquad \varphi_6(x) = \begin{cases} 2 & if \quad x_\downarrow \in H_0, \quad x \in H_0, \\ 1 & if \quad x_\downarrow \in H_0, \quad x \in H_1, \\ 3 & if \quad x_\downarrow \in H_1, \quad x \in H_0, \\ 2 & if \quad x_\downarrow \in H_1, \quad x \in H_1, \end{cases}$$

$$\varphi_7(x) = \begin{cases} 2 & if \quad x_\downarrow \in H_0, \quad x \in H_0, \\ 2 & if \quad x_\downarrow \in H_0, \quad x \in H_1, \\ 1 & if \quad x_\downarrow \in H_1, \quad x \in H_0, \\ 2 & if \quad x_\downarrow \in H_1, \quad x \in H_1, \end{cases} \qquad \varphi_8(x) = \begin{cases} 2 & if \quad x_\downarrow \in H_0, \quad x \in H_0, \\ 2 & if \quad x_\downarrow \in H_0, \quad x \in H_1, \\ 2 & if \quad x_\downarrow \in H_1, \quad x \in H_0, \\ 1 & if \quad x_\downarrow \in H_1, \quad x \in H_1, \end{cases}$$

$$\varphi_9(x) = \begin{cases} 2 & if \quad x_\downarrow \in H_0, \quad x \in H_0, \\ 2 & if \quad x_\downarrow \in H_0, \quad x \in H_1, \\ 2 & if \quad x_\downarrow \in H_1, \quad x \in H_0, \\ 3 & if \quad x_\downarrow \in H_1, \quad x \in H_1, \end{cases} \qquad \varphi_{10}(x) = \begin{cases} 2 & if \quad x_\downarrow \in H_0, \quad x \in H_0, \\ 2 & if \quad x_\downarrow \in H_0, \quad x \in H_1, \\ 3 & if \quad x_\downarrow \in H_1, \quad x \in H_0, \\ 2 & if \quad x_\downarrow \in H_1, \quad x \in H_1, \end{cases}$$

$$\varphi_{11}(x) = \begin{cases} 2 & if \quad x_\downarrow \in H_0, \quad x \in H_0, \\ 2 & if \quad x_\downarrow \in H_0, \quad x \in H_1, \\ 1 & if \quad x_\downarrow \in H_1, \quad x \in H_0, \\ 1 & if \quad x_\downarrow \in H_1, \quad x \in H_1, \end{cases} \qquad \varphi_{12}(x) = \begin{cases} 3 & if \quad x_\downarrow \in H_0, \quad x \in H_0, \\ 3 & if \quad x_\downarrow \in H_0, \quad x \in H_1, \\ 3 & if \quad x_\downarrow \in H_1, \quad x \in H_0, \\ 2 & if \quad x_\downarrow \in H_1, \quad x \in H_1, \end{cases}$$

$$\varphi_{13}(x) = \begin{cases} 1 & if \quad x_\downarrow \in H_0, \quad x \in H_0, \\ 1 & if \quad x_\downarrow \in H_0, \quad x \in H_1, \\ 1 & if \quad x_\downarrow \in H_1, \quad x \in H_0, \\ 2 & if \quad x_\downarrow \in H_1, \quad x \in H_1, \end{cases} \qquad \varphi_{14}(x) = \begin{cases} 1 & if \quad x_\downarrow \in H_0, \quad x \in H_0, \\ 2 & if \quad x_\downarrow \in H_0, \quad x \in H_1, \\ 2 & if \quad x_\downarrow \in H_1, \quad x \in H_0, \\ 2 & if \quad x_\downarrow \in H_1, \quad x \in H_1, \end{cases}$$

$$\varphi_{15}(x)=\begin{cases}1 & if\quad x_\downarrow\in H_0,\quad x\in H_0,\\ 3 & if\quad x_\downarrow\in H_0,\quad x\in H_1,\\ 3 & if\quad x_\downarrow\in H_1,\quad x\in H_0,\\ 3 & if\quad x_\downarrow\in H_1,\quad x\in H_1,\end{cases}\qquad \varphi_{16}(x)=\begin{cases}3 & if\quad x_\downarrow\in H_0,\quad x\in H_0,\\ 1 & if\quad x_\downarrow\in H_0,\quad x\in H_1,\\ 1 & if\quad x_\downarrow\in H_1,\quad x\in H_0,\\ 3 & if\quad x_\downarrow\in H_1,\quad x\in H_1,\end{cases}$$

$$\varphi_{17}(x)=\begin{cases}3 & if\quad x_\downarrow\in H_0,\quad x\in H_0,\\ 3 & if\quad x_\downarrow\in H_0,\quad x\in H_1,\\ 1 & if\quad x_\downarrow\in H_1,\quad x\in H_0,\\ 1 & if\quad x_\downarrow\in H_1,\quad x\in H_1,\end{cases}\qquad \varphi_{18}(x)=\begin{cases}3 & if\quad x_\downarrow\in H_0,\quad x\in H_0,\\ 3 & if\quad x_\downarrow\in H_0,\quad x\in H_1,\\ 1 & if\quad x_\downarrow\in H_1,\quad x\in H_0,\\ 3 & if\quad x_\downarrow\in H_1,\quad x\in H_1,\end{cases}$$

$$\varphi_{19}(x)=\begin{cases}3 & if\quad x_\downarrow\in H_0,\quad x\in H_0,\\ 3 & if\quad x_\downarrow\in H_0,\quad x\in H_1,\\ 3 & if\quad x_\downarrow\in H_1,\quad x\in H_0,\\ 1 & if\quad x_\downarrow\in H_1,\quad x\in H_1,\end{cases}\qquad \varphi_{20}(x)=\begin{cases}1 & if\quad x_\downarrow\in H_0,\quad x\in H_0,\\ 3 & if\quad x_\downarrow\in H_0,\quad x\in H_1,\\ 3 & if\quad x_\downarrow\in H_1,\quad x\in H_0,\\ 1 & if\quad x_\downarrow\in H_1,\quad x\in H_1,\end{cases}$$

Theorem 5.1. *Let $k=2$ and $|A|=1$. Then the following statements hold for the λ-model given by (2):*

I. There exist fourteen H_A-weakly periodic ground states if $(\overline{a},\overline{b},\overline{c})\in A_9$ that are not H_A-periodic, and they are described by φ_i, $i=1,\dots,14$. Moreover, they are not ground states on the set $\mathbb{R}^3\setminus A_9$.

II. There exist six H_A-weakly periodic ground states if $(\overline{a},\overline{b},\overline{c})\in A_5$ that are not H_A-periodic, and they are described by φ_i, $i=15,\dots,20$, and they are not ground states on the set $\mathbb{R}^3\setminus A_5$.

III. All H_A-weakly periodic (non periodic, in particular non translation-invariant) configurations (not mentioned in assertions I and II) are not weakly periodic ground states.

Proof. 1) Consider

$$\varphi_1(x)=\begin{cases}1 & if\quad x_\downarrow\in H_0,\quad x\in H_0,\\ 1 & if\quad x_\downarrow\in H_0,\quad x\in H_1,\\ 2 & if\quad x_\downarrow\in H_1,\quad x\in H_0,\\ 2 & if\quad x_\downarrow\in H_1,\quad x\in H_1.\end{cases}$$

Let $C_b\in H_0$, we have the following possible cases

a) $C_{b_\downarrow}\in H_0$ and $\varphi_{1,b}(C_{b_\downarrow})=1$, then $\varphi_{1,b}(C_b)=1$, $B^{(1)}=3$, $\varphi_{1,b}\in C_3$,

b) $C_{b_\downarrow}\in H_0$ and $\varphi_{1,b}(C_{b_\downarrow})=2$, then $\varphi_{1,b}(C_b)=1$, $B^{(1)}=2$, $B^{(2)}=1$, $\varphi_{1,b}\in C_7$,

c) $C_{b_\downarrow}\in H_1$ and $\varphi_{1,b}(C_{b_\downarrow})=1$, this case is not exist, if $C_{b_\downarrow}\in H_1$, then $\varphi_{1,b}(C_{b_\downarrow})=2$,

d) $C_{b_\downarrow}\in H_1$ and $\varphi_{1,b}(C_{b_\downarrow})=2$, then $\varphi_{1,b}(C_b)=2$, $B^{(1)}=2$, $B^{(2)}=1$, $\varphi_{1,b}\in C_9$.

Let $C_b\in H_1$, then the possible cases are

a) $C_{b_\downarrow}\in H_0$ and $\varphi_{1,b}(C_{b_\downarrow})=1$, then $\varphi_{1,b}(C_b)=1$, $B^{(1)}=1$, $B^{(2)}=2$, $\varphi_{1,b}\in C_9$,

b) $C_{b_\downarrow}\in H_0$ and $\varphi_{1,b}(C_{b_\downarrow})=2$, this case is not exist, if $C_{b_\downarrow}\in H_0$, then $\varphi_{1,b}(C_{b_\downarrow})=1$,

c) $C_{b_\downarrow}\in H_1$ and $\varphi_{1,b}(C_{b_\downarrow})=1$, then $\varphi_{1,b}(C_b)=2$, $B^{(1)}=1$, $B^{(2)}=2$, $\varphi_{1,b}\in C_7$,

d) $C_{b_\downarrow}\in H_1$ and $\varphi_{1,b}(C_{b_\downarrow})=2$, then $\varphi_{1,b}(C_b)=2$, $B^{(2)}=3$, $\varphi_{1,b}\in C_3$.

Therefore, the configuration φ_1 is a weakly periodic ground state on the set $A_3\cap A_7\cap A_9=A_7=A_9$.

2) Consider

$$\varphi_2(x)=\begin{cases}1 & if\quad x_\downarrow\in H_0,\quad x\in H_0,\\ 2 & if\quad x_\downarrow\in H_0,\quad x\in H_1,\\ 2 & if\quad x_\downarrow\in H_1,\quad x\in H_0,\\ 1 & if\quad x_\downarrow\in H_1,\quad x\in H_1.\end{cases}$$

Let $C_b \in H_0$, the possible cases are

a) $C_{b_\downarrow} \in H_0$ and $\varphi_{2,b}(C_{b_\downarrow}) = 1$, then $\varphi_{2,b}(C_b) = 2$, $B^{(1)} = 2$, $B^{(2)} = 1$, $\varphi_{2,b} \in C_7$,

b) $C_{b_\downarrow} \in H_0$ and $\varphi_{2,b}(C_{b_\downarrow}) = 2$, then $\varphi_{2,b}(C_b) = 2$, $B^{(1)} = 1$, $B^{(2)} = 2$, $\varphi_{2,b} \in C_9$,

c) $C_{b_\downarrow} \in H_1$ and $\varphi_{2,b}(C_{b_\downarrow}) = 1$, then $\varphi_{2,b}(C_b) = 2$, $B^{(1)} = 3$, $\varphi_{2,b} \in C_2$,

d) $C_{b_\downarrow} \in H_1$ and $\varphi_{2,b}(C_{b_\downarrow}) = 2$, this case is not exist, if $C_{b_\downarrow} \in H_1$, then $\varphi_{2,b}(C_{b_\downarrow}) = 1$.

Let $C_b \in H_1$, the possible cases are

a) $C_{b_\downarrow} \in H_0$ and $\varphi_{2,b}(C_{b_\downarrow}) = 1$, then $\varphi_{2,b}(C_b) = 2$, $B^{(1)} = 3$, $\varphi_{2,b} \in C_2$,

b) $C_{b_\downarrow} \in H_0$ and $\varphi_{2,b}(C_{b_\downarrow}) = 2$, this case is not exist, if $C_{b_\downarrow} \in H_0$, then $\varphi_{2,b}(C_{b_\downarrow}) = 1$,

c) $C_{b_\downarrow} \in H_1$ and $\varphi_{2,b}(C_{b_\downarrow}) = 1$, then $\varphi_{2,b}(C_b) = 1$, $B^{(1)} = 2$, $B^{(2)} = 1$, $\varphi_{2,b} \in C_7$,

d) $C_{b_\downarrow} \in H_1$ and $\varphi_{2,b}(C_{b_\downarrow}) = 2$, then $\varphi_{2,b}(C_b) = 1$, $B^{(1)} = 1$, $B^{(2)} = 2$, $\varphi_{2,b} \in C_9$.

Therefore, the configuration φ_2 is a weakly periodic ground state on the set $A_2 \cap A_7 \cap A_9 = A_7 = A_9$.

3) Consider

$$\varphi_3(x) = \begin{cases} 1 & if \quad x_\downarrow \in H_0, \quad x \in H_0, \\ 2 & if \quad x_\downarrow \in H_0, \quad x \in H_1, \\ 2 & if \quad x_\downarrow \in H_1, \quad x \in H_0, \\ 3 & if \quad x_\downarrow \in H_1, \quad x \in H_1. \end{cases}$$

Let $C_b \in H_0$, the possible cases are

a) $C_{b_\downarrow} \in H_0$ and $\varphi_{3,b}(C_{b_\downarrow}) = 1$, then $\varphi_{3,b}(C_b) = 1$, $B^{(1)} = 2$, $B^{(2)} = 1$, $\varphi_{3,b} \in C_7$,

b) $C_{b_\downarrow} \in H_0$ and $\varphi_{3,b}(C_{b_\downarrow}) = 2$, then $\varphi_{3,b}(C_b) = 2$, $B^{(1)} = 1$, $B^{(2)} = 2$, $\varphi_{3,b} \in C_9$,

c) $C_{b_\downarrow} \in H_1$ and $\varphi_{3,b}(C_{b_\downarrow}) = 2$, this case is not exist, if $C_{b_\downarrow} \in H_1$, then $\varphi_{3,b}(C_{b_\downarrow}) = 3$,

d) $C_{b_\downarrow} \in H_1$ and $\varphi_{3,b}(C_{b_\downarrow}) = 3$, then $\varphi_{3,b}(C_b) = 2$, $B^{(1)} = 2$, $B^{(3)} = 1$, $\varphi_{3,b} \in C_9$.

Let $C_b \in H_1$, the possible cases are

a) $C_{b_\downarrow} \in H_0$ and $\varphi_{3,b}(C_{b_\downarrow}) = 1$, then $\varphi_{3,b}(C_b) = 2$, $B^{(1)} = 1$, $B^{(3)} = 2$, $\varphi_{3,b} \in C_2$,

b) $C_{b_\downarrow} \in H_0$ and $\varphi_{3,b}(C_{b_\downarrow}) = 2$, this case is not exist, if $C_{b_\downarrow} \in H_0$, then $\varphi_{3,b}(C_{b_\downarrow}) = 1$,

c) $C_{b_\downarrow} \in H_1$ and $\varphi_{3,b}(C_{b_\downarrow}) = 2$, then $\varphi_{3,b}(C_b) = 3$, $B^{(2)} = 2$, $B^{(3)} = 1$, $\varphi_{3,b} \in C_9$,

d) $C_{b_\downarrow} \in H_1$ and $\varphi_{3,b}(C_{b_\downarrow}) = 3$, then $\varphi_{3,b}(C_b) = 3$, $B^{(2)} = 1$, $B^{(3)} = 2$, $\varphi_{3,b} \in C_9$.

Therefore, configuration φ_3 a weakly periodic ground state on the set $A_2 \cap A_7 \cap A_9 = A_7 = A_9$.

4) consider

$$\varphi_4(x) = \begin{cases} 2 & if \quad x_\downarrow \in H_0, \quad x \in H_0, \\ 1 & if \quad x_\downarrow \in H_0, \quad x \in H_1, \\ 1 & if \quad x_\downarrow \in H_1, \quad x \in H_0, \\ 1 & if \quad x_\downarrow \in H_1, \quad x \in H_1. \end{cases}$$

Let $C_b \in H_0$, the possible cases are

a) $C_{b_\downarrow} \in H_0$ and $\varphi_{4,b}(C_{b_\downarrow}) = 1$, then $\varphi_{4,b}(C_b) = 2$, $B^{(1)} = 2$, $B^{(2)} = 1$, $\varphi_{4,b} \in C_9$,

b) $C_{b_\downarrow} \in H_0$ and $\varphi_{4,b}(C_{b_\downarrow}) = 2$, then $\varphi_{4,b}(C_b) = 2$, $B^{(1)} = 1$, $B^{(2)} = 2$, $\varphi_{4,b} \in C_7$,

c) $C_{b_\downarrow} \in H_1$ and $\varphi_{4,b}(C_{b_\downarrow}) = 1$, then $\varphi_{4,b}(C_b) = 1$, $B^{(1)} = 1$, $B^{(2)} = 2$, $\varphi_{4,b} \in C_9$.

Let $C_b \in H_1$, the possible cases are

a) $C_{b_\downarrow} \in H_0$ and $\varphi_{4,b}(C_{b_\downarrow}) = 1$, then $\varphi_{4,b}(C_b) = 1$, $B^{(1)} = 3$, $\varphi_{4,b} \in C_3$,

b) $C_{b_\downarrow} \in H_0$ and $\varphi_{4,b}(C_{b_\downarrow}) = 2$, then $\varphi_{4,b}(C_b) = 1$, $B^{(1)} = 2$, $B^{(2)} = 1$, $\varphi_{4,b} \in C_7$,

c) $C_{b_\downarrow} \in H_1$ and $\varphi_{4,b}(C_{b_\downarrow}) = 1$, then $\varphi_{4,b}(C_b) = 1$, $B^{(1)} = 3$, $\varphi_{4,b} \in C_9$.

Therefore, configuration φ_4 a weakly periodic ground state on the set $A_3 \cap A_7 \cap A_9 = A_7 = A_9$.

5) Consider

$$\varphi_5(x) = \begin{cases} 2 & if \quad x_\downarrow \in H_0, \quad x \in H_0, \\ 1 & if \quad x_\downarrow \in H_0, \quad x \in H_1, \\ 2 & if \quad x_\downarrow \in H_1, \quad x \in H_0, \\ 2 & if \quad x_\downarrow \in H_1, \quad x \in H_1. \end{cases}$$

Let $C_b \in H_0$, the possible cases are

a) $C_{b_\downarrow} \in H_0$ and $\varphi_{5,b}(C_{b_\downarrow}) = 2$, then $\varphi_{5,b}(C_b) = 2,\ B^{(1)} = 1,\ B^{(2)} = 2,\ \varphi_{5,b} \in C_7$,
b) $C_{b_\downarrow} \in H_1$ and $\varphi_{5,b}(C_{b_\downarrow}) = 1$, this case is not exist, if $C_{b_\downarrow} \in H_1$, then $\varphi_{5,b}(C_{b_\downarrow}) = 2$,
c) $C_{b_\downarrow} \in H_1$ and $\varphi_{5,b}(C_{b_\downarrow}) = 2$, then $\varphi_{5,b}(C_b) = 2,\ B^{(2)} = 3,\ \varphi_{5,b} \in C_3$.

Let $C_b \in H_1$, the possible cases are

a) $C_{b_\downarrow} \in H_0$ and $\varphi_{5,b}(C_{b_\downarrow}) = 2$, then $\varphi_{5,b}(C_b) = 1,\ B^{(2)} = 3,\ \varphi_{5,b} \in C_2$,
b) $C_{b_\downarrow} \in H_1$ and $\varphi_{5,b}(C_{b_\downarrow}) = 1$, then $\varphi_{5,b}(C_b) = 2,\ B^{(1)} = 2,\ B^{(2)} = 1,\ \varphi_{5,b} \in C_9$,
c) $C_{b_\downarrow} \in H_1$ and $\varphi_{5,b}(C_{b_\downarrow}) = 2$, then $\varphi_{5,b}(C_b) = 2,\ B^{(2)} = 3,\ \varphi_{5,b} \in C_3$.

Therefore, the configuration φ_5 is a weakly periodic ground state on the set $A_2 \cap A_3 \cap A_7 \cap A_9 = A_7 = A_9$.

6) Consider

$$\varphi_6(x) = \begin{cases} 2 & if \quad x_\downarrow \in H_0, \quad x \in H_0, \\ 1 & if \quad x_\downarrow \in H_0, \quad x \in H_1, \\ 3 & if \quad x_\downarrow \in H_1, \quad x \in H_0, \\ 2 & if \quad x_\downarrow \in H_1, \quad x \in H_1. \end{cases}$$

Let $C_b \in H_0$, the possible cases are

a) $C_{b_\downarrow} \in H_0$ and $\varphi_{6,b}(C_{b_\downarrow}) = 2$, then $\varphi_{6,b}(C_b) = 2,\ B^{(2)} = 2,\ B^{(3)} = 3,\ \varphi_{6,b} \in C_7$,
b) $C_{b_\downarrow} \in H_0$ and $\varphi_{6,b}(C_{b_\downarrow}) = 3$, then $\varphi_{6,b}(C_b) = 2,\ B^{(1)} = B^{(2)} = B^{(3)} = 1,\ \varphi_{6,b} \in C_9$,
c) $C_{b_\downarrow} \in H_1$ and $\varphi_{6,b}(C_{b_\downarrow}) = 1$, this case is not exist, if $C_{b_\downarrow} \in H_1$, then $\varphi_{6,b}(C_{b_\downarrow}) = 2$,
d) $C_{b_\downarrow} \in H_1$ and $\varphi_{6,b}(C_{b_\downarrow}) = 2$, then $\varphi_{6,b}(C_b) = 3,\ B^{(2)} = 3,\ \varphi_{6,b} \in C_2$.

Let $C_b \in H_1$, the possible cases are

a) $C_{b_\downarrow} \in H_0$ and $\varphi_{6,b}(C_{b_\downarrow}) = 2$, then $\varphi_{6,b}(C_b) = 1,\ B^{(2)} = 3,\ \varphi_{6,b} \in C_2$,
b) $C_{b_\downarrow} \in H_0$ and $\varphi_{6,b}(C_{b_\downarrow}) = 3$, this case is not exist, if $C_{b_\downarrow} \in H_0$, then $\varphi_{6,b}(C_{b_\downarrow}) = 2$,
c) $C_{b_\downarrow} \in H_1$ and $\varphi_{6,b}(C_{b_\downarrow}) = 1$, then $\varphi_{6,b}(C_b) = 2,\ B^{(1)} = B^{(2)} = B^{(3)} = 1,\ \varphi_{6,b} \in C_9$,
d) $C_{b_\downarrow} \in H_1$ and $\varphi_{6,b}(C_{b_\downarrow}) = 2$, then $\varphi_{6,b}(C_b) = 2,\ B^{(2)} = 2,\ B^{(3)} = 1,\ \varphi_{6,b} \in C_7$.

Therefore, the configuration φ_6 is a weakly periodic ground state on the set $A_2 \cap A_7 \cap A_9 = A_7 = A_9$.

7) Consider

$$\varphi_7(x) = \begin{cases} 2 & if \quad x_\downarrow \in H_0, \quad x \in H_0, \\ 2 & if \quad x_\downarrow \in H_0, \quad x \in H_1, \\ 1 & if \quad x_\downarrow \in H_1, \quad x \in H_0, \\ 2 & if \quad x_\downarrow \in H_1, \quad x \in H_1. \end{cases}$$

Let $C_b \in H_0$, the possible cases are

a) $C_{b_\downarrow} \in H_0$ and $\varphi_{7,b}(C_{b_\downarrow}) = 1$, then $\varphi_{7,b}(C_b) = 2,\ B^{(1)} = 1,\ B^{(2)} = 2,\ \varphi_{7,b} \in C_7$,
b) $C_{b_\downarrow} \in H_0$ and $\varphi_{7,b}(C_{b_\downarrow}) = 2$, then $\varphi_{7,b}(C_b) = 2,\ B^{(2)} = 3,\ \varphi_{7,b} \in C_3$,
c) $C_{b_\downarrow} \in H_1$ and $\varphi_{7,b}(C_{b_\downarrow}) = 2$, then $\varphi_{7,b}(C_b) = 1,\ B^{(2)} = 3,\ \varphi_{7,b} \in C_2$.

Let $C_b \in H_1$, the possible cases are

a) $C_{b_\downarrow} \in H_0$ and $\varphi_{7,b}(C_{b_\downarrow}) = 1$, then $\varphi_{7,b}(C_b) = 2,\ B^{(1)} = 1,\ B^{(2)} = 2,\ \varphi_{7,b} \in C_7$,
b) $C_{b_\downarrow} \in H_0$ and $\varphi_{7,b}(C_{b_\downarrow}) = 2$, then $\varphi_{7,b}(C_b) = 2,\ B^{(2)} = 3,\ \varphi_{7,b} \in C_3$,

c) $C_{b_\downarrow} \in H_1$ and $\varphi_{7,b}(C_{b_\downarrow}) = 2$, then $\varphi_{7,b}(C_b) = 2,\ B^{(1)} = 1,\ B^{(2)} = 2,\ \varphi_{7,b} \in C_7$.

Therefore, the configuration φ_7 is a weakly periodic ground state on the set $A_2 \cap A_3 \cap A_7 = A_7$.

8) Consider

$$\varphi_8(x) = \begin{cases} 2 & if \quad x_\downarrow \in H_0, \quad x \in H_0, \\ 2 & if \quad x_\downarrow \in H_0, \quad x \in H_1, \\ 2 & if \quad x_\downarrow \in H_1, \quad x \in H_0, \\ 1 & if \quad x_\downarrow \in H_1, \quad x \in H_1. \end{cases}$$

Let $C_b \in H_0$, the possible cases are

a) $C_{b_\downarrow} \in H_0$ and $\varphi_{8,b}(C_{b_\downarrow}) = 2$, then $\varphi_{8,b}(C_b) = 2,\ B^{(2)} = 3,\ \varphi_{8,b} \in C_3$,

b) $C_{b_\downarrow} \in H_1$ and $\varphi_{8,b}(C_{b_\downarrow}) = 1$, then $\varphi_{8,b}(C_b) = 2,\ B^{(1)} = 1,\ B^{(2)} = 2,\ \varphi_{8,b} \in C_7$,

c) $C_{b_\downarrow} \in H_1$ and $\varphi_{8,b}(C_{b_\downarrow}) = 2$, this case is not exist, if $C_{b_\downarrow} \in H_1$, then $\varphi_{8,b}(C_{b_\downarrow}) = 1$.

Let $C_b \in H_1$, the possible cases are

a) $C_{b_\downarrow} \in H_0$ and $\varphi_{8,b}(C_{b_\downarrow}) = 2$, then we can have two configuration:

(i) $\varphi_{8,b}(C_b) = 2,\ B^{(1)} = 1,\ B^{(2)} = 2,\ \varphi_{8,b} \in C_7$,

(ii) $\varphi_{8,b}(C_b) = 2,\ B^{(1)} = 3,\ \varphi_{8,b} \in C_1$,

b) $C_{b_\downarrow} \in H_1$ and $\varphi_{8,b}(C_{b_\downarrow}) = 1$, then $\varphi_{8,b}(C_b) = 1,\ B^{(1)} = 2,\ B^{(2)} = 1,\ \varphi_{8,b} \in C_7$,

c) $C_{b_\downarrow} \in H_1$ and $\varphi_{8,b}(C_{b_\downarrow}) = 2$, then $\varphi_{8,b}(C_b) = 1,\ B^{(1)} = 1,\ B^{(2)} = 2,\ \varphi_{8,b} \in C_9$.

Therefore, the configuration φ_8 is a weakly periodic ground state on the set $A_2 \cap A_3 \cap A_7 \cap A_9 = A_7 = A_9$.

9) Consider

$$\varphi_9(x) = \begin{cases} 2 & if \quad x_\downarrow \in H_0, \quad x \in H_0, \\ 2 & if \quad x_\downarrow \in H_0, \quad x \in H_1, \\ 2 & if \quad x_\downarrow \in H_1, \quad x \in H_0, \\ 3 & if \quad x_\downarrow \in H_1, \quad x \in H_1. \end{cases}$$

Let $C_b \in H_0$, the possible cases are

a) $C_{b_\downarrow} \in H_0$ and $\varphi_{9,b}(C_{b_\downarrow}) = 2$, then $\varphi_{9,b}(C_b) = 2,\ B^{(2)} = 3,\ \varphi_{9,b} \in C_3$,

b) $C_{b_\downarrow} \in H_1$ and $\varphi_{9,b}(C_{b_\downarrow}) = 2$, this case is not exist, if $C_{b_\downarrow} \in H_1$, then $\varphi_{9,b}(C_{b_\downarrow}) = 3$,

c) $C_{b_\downarrow} \in H_1$ and $\varphi_{9,b}(C_{b_\downarrow}) = 3$, then $\varphi_{9,b}(C_b) = 2,\ B^{(2)} = 2,\ B^{(3)} = 1,\ \varphi_{9,b} \in C_7$.

Let $C_b \in H_1$, the possible cases are

a) $C_{b_\downarrow} \in H_0$ and $\varphi_{9,b}(C_{b_\downarrow}) = 2$, then $\varphi_{9,b}(C_b) = 2,\ B^{(2)} = 1,\ B^{(3)} = 2,\ \varphi_{9,b} \in C_9$,

b) $C_{b_\downarrow} \in H_1$ and $\varphi_{9,b}(C_{b_\downarrow}) = 2$, then $\varphi_{9,b}(C_b) = 3,\ B^{(2)} = 2,\ B^{(3)} = 1,\ \varphi_{9,b} \in C_9$,

c) $C_{b_\downarrow} \in H_1$ and $\varphi_{9,b}(C_{b_\downarrow}) = 3$, then $\varphi_{9,b}(C_b) = 3,\ B^{(2)} = 1,\ B^{(3)} = 2,\ \varphi_{9,b} \in C_7$.

Therefore, the configuration φ_9 is a weakly periodic ground state on the set $A_3 \cap A_7 \cap A_9 = A_7 = A_9$.

10) Consider

$$\varphi_{10}(x) = \begin{cases} 2 & if \quad x_\downarrow \in H_0, \quad x \in H_0, \\ 2 & if \quad x_\downarrow \in H_0, \quad x \in H_1, \\ 3 & if \quad x_\downarrow \in H_1, \quad x \in H_0, \\ 2 & if \quad x_\downarrow \in H_1, \quad x \in H_1. \end{cases}$$

Let $C_b \in H_0$, the possible cases are

a) $C_{b_\downarrow} \in H_0$ and $\varphi_{10,b}(C_{b_\downarrow}) = 2$, then $\varphi_{10,b}(C_b) = 2,\ B^{(2)} = 3,\ \varphi_{10,b} \in C_3$,

b) $C_{b_\downarrow} \in H_0$ and $\varphi_{10,b}(C_{b_\downarrow}) = 3$, then $\varphi_{10,b}(C_b) = 2,\ B^{(2)} = 2,\ B^{(3)} = 1,\ \varphi_{10,b} \in C_7$,

c) $C_{b_\downarrow} \in H_1$ and $\varphi_{10,b}(C_{b_\downarrow}) = 2$, then $\varphi_{10,b}(C_b) = 2,\ B^{(2)} = 3,\ \varphi_{10,b} \in C_7$.

Let $C_b \in H_1$, the possible cases are

a) $C_{b_\downarrow} \in H_0$ and $\varphi_{10,b}(C_{b_\downarrow}) = 2$, then $\varphi_{10,b}(C_b) = 2,\ B^{(2)} = 3,\ \varphi_{10,b} \in C_3$,

b) $C_{b_\downarrow} \in H_0$ and $\varphi_{10,b}(C_{b_\downarrow}) = 3$, this case is not exist, if $C_{b_\downarrow} \in H_0$, then $\varphi_{10,b}(C_{b_\downarrow}) = 2$,

c) $C_{b_\downarrow} \in H_1$ and $\varphi_{10,b}(C_{b_\downarrow}) = 3$, then $\varphi_{10,b}(C_b) = 2,\ B^{(2)} = 2,\ B^{(3)} = 1,\ \varphi_{10,b} \in C_7$.

Therefore, the configuration φ_{10} is a weakly periodic ground state on the set $A_2 \cap A_3 \cap A_7 = A_7$.

11) Consider

$$\varphi_{11}(x) = \begin{cases} 2 & if \quad x_\downarrow \in H_0, \quad x \in H_0, \\ 2 & if \quad x_\downarrow \in H_0, \quad x \in H_1, \\ 1 & if \quad x_\downarrow \in H_1, \quad x \in H_0, \\ 1 & if \quad x_\downarrow \in H_1, \quad x \in H_1. \end{cases}$$

Let $C_b \in H_0$, the possible cases are

a) $C_{b_\downarrow} \in H_0$ and $\varphi_{11,b}(C_{b_\downarrow}) = 1$, then $\varphi_{11,b}(C_b) = 2,\ B^{(1)} = 1,\ B^{(2)} = 2,\ \varphi_{11,b} \in C_7$,

b) $C_{b_\downarrow} \in H_0$ and $\varphi_{11,b}(C_{b_\downarrow}) = 2$, then $\varphi_{11,b}(C_b) = 2,\ B^{(2)} = 3,\ \varphi_{11,b} \in C_3$,

c) $C_{b_\downarrow} \in H_1$ and $\varphi_{11,b}(C_{b_\downarrow}) = 1$, then $\varphi_{11,b}(C_b) = 1,\ B^{(1)} = 1,\ B^{(2)} = 2\ \varphi_{11,b} \in C_9$,

d) $C_{b_\downarrow} \in H_1$ and $\varphi_{11,b}(C_{b_\downarrow}) = 2$, then $\varphi_{11,b}(C_b) = 1,\ B^{(1)} = 2,\ B^{(2)} = 1\ \varphi_{11,b} \in C_7$.

Let $C_b \in H_1$, the possible cases are

a) $C_{b_\downarrow} \in H_0$ and $\varphi_{11,b}(C_{b_\downarrow}) = 1$, then $\varphi_{11,b}(C_b) = 2,\ B^{(1)} = 3,\ \varphi_{11,b} \in C_2$,

b) $C_{b_\downarrow} \in H_0$ and $\varphi_{11,b}(C_{b_\downarrow}) = 2$, then $\varphi_{11,b}(C_b) = 2,\ B^{(1)} = 2,\ B^{(2)} = 1,\ \varphi_{11,b} \in C_9$,

c) $C_{b_\downarrow} \in H_1$ and $\varphi_{11,b}(C_{b_\downarrow}) = 1$, then $\varphi_{11,b}(C_b) = 1,\ B^{(1)} = 3,\ \varphi_{11,b} \in C_3$,

d) $C_{b_\downarrow} \in H_1$ and $\varphi_{11,b}(C_{b_\downarrow}) = 2$, then $\varphi_{11,b}(C_b) = 1,\ B^{(1)} = 2,\ B^{(2)} = 1,\ \varphi_{11,b} \in C_7$.

Therefore, the configuration φ_{11} is a weakly periodic ground state on the set $A_2 \cap A_3 \cap A_7 \cap A_9 = A_7 = A_9$.

12) Consider

$$\varphi_{12}(x) = \begin{cases} 3 & if \quad x_\downarrow \in H_0, \quad x \in H_0, \\ 3 & if \quad x_\downarrow \in H_0, \quad x \in H_1, \\ 3 & if \quad x_\downarrow \in H_1, \quad x \in H_0, \\ 2 & if \quad x_\downarrow \in H_1, \quad x \in H_1. \end{cases}$$

Let $C_b \in H_0$, the possible cases are

a) $C_{b_\downarrow} \in H_0$ and $\varphi_{12,b}(C_{b_\downarrow}) = 3$, then $\varphi_{12,b}(C_b) = 3,\ B^{(3)} = 3,\ \varphi_{12,b} \in C_3$,

b) $C_{b_\downarrow} \in H_1$ and $\varphi_{12,b}(C_{b_\downarrow}) = 2$, then $\varphi_{12,b}(C_b) = 3,\ B^{(2)} = 1,\ B^{(3)} = 2,\ \varphi_{12,b} \in C_7$,

c) $C_{b_\downarrow} \in H_1$ and $\varphi_{12,b}(C_{b_\downarrow}) = 3$, this case is not exist, if $C_{b_\downarrow} \in H_1$, then $\varphi_{12,b}(C_{b_\downarrow}) = 2$.

Let $C_b \in H_1$, the possible cases are

a) $C_{b_\downarrow} \in H_0$ and $\varphi_{12,b}(C_{b_\downarrow}) = 3$, then $\varphi_{12,b}(C_b) = 3,\ B^{(2)} = 2,\ B^{(3)} = 1,\ \varphi_{12,b} \in C_9$,

b) $C_{b_\downarrow} \in H_1$ and $\varphi_{12,b}(C_{b_\downarrow}) = 2$, then $\varphi_{12,b}(C_b) = 2,\ B^{(2)} = 2,\ B^{(3)} = 1,\ \varphi_{12,b} \in C_7$,

c) $C_{b_\downarrow} \in H_1$ and $\varphi_{12,b}(C_{b_\downarrow}) = 3$, then $\varphi_{12,b}(C_b) = 2,\ B^{(2)} = 1,\ B^{(3)} = 2,\ \varphi_{12,b} \in C_9$.

Therefore, the configuration φ_{12} is a weakly periodic ground state on the set $A_1 \cap A_3 \cap A_7 \cap A_9 = A_7 = A_9$.

13) Consider

$$\varphi_{13}(x) = \begin{cases} 1 & if \quad x_\downarrow \in H_0, \quad x \in H_0, \\ 1 & if \quad x_\downarrow \in H_0, \quad x \in H_1, \\ 1 & if \quad x_\downarrow \in H_1, \quad x \in H_0, \\ 2 & if \quad x_\downarrow \in H_1, \quad x \in H_1. \end{cases}$$

Let $C_b \in H_0$, the possible cases are

a) $C_{b_\downarrow} \in H_0$ and $\varphi_{13,b}(C_{b_\downarrow}) = 1$, then $\varphi_{13,b}(C_b) = 1,\ B^{(1)} = 3,\ \varphi_{13,b} \in C_3$,
b) $C_{b_\downarrow} \in H_1$ and $\varphi_{13,b}(C_{b_\downarrow}) = 1$, this case is not exist, if $C_{b_\downarrow} \in H_1$, then $\varphi_{13,b}(C_{b_\downarrow}) = 2$,
c) $C_{b_\downarrow} \in H_1$ and $\varphi_{13,b}(C_{b_\downarrow}) = 2$, then $\varphi_{3,b}(C_b) = 2,\ B^{(1)} = 2,\ B^{(2)} = 1,\ \varphi_{13,b} \in C_7$.

Let $C_b \in H_1$, the possible cases are

a) $C_{b_\downarrow} \in H_0$ and $\varphi_{13,b}(C_{b_\downarrow}) = 1$, then $\varphi_{13,b}(C_b) = 1,\ B^{(1)} = 1,\ B^{(2)} = 2,\ \varphi_{13,b} \in C_9$,
b) $C_{b_\downarrow} \in H_1$ and $\varphi_{13,b}(C_{b_\downarrow}) = 1$, then $\varphi_{13,b}(C_b) = 2,\ B^{(1)} = 2,\ B^{(2)} = 1,\ \varphi_{13,b} \in C_9$,
c) $C_{b_\downarrow} \in H_1$ and $\varphi_{13,b}(C_{b_\downarrow}) = 2$, then $\varphi_{13,b}(C_b) = 2,\ B^{(1)} = 1,\ B^{(2)} = 2,\ \varphi_{13,b} \in C_7$.

Therefore, the configuration φ_{13} is a weakly periodic ground state on the set $A_3 \cap A_7 \cap A_9 = A_7 = A_9$.

14) Consider

$$\varphi_{14}(x) = \begin{cases} 1 & if \quad x_\downarrow \in H_0, \quad x \in H_0, \\ 2 & if \quad x_\downarrow \in H_0, \quad x \in H_1, \\ 2 & if \quad x_\downarrow \in H_1, \quad x \in H_0, \\ 2 & if \quad x_\downarrow \in H_1, \quad x \in H_1. \end{cases}$$

Let $C_b \in H_0$, the possible cases are

a) $C_{b_\downarrow} \in H_0$ and $\varphi_{14,b}(C_{b_\downarrow}) = 1$, then $\varphi_{14,b}(C_b) = 1,\ B^{(1)} = 2,\ B^{(2)} = 1,\ \varphi_{14,b} \in C_7$,
b) $C_{b_\downarrow} \in H_0$ and $\varphi_{14,b}(C_{b_\downarrow}) = 2$, then $\varphi_{14,b}(C_b) = 1,\ B^{(1)} = 1,\ B^{(2)} = 2,\ \varphi_{14,b} \in C_9$,
c) $C_{b_\downarrow} \in H_1$ and $\varphi_{14,b}(C_{b_\downarrow}) = 2$, then $\varphi_{14,b}(C_b) = 1,\ B^{(1)} = 1,\ B^{(2)} = 2,\ \varphi_{14,b} \in C_9$.

Let $C_b \in H_1$, the possible cases are

a) $C_{b_\downarrow} \in H_0$ and $\varphi_{14,b}(C_{b_\downarrow}) = 1$, then $\varphi_{14,b}(C_b) = 2,\ B^{(1)} = 1,\ B^{(2)} = 2,\ \varphi_{14,b} \in C_7$,
b) $C_{b_\downarrow} \in H_1$ and $\varphi_{14,b}(C_{b_\downarrow}) = 2$, this case is not exist, if $C_{b_\downarrow} \in H_1$, then $\varphi_{14,b}(C_{b_\downarrow}) = 1$,
c) $C_{b_\downarrow} \in H_1$ and $\varphi_{14,b}(C_{b_\downarrow}) = 2$, then $\varphi_{14,b}(C_b) = 2,\ B^{(2)} = 3,\ \varphi_{14,b} \in C_3$.

Therefore, the configuration φ_{14} is a weakly periodic ground state on the set $A_3 \cap A_7 \cap A_9 = A_7 = A_9$.

15) Consider

$$\varphi_6(x) = \begin{cases} 1 & if \quad x_\downarrow \in H_0, \quad x \in H_0, \\ 3 & if \quad x_\downarrow \in H_0, \quad x \in H_1, \\ 3 & if \quad x_\downarrow \in H_1, \quad x \in H_0, \\ 3 & if \quad x_\downarrow \in H_1, \quad x \in H_1. \end{cases}$$

Let $C_b \in H_0$, the possible cases are

a) $C_{b_\downarrow} \in H_0$ and $\varphi_{15,b}(C_{b_\downarrow}) = 1$, then $\varphi_{15,b}(C_b) = 1,\ B^{(1)} = 2,\ B^{(3)} = 1,\ \varphi_{15,b} \in C_5$,
b) $C_{b_\downarrow} \in H_0$ and $\varphi_{15,b}(C_{b_\downarrow}) = 3$, then $\varphi_{15,b}(C_b) = 1,\ B^{(1)} = 1,\ B^{(3)} = 2,\ \varphi_{15,b} \in C_8$,
c) $C_{b_\downarrow} \in H_1$ and $\varphi_{15,b}(C_{b_\downarrow}) = 3$, then $\varphi_{15,b}(C_b) = 3,\ B^{(1)} = 2,\ B^{(3)} = 1,\ \varphi_{15,b} \in C_8$.

Let $C_b \in H_1$, the possible cases are

a) $C_{b_\downarrow} \in H_0$ and $\varphi_{15,b}(C_{b_\downarrow}) = 1$, then $\varphi_{15,b}(C_b) = 3,\ B^{(1)} = 1,\ B^{(3)} = 2,\ \varphi_{15,b} \in C_5$,
b) $C_{b_\downarrow} \in H_0$ and $\varphi_{15,b}(C_{b_\downarrow}) = 3$, this case is not exist, if $C_{b_\downarrow} \in H_0$, then $\varphi_{15,b}(C_{b_\downarrow}) = 1$,
c) $C_{b_\downarrow} \in H_1$ and $\varphi_{15,b}(C_{b_\downarrow}) = 3$, then $\varphi_{15,b}(C_b) = 3,\ B^{(3)} = 3,\ \varphi_{15,b} \in C_3$.

Therefore, configuration φ_{15} a weakly periodic ground state on the set $A_3 \cap A_5 \cap A_8 = A_5 = A_8$.

16) Consider

$$\varphi_{16}(x) = \begin{cases} 3 & if \quad x_\downarrow \in H_0, \quad x \in H_0, \\ 1 & if \quad x_\downarrow \in H_0, \quad x \in H_1, \\ 1 & if \quad x_\downarrow \in H_1, \quad x \in H_0, \\ 3 & if \quad x_\downarrow \in H_1, \quad x \in H_1. \end{cases}$$

Let $C_b \in H_0$, the possible cases are

a) $C_{b_\downarrow} \in H_0$ and $\varphi_{16,b}(C_{b_\downarrow}) = 1$, then $\varphi_{16,b}(C_b) = 3,\ B^{(1)} = 2,\ B^{(3)} = 1,\ \varphi_{16,b} \in C_8$,
b) $C_{b_\downarrow} \in H_0$ and $\varphi_{16,b}(C_{b_\downarrow}) = 3$, then $\varphi_{16,b}(C_b) = 3,\ B^{(1)} = 1,\ B^{(3)} = 2,\ \varphi_{16,b} \in C_5$,
c) $C_{b_\downarrow} \in H_1$ and $\varphi_{16,b}(C_{b_\downarrow}) = 1$, this case is not exist, if $C_{b_\downarrow} \in H_1$, then $\varphi_{16,b}(C_{b_\downarrow}) = 3$,
d) $C_{b_\downarrow} \in H_1$ and $\varphi_{16,b}(C_{b_\downarrow}) = 3$, then $\varphi_{16,b}(C_b) = 1,\ B^{(3)} = 3,\ \varphi_{16,b} \in C_1$.

Let $C_b \in H_1$, the possible cases are
a) $C_{b_\downarrow} \in H_0$ and $\varphi_{16,b}(C_{b_\downarrow}) = 1$, then $\varphi_{16,b}(C_b) = 1,\ B^{(1)} = 1,\ B^{(3)} = 2,\ \varphi_{16,b} \in C_8$,
b) $C_{b_\downarrow} \in H_0$ and $\varphi_{16,b}(C_{b_\downarrow}) = 3$, then $\varphi_{16,b}(C_b) = 1,\ B^{(3)} = 3,\ \varphi_{16,b} \in C_2$,
c) $C_{b_\downarrow} \in H_1$ and $\varphi_{16,b}(C_{b_\downarrow}) = 1$, then $\varphi_{16,b}(C_b) = 3,\ B^{(1)} = 2,\ B^{(3)} = 1,\ \varphi_{16,b} \in C_8$,
d) $C_{b_\downarrow} \in H_1$ and $\varphi_{16,b}(C_{b_\downarrow}) = 3$, then $\varphi_{16,b}(C_b) = 3,\ B^{(1)} = 1,\ B^{(3)} = 2,\ \varphi_{16,b} \in C_5$.

Therefore, the configuration φ_{16} is a weakly periodic ground state on the set $A_1 \cap A_5 \cap A_8 = A_5 = A_8$.

17) Consider

$$\varphi_{17}(x) = \begin{cases} 3 & if \quad x_\downarrow \in H_0, \quad x \in H_0, \\ 3 & if \quad x_\downarrow \in H_0, \quad x \in H_1, \\ 1 & if \quad x_\downarrow \in H_1, \quad x \in H_0, \\ 1 & if \quad x_\downarrow \in H_1, \quad x \in H_1. \end{cases}$$

Let $C_b \in H_0$, the possible cases are
a) $C_{b_\downarrow} \in H_0$ and $\varphi_{17,b}(C_{b_\downarrow}) = 1$, then $\varphi_{17,b}(C_b) = 3,\ B^{(1)} = 1,\ B^{(3)} = 2,\ \varphi_{17,b} \in C_5$,
b) $C_{b_\downarrow} \in H_0$ and $\varphi_{17,b}(C_{b_\downarrow}) = 3$, then $\varphi_{17,b}(C_b) = 3,\ B^{(3)} = 3,\ \varphi_{17,b} \in C_3$,
c) $C_{b_\downarrow} \in H_1$ and $\varphi_{17,b}(C_{b_\downarrow}) = 1$, then $\varphi_{17,b}(C_b) = 1,\ B^{(1)} = 1,\ B^{(3)} = 2,\ \varphi_{17,b} \in C_8$,
d) $C_{b_\downarrow} \in H_1$ and $\varphi_{17,b}(C_{b_\downarrow}) = 3$, this case is not exist, if $C_{b_\downarrow} \in H_1$, then $\varphi_{17,b}(C_{b_\downarrow}) = 1$.

Let $C_b \in H_1$, the possible cases are
a) $C_{b_\downarrow} \in H_0$ and $\varphi_{17,b}(C_{b_\downarrow}) = 1$, then $\varphi_{17,b}(C_b) = 3,\ B^{(1)} = 3,\ \varphi_{17,b} \in C_1$,
b) $C_{b_\downarrow} \in H_0$ and $\varphi_{17,b}(C_{b_\downarrow}) = 3$, then $\varphi_{17,b}(C_b) = 3,\ B^{(1)} = 2,\ B^{(3)} = 1,\ \varphi_{17,b} \in C_8$,
c) $C_{b_\downarrow} \in H_1$ and $\varphi_{17,b}(C_{b_\downarrow}) = 1$, then $\varphi_{17,b}(C_b) = 1,\ B^{(1)} = 3,\ \varphi_{17,b} \in C_3$,
d) $C_{b_\downarrow} \in H_1$ and $\varphi_{17,b}(C_{b_\downarrow}) = 3$, then $\varphi_{17,b}(C_b) = 1,\ B^{(1)} = 2,\ B^{(3)} = 1,\ \varphi_{17,b} \in C_5$.

Therefore, the configuration φ_{17} is a weakly periodic ground state on the set $A_1 \cap A_3 \cap A_5 \cap A_8 = A_5 = A_8$.

18) Consider

$$\varphi_{18}(x) = \begin{cases} 3 & if \quad x_\downarrow \in H_0, \quad x \in H_0, \\ 3 & if \quad x_\downarrow \in H_0, \quad x \in H_1, \\ 1 & if \quad x_\downarrow \in H_1, \quad x \in H_0, \\ 3 & if \quad x_\downarrow \in H_1, \quad x \in H_1. \end{cases}$$

Let $C_b \in H_0$, the possible cases are
a) $C_{b_\downarrow} \in H_0$ and $\varphi_{18,b}(C_{b_\downarrow}) = 1$, then $\varphi_{18,b}(C_b) = 3,\ B^{(1)} = 1,\ B^{(3)} = 2,\ \varphi_{18,b} \in C_5$,
b) $C_{b_\downarrow} \in H_0$ and $\varphi_{18,b}(C_{b_\downarrow}) = 3$, then $\varphi_{18,b}(C_b) = 3,\ B^{(3)} = 3,\ \varphi_{18,b} \in C_3$,
c) $C_{b_\downarrow} \in H_1$ and $\varphi_{18,b}(C_{b_\downarrow}) = 3$, then $\varphi_{18,b}(C_b) = 1,\ B^{(3)} = 3,\ \varphi_{18,b} \in C_1$.

Let $C_b \in H_1$, the possible cases are
a) $C_{b_\downarrow} \in H_0$ and $\varphi_{18,b}(C_{b_\downarrow}) = 1$, then $\varphi_{18,b}(C_b) = 3,\ B^{(1)} = 1,\ B^{(3)} = 2,\ \varphi_{18,b} \in C_5$,
b) $C_{b_\downarrow} \in H_0$ and $\varphi_{18,b}(C_{b_\downarrow}) = 3$, then $\varphi_{18,b}(C_b) = 3,\ B^{(3)} = 3,\ \varphi_{18,b} \in C_3$,
c) $C_{b_\downarrow} \in H_1$ and $\varphi_{18,b}(C_{b_\downarrow}) = 3$, then $\varphi_{18,b}(C_b) = 3,\ B^{(1)} = 1,\ B^{(3)} = 2,\ \varphi_{18,b} \in C_5$.

Therefore, the configuration φ_{18} is a weakly periodic ground state on the set $A_1 \cap A_3 \cap A_5 = A_5$.

19) Consider

$$\varphi_{19}(x)=\begin{cases}3 & if \quad x_{\downarrow}\in H_0, \quad x\in H_0,\\ 3 & if \quad x_{\downarrow}\in H_0, \quad x\in H_1,\\ 3 & if \quad x_{\downarrow}\in H_1, \quad x\in H_0,\\ 1 & if \quad x_{\downarrow}\in H_1, \quad x\in H_1.\end{cases}$$

Let $C_b\in H_0$, the possible cases are

a) $C_{b_\downarrow}\in H_0$ and $\varphi_{19,b}(C_{b_\downarrow})=3$, then $\varphi_{19,b}(C_b)=3,\ B^{(3)}=3,\ \varphi_{19,b}\in C_3$,

b) $C_{b_\downarrow}\in H_1$ and $\varphi_{19,b}(C_{b_\downarrow})=1$, then $\varphi_{19,b}(C_b)=3,\ B^{(1)}=1,\ B^{(3)}=2,\ \varphi_{19,b}\in C_5$,

c) $C_{b_\downarrow}\in H_1$ and $\varphi_{19,b}(C_{b_\downarrow})=3$, then $\varphi_{19,b}(C_b)=3,\ B^{(1)}=1,\ B^{(3)}=2,\ \varphi_{19,b}\in C_5$.

Let $C_b\in H_1$, the possible cases are

a) $C_{b_\downarrow}\in H_0$ and $\varphi_{19,b}(C_{b_\downarrow})=3$, then $\varphi_{19,b}(C_b)=3,\ B^{(1)}=2,\ B^{(3)}=1,\ \varphi_{19,b}\in C_8$,

b) $C_{b_\downarrow}\in H_1$ and $\varphi_{19,b}(C_{b_\downarrow})=1$, then $\varphi_{19,b}(C_b)=1,\ B^{(1)}=2,\ B^{(3)}=1,\ \varphi_{19,b}\in C_5$,

c) $C_{b_\downarrow}\in H_1$ and $\varphi_{19,b}(C_{b_\downarrow})=3$, then $\varphi_{19,b}(C_b)=3,\ B^{(1)}=1,\ B^{(3)}=2,\ \varphi_{19,b}\in C_8$.

Therefore, configuration φ_{19} a weakly periodic ground state on the set $A_3\cap A_5\cap A_8=A_5=A_8$.

20) Consider

$$\varphi_{20}(x)=\begin{cases}1 & if \quad x_{\downarrow}\in H_0, \quad x\in H_0,\\ 3 & if \quad x_{\downarrow}\in H_0, \quad x\in H_1,\\ 3 & if \quad x_{\downarrow}\in H_1, \quad x\in H_0,\\ 1 & if \quad x_{\downarrow}\in H_1, \quad x\in H_1.\end{cases}$$

Let $C_b\in H_0$, the possible cases are

a) $C_{b_\downarrow}\in H_1$ and $\varphi_{20,b}(C_{b_\downarrow})=1$, then $\varphi_{20,b}(C_b)=3,\ B^{(1)}=3,\ \varphi_{20,b}\in C_3$,

b) $C_{b_\downarrow}\in H_0$ and $\varphi_{20,b}(C_{b_\downarrow})=1$, then $\varphi_{20,b}(C_b)=1,\ B^{(1)}=2,\ B^{(3)}=1,\ \varphi_{20,b}\in C_5$,

c) $C_{b_\downarrow}\in H_0$ and $\varphi_{20,b}(C_{b_\downarrow})=3$, then $\varphi_{20,b}(C_b)=1,\ B^{(1)}=2,\ B^{(3)}=1,\ \varphi_{20,b}\in C_5$.

Let $C_b\in H_1$, the possible cases are

a) $C_{b_\downarrow}\in H_0$ and $\varphi_{20,b}(C_{b_\downarrow})=1$, then $\varphi_{20,b}(C_b)=3,\ B^{(1)}=3,\ B^{(3)}=0,\ \varphi_{20,b}\in C_3$,

b) $C_{b_\downarrow}\in H_1$ and $\varphi_{20,b}(C_{b_\downarrow})=1$, then $\varphi_{20,b}(C_b)=3,\ B^{(1)}=2,\ B^{(3)}=1,\ \varphi_{20,b}\in C_8$,

c) $C_{b_\downarrow}\in H_1$ and $\varphi_{20,b}(C_{b_\downarrow})=3$, then $\varphi_{20,b}(C_b)=1,\ B^{(1)}=1,\ B^{(3)}=2,\ \varphi_{20,b}\in C_5$.

Therefore, the configuration φ_{20} is a weakly periodic ground state on the set $A_3\cap A_5\cap A_8=A_5=A_8$.

Let we prove assertion III. We consider the remaining configurations. For example, let:

$$\varphi(x)=\begin{cases}3 & if \quad x_{\downarrow}\in H_0, \quad x\in H_0,\\ 1 & if \quad x_{\downarrow}\in H_0, \quad x\in H_1,\\ 3 & if \quad x_{\downarrow}\in H_1, \quad x\in H_0,\\ 3 & if \quad x_{\downarrow}\in H_1, \quad x\in H_1.\end{cases}$$

We show that $\varphi(x)$ is not a ground state.

Let $C_b\in H_0$, the possible cases are

a) $C_{b_\downarrow}\in H_0$ and $\varphi_b(C_{b_\downarrow})=3$, then $\varphi_b(C_b)=3,\ B^{(1)}=1,\ B^{(3)}=2,\ \varphi_b\in C_8$,

b) $C_{b_\downarrow}\in H_1$ and $\varphi_b(C_{b_\downarrow})=3$, then $\varphi_b(C_b)=3,\ B^{(1)}=0,\ B^{(3)}=3,\ \varphi_b\in C_1$.

Let $C_b\in H_1$, the possible cases are

a) $C_{b_\downarrow}\in H_0$ and $\varphi_b(C_{b_\downarrow})=3$, then $\varphi_b(C_b)=1,\ B^{(1)}=0,\ B^{(3)}=3,\ \varphi_b\in C_3$,

b) $C_{b_\downarrow}\in H_1$ and $\varphi_b(C_{b_\downarrow})=1$, then $\varphi_b(C_b)=3,\ B^{(1)}=1,\ B^{(3)}=2,\ \varphi_b\in C_8$,

c) $C_{b_\downarrow} \in H_1$ and $\varphi_b(C_{b_\downarrow}) = 3$, then $\varphi_b(C_b) = 3$, $B^{(1)} = 0$, $B^{(3)} = 3$, $\varphi_b \in C_1$.

We conclude that the configuration $\varphi(x)$ is a ground state on the set $\{(a,b,c) \in R^3 : a = b = c\}$. Therefore, if $(a,b,c) \notin \{(a,b,c) \in R^3 : a = b = c\}$, then the weakly periodic configuration $\varphi(x)$ is not a weakly periodic ground state.

As above, we analyze all possible cases, except for periodic (in particular translation-invariant) configurations and the configurations mentioned in the assertions I and II we find that none of them is a ground state. This completes the proof of Theorem 5.1. □

Remark 5.1. *Notice that all the configurations on the set $\{(a,b,c) \in R^3 : a = b = c\}$ are ground states.*

6. References

[1] Y.G.Sinai, *Theory of Phase Transitions:Rigorous Results*, (Pergamon Press, Oxford, 1982).
[2] R. A. Minlos, *Introduction to Mathematical Statistical Physics*, (Amer. Math. Soc., Providence, RI, 2000).
[3] N.N. Ganikhodjaev, F.M. Mukhamedov, J.F.F. Mendes, *Jour. Stat. Mech.* 2006, P08012.
[4] G. I. Botirov, U. A. Rozikov, *Theo. Math. Phys.* **153**, p. 1423 (2007).
[5] F.M. Mukhamedov, U.A. Rozikov U.A., J.F.F. Mendes,*Jour. Math. Phys.* **48**, 013301 (2007).
[6] U. A. Rozikov, *Gibbs Measures on Cayley Trees,* (World Scientific, Hackensack, 2013).
[7] U. A. Rozikov and M. M. Rakhmatullaev,*Theo. Math. Phys.* **160**, p. 1292 (2009).
[8] M. M. Rakhmatullaev, *Appl. Math. Inf. Sci.* **4**, p. 237 (2010).
[9] M. M. Rakhmatullaev, *Theo. Math. Phys.* **176**, p. 1236 (2013).
[10] M.M. Rahmatullaev, M.A. Rasulova,*Siberian Adv. Math.*, **26**, p. 215-229 (2016) .
[11] D. Gandolfo, J. Ruiz, S. Shlosman, *Jour. Stat. Phys.* **148**, p. 999-1005 (2012).
[12] F.M. Mukhamedov,*Rep. Math. Phys.***53**, p. 1-18 (2004).
[13] U.A. Rozikov,*Siberan Math. Jour.* **39**, p. 373-380, (1998).
[14] F.M.Mukhamedov, U.A.Rozikov,*Theo. Math. Phys.* **124**, p. 1202-1210 (2000)
[15] F. Mukhamedov, C. H. Pah, H. Jamil, `arXiv:1610.07019`
[16] U.A. Rozikov, R.M. Khakimov,*Theor. Math. Phys.* **175**, p. 699-709 (2013).
[17] M. I. Kargapolov, Yu. I. Merzlyakov, *Fundamentals of the Theory of Groups*, (Springer-Verlag, Berlin, 1979).
[18] N. N. Ganikhodzhaev, *Dokl. Akad. nauk Resp. Uzbekistan,* **4** , 3 (1994) [in Russian].
[19] O. Melnikov, R. I. Tyshkevich, V. A. Yemelichev, V. I. Sarvanov, *Lectures on Graph Theory* (B. I. Wissenschaftsverlag, Mannheim ,1994).

On Lyapunov Functions for Infinite Dimensional Volterra Quadratic Stochastic Operators

Farrukh Mukhamedov[1,*], **Ahmad Fadillah Embong**[2,3,**]

[1], Department of Mathematical Sciences, College of Science, The United Arab Emirates University, Al Ain, Abu Dhabi, 15551, UAE

[2], Department of Computational & Theoretical Sciences, Faculty of Science, International Islamic University Malaysia, 25200, Kuantan, Pahang, Malaysia

[3]Corresponding author

E-mail: *farrukh.m@uaeu.ac.ae, **ahmadfadillah.90@gmail.com

Abstract. In the present paper, we study the existence of Lyapunov functions for Volterra qudratic stochastic operators(QSO) on infinite dimensional simplex. We provide a construction of Lyapunov functions for such kind of operators, which allows us to estimate their limiting points.

1. Introduction

An original work on quadratic stochastic operators (in short QSOs) was done by Bernstein [1] where such kind of operators appeared from the problems of population genetics (see also [2]). These operator appear to have tremendous applications especially in modelings in many different fields such as biology [3] (population and disease dynamics), physics [4, 5](non-equilibrium statistical mechanics) , economics and mathematics [2, 5] (replicator dynamics and games). A quadratic stochastic operator is usually used to present the time evolution of species in biology, which arises as follows. Consider evolution of species in biology as given in the following situation. Let $I = \{1, 2, \dots, n\}$ be the n type of species (or traits) in a population and we denote $x^{(0)} = (x_1^{(0)}, \dots, x_n^{(0)})$ to be the probability distribution of the species in an early state of that population. By $P_{ij,k}$ we mean the probability of an individual in the ith species and j^{th} species to cross-fertilize and produce an individual from k^{th} species (trait). Given $x^{(0)} = (x_1^{(0)}, \dots, x_n^{(0)})$, we can find the probability distribution of the first generation, $x^{(1)} = (x_1^{(1)}, \dots, x_n^{(1)})$ by using a total probability, i.e.,

$$x_k^{(1)} = \sum_{i,j=1}^{n} P_{ij,k} x_i^{(0)} x_j^{(0)}, \quad k \in \{1, \dots, n\}.$$

This relation defines an operator which is denoted by V and it is called *quadratic stochastic operator (QSO)*. Each QSO can be reinterprate as evolutionary operator that describes the sequence of generations in terms of probabilistic distribution if the values of $P_{ij,k}$ and the distribution of the current generation are given.

Currently, there are only a small number of studies on dynamical phenomena on higher dimensional systems, even though they are very important. Note that, most of the studies in

IOP Conf. Series: Journal of Physics: Conf. Series **949** (2017) 012022 doi:10.1088/1742-6596/949/1/012022

theory of QSO were done on finite set of all probabilistic distributions. However, there are models where the probability distribution is given on countable set, which means that the corresponding QSO is defined on infinite-dimensional space. The simplest case, the infinite-dimensional space should be the Banach space ℓ_1 of absolutely summable sequences. It is worth mentioning that Volterra QSOs, orthogonal preserving QSO on infinite-dimensional space were studied in [6], [7]. For more information on recent achievements and open problems in the theory of QSO, one may refer to [8, 9].

One of the main problems in mathematical biology is to study the asymptotic behavior of the trajectories. In [10–13] this problem was solved for the finite dimensional Volterra QSO's by using the theories of the Lyapunov function and tournaments. The functions constructed were based on finite dimensional case. Therefore, it is natural to continue this study by constructing Lyapunov function for infinite dimensional Volterra QSO which allows us to estimate their limit points for future study. Note that, up to now there is no paper yet is devoted to construct Lyapunov function for infinite dimensional Volterra QSOs.

2. Preliminaries

In this paper we are going to consider the set of absolutely summable sequences i.e.,

$$\ell^1 = \left\{ \mathbf{x} = \{x_k\} : \ \|\mathbf{x}\|_1 = \sum_{k=1}^{\infty} |x_k| < \infty \right\}. \tag{2.1}$$

Denote

$$S = \{\mathbf{x} = \{x_k\} \in \ell_1 : \ x_k \geq 0 \text{ for all } k \in \mathbb{N}, \ \|\mathbf{x}\|_1 = 1\}. \tag{2.2}$$

It is known that $S = convh(ExtrS)$, where $Extr(S)$ is the extremal points of S and $convh(A)$ is the convex hall of a set A. Any extremal point of S has the following form:

$$\mathbf{e}_k = (0, \dots, 0, 1, 0, \dots), \quad k \in \mathbb{N}$$

where 1 stands in kth position. That is, vertices of the simplex are extreme points of the simplex. Here and henceforth we denote

$$riS = \{\mathbf{x} = (x_1, x_2, \dots, x_k, \dots) \in S : \ x_k > 0, \text{ for all } k \in \mathbb{N}\}, \quad \partial S = S \backslash riS$$

Next, let V be a mapping defined by

$$V(\mathbf{x})_k = \sum_{i,j=1}^{\infty} P_{ij,k} x_i x_j \quad , k \in \mathbb{N}. \tag{2.3}$$

Here, $P_{ij,k}$ are hereditary coefficients and satisfy

$$P_{ij,k} \geq 0, \quad P_{ij,k} = P_{ji,k}, \quad \sum_{k=1}^{\infty} P_{ij,k} = 1, \quad i, j, k \in \mathbb{N}. \tag{2.4}$$

It is important to see that the mapping V is well-defined i.e., $V(\mathbf{x})_k < \infty$ for any $k \in \mathbb{N}$. Moreover one can check that V maps S into itself. Such kind of mapping V is called *Quadratic Stochastic Operator (QSO)*.

Recall that a QSO $V : S \to S$ is called Volterra if one has

$$P_{ij,k} = 0 \ \text{ when } \ k \notin \{i, j\} \tag{2.5}$$

The condition (2.5) biologically means that each individual can inherit only the species of the parents.

Taking into account (2.3), one easily can check that (2.5) is equivalent to the following canonical form of V

$$V(\mathbf{x})_k = x_k \left(1 + \sum_{i=1}^{\infty} a_{ki} x_i\right) \text{ for all } k \in \mathbb{N}, \tag{2.6}$$

where $a_{ki} = 1 - 2P_{ik,k}$. From (2.6), one can deduce that all faces of the simplex S are invariant with respect to V. In particular, all the vertices of S are the fixed points. Several properties of infinite dimensional Volterra operators have been investigated in [6].

Next we recall Young's inequality:

$$b_1^{p_1} \cdots b_n^{p_n} \leq b_1^{p_1} + \cdots + b_n^{p_n},$$

for any $b_i > 0$ and $p_i \geq 0$, $\sum_{i=1}^{n} p_i = 1$.

Here and henceforth, we use $V^{(n)}(\mathbf{x})$ to denote the iterations of the given operator V at $\mathbf{x} \in S$ i.e.,

$$V^{(n+1)}(\mathbf{x}) = V(V^{(n)}(\mathbf{x})), \ n \in \mathbb{N}.$$

A continuous function $\varphi : S \to \mathbb{R}$ is called a *Lyapunov function* for V if the limit $\lim_{n\to\infty} \varphi(V^{(n)}(\mathbf{x}))$ exists for any initial point $\mathbf{x} \in S$.

Let $\{V^{(n)}(\mathbf{x})\}_{n=1}^{\infty}$ be the trajectory of the point $\mathbf{x} \in S$ under QSO V. By $\omega_V(\mathbf{x})$ we denote the set of limit points with respect to ℓ^1-norm of the trajectory. Namely, for $\mathbf{x}^* \in \omega_V(\mathbf{x})$ means, there exist a subsequence $\{n_k\}$ such that

$$V^{(n_k)}(\mathbf{x}) \to \mathbf{x}^*$$

in ℓ^1-norm as n_k goes to infinity.

Obviously, if $\omega_V(\mathbf{x})$ consists of a single point, then the trajectory converges, to a fixed point of (2.5). However, looking ahead, we remark that convergence of the trajectories is not the typical case for the dynamical systems (2.5). Therefore, it is of particular interest to obtain an upper bound for $\mathbf{x}_0 \in S$, i.e., to determine a sufficiently "small" set containing $\mathbf{x}_0 \in S$.

Obviously, if $\lim_{n\to\infty} \varphi(V^{(n)}(\mathbf{x}_0)) = c$, then $\omega_V(\mathbf{x}_0) \subset \varphi^{-1}(c)$. Consequently, to estimate of $\omega_V(\mathbf{x}_0)$ we should construct the set of Lyapunov functions that is as large as possible.

3. Finite Dimensional Volterra QSO

Before, considering infinite dimensional Volterra QSO, let us observe some facts on finite dimensional Volterra QSO which turns to be very useful in the investigation one of the classes of infinite dimensional Volterra QSO. Denote

$$S_r^{n-1} = \left\{\mathbf{x} = (x_1, \ldots, x_n) : \ x_i \geq 0 \ \forall i = \overline{1,n}; \ \sum_{i=1}^{n} x_i = r\right\},$$

for any $r \in (0, 1]$. Note that, a huge number of papers on the theory of QSO were done on S_1^{n-1} (or simply we write S^{n-1}). Therefore, it is self-interest to examine Volterra QSO defined on S_r^{n-1} for any $r \in (0, 1)$. We need an auxiliary result to further in this topic. Denote

$$\mathbb{P}^{n-1} = \left\{\mathbf{x} \in S^{n-1} : \ \sum_{i=1}^{n} a_{ki} x_i \geq 0 \text{ for every } k \in \{1, \ldots, n\}\right\}.$$

IOP Conf. Series: Journal of Physics: Conf. Series **949** (2017) 012022 doi:10.1088/1742-6596/949/1/012022

Analogously,

$$\mathbb{P} = \left\{ \mathbf{x} \in S : \sum_{i=1}^{\infty} a_{ki} x_i \geq 0 \text{ for every } k \in \mathbb{N} \right\}.$$

We stress that the set $\mathbb{P}^{m-1}$ is non-empty (see [10]), in contrast to the set $\mathbb{P}$ could be empty [6]. Now, we recall some of known facts on Volterra QSO over a finite dimensional simplex.

Theorem 3.1 ([10]). *Let V be a Volterra QSO. If $p = (p_1, \ldots, p_n) \in \mathbb{P}^{n-1}$, then*

$$\varphi_p(\mathbf{x}) = x_1^{p_1} \cdots x_n^{p_n} \tag{3.1}$$

is a Lyapunov function for V for $\mathbf{x} \in riS$.

Using the constructed Lyapunov function, it was proved the following result

Theorem 3.2 ([10]). *If $\mathbf{x} \in riS^{n-1}$ is not a fixed point, then $\omega_V(\mathbf{x}) \subset \partial S^{n-1}$ where ∂S^{n-1} is the boundary of the simplex S^{n-1}.*

Proposition 3.3. *Let $V : S_r^{n-1} \to S_r^{n-1}$ (for any $r \in (0, 1]$) be a Volterra QSO associated with a skew-symmetric matrix $A = (a_{ki})_{k,i=1}^n$, then its dynamics is the same as $\widetilde{V} : S^{n-1} \to S^{n-1}$ where the associated skew-symmetric matrix for $\widetilde{V}$ is given by $\widetilde{A} = rA$ i.e.,*

$$rA = (ra_{ki})_{k,i=1}^n$$

Proof. Let us prove the statement by showing V and $\widetilde{V}$ are conjugate to each other i.e., there exist a homeomorphism function $T : S^{n-1} \to S_r^{n-1}$ such that

$$T_r^{-1} \circ V \circ T_r(\mathbf{y}) = \widetilde{V}$$

Indeed, V is well-defined on S_r^{n-1} i.e., it maps from S_r^{n-1} to itself. Next, we define a linear operator by

$$T_r(\mathbf{y}) := r\,(\mathbf{y}) = (ry_1, ry_2, \ldots, ry_n)$$

for any $\mathbf{y} \in S^{n-1}$. One can check that $T_r(S^{n-1}) = S_r^{n-1}$ and $T^{-1}(S_r^{n-1}) = S^{n-1}$, where $T^{-1}(\mathbf{x})$ is the pre-image of $\mathbf{x}$ under mapping T. Moreover, for any $\mathbf{y} \in S^{n-1}$ we have

$$\begin{aligned} T_r^{-1} \circ V \circ T_r(\mathbf{y}) &= T_r^{-1} \circ V(r(\mathbf{y})) \\ &= T_r^{-1} \circ \left(\left\{ ry_k \left(1 + r \sum_{i=1}^{n} a_{ki} y_i \right) \right\}_{k=1}^{n} \right) \\ &= \widetilde{V}(\mathbf{y}) \end{aligned}$$

The last statement means $V : S_r^{n-1} \to S_r^{n-1}$ is conjugate to $\widetilde{V} : S^{n-1} \to S^{n-1}$. Hence it proves the statement. □

In [6], it has been proved that every infinite dimensional Volterra QSO can be constructed by means of compatible sequence of finite dimensional Volterra QSO. Therefore, it is interesting to know what is the relationship between the sequence of compatible Volterra QSO and the corresponding Lyapunov functions. Let us consider a sequence $V_{n]} : S^{n-1} \to S^{n-1}$ finite dimensional Volterra operators, i.e.

$$(V_{n]}(x))_k = x_k \left(1 + \sum_{i=1}^{n} a_{ki}^{n]} x_i \right) \quad k = 1, \cdots, n, \quad (n \in \mathbb{N}), \tag{3.2}$$

here $(a_{ki}^{n]})$ is a skew-symmetric matrix. We recall that the sequence $\{V_{n]}\}$ of Volterra operators is *compatible* if

$$V_{n+1]} \upharpoonright S^{n-1} = V_{n]} \tag{3.3}$$

for every $n \in \mathbb{N}$. Here, $V_{n+1]} \upharpoonright S^{n-1}$ means restriction of $V_{n+1]}$ to the simplex S^{n-1}.

Remark 3.4. Let $\{V_{n]}(\mathbf{x})\}$ be a sequence of compatible Volterra QSO and let $p = (p_1, \dots, p_m) \in \mathbb{P}^{m-1}$ for $m < \infty$. If there exists $l_0 < m$ such that

$$\sum_{i=1}^{l_0} p_i = 1$$

then $\varphi(\mathbf{x}) = \prod_{i=1}^{l_0} x_i^{p_i}$ is a Lyapunov function for $V_{l_0]}(\mathbf{x})$.

4. Infinite Dimensional Volterra QSO and Lyapunov Function

From Proposition 3.3 one can deduce the following remark.

Remark 4.1. Let V be a Volterra QSO defined on S associated with a skew-symmetric matrix $A = (a_{ij})$ such that we have finitely many $a_{ij} > 0$ i.e., there exist n such that

$$A = \begin{bmatrix} a_{11} & a_{12} & \cdots & a_{1n} & 0 & \cdots \\ a_{21} & a_{22} & \cdots & a_{2n} & 0 & \cdots \\ \vdots & \vdots & \cdots & \vdots & \vdots & \cdots \\ a_{n1} & a_{n2} & \cdots & a_{nn} & 0 & \cdots \\ 0 & 0 & \cdots & 0 & 0 & \cdots \\ \vdots & \vdots & \cdots & \vdots & \vdots & \cdots \end{bmatrix}$$

We emphasize that $|a_{ij}| \leq 1$ for any $i, j \in \{1, \dots, n\}$. Then, for any $\mathbf{x} \in S$, the operator V can be written as follows

$$\begin{aligned} V(\mathbf{x})_k &= x_k \left(1 + \sum_{i=1}^{n} a_{ki} x_i\right) \quad \text{for } 1 \leq k \leq n, \\ V(\mathbf{x})_k &= x_k \quad \text{for } k \geq n+1 \end{aligned} \tag{4.1}$$

Thus, the investigation of dynamics on $V(\mathbf{x})$ reduces to $V(\mathbf{x})_k$ for $1 \leq k \leq n$ which defines a finite Volterra QSO $\widetilde{V}$ on S_r^{n-1} for any $r \in (0,1]$. Here $\widetilde{V}$ is associated to skew-syammetric matrix $A = (a_{ij})_{i,j=1}^{n}$. Since all faces of the simplex S_r^{n-1} are invariant with respect to V, therefore it is enough to study $\widetilde{V}$ on riS_r^{n-1}. Thanks to Proposition 3.3 and Theorem 3.2 we conclude that for any $\mathbf{x}_0 = (x_1^0, x_2^0, \dots, x_n^0) \in riS_r^{n-1}$ and $\mathbf{x}_0 \notin Fix(\widetilde{V})$,

$$\omega_{\widetilde{V}}(\mathbf{x}_0) \in \partial S_r^{n-1} \tag{4.2}$$

Define $\mathbf{y}_0 = (x_1^0, x_2^0, \dots, x_n^0, y_{n+1}^0, \dots) \in riS$ such that $\sum_{k=n+1}^{\infty} y_k^0 = 1 - r$. Due to (4.1) and (4.2) one has

$$\omega_V(\mathbf{y}_0) \in \partial S$$

Note that, the limiting points with respect to ℓ^1 and point-wise are the same because of (4.1).

Now, we want to establish Lyapunov function for infinite dimensional Volterra QSO which has the same form as given by (3.1).

Theorem 4.2. *Let V be an infinite dimensional Volterra QSO such that $\mathbb{P}$ is non-empty. If there exist finitely many $p_k > 0$ i.e., there is m such that*

$$p = (p_1, p_2, \dots, p_m, 0, \dots) \in \mathbb{P};\ p_i \geq 0 \ \text{for} \ 1 \leq i \leq m \tag{4.3}$$

then the functional

$$\varphi_p(\mathbf{x}) = \prod_{k=1}^{\infty} x_k^{p_k}$$

is a Lyapunov function for V.

Proof. Using the same argument as before, we can choose $\mathbf{x} \in riS$ without the loss of generality. Next, one sees that

$$\begin{aligned} \varphi_p(V(\mathbf{x})) &= \prod_{k=1}^{\infty}\left(x_k\left(1+\sum_{i=1}^{\infty} a_{ki}x_i\right)\right)^{p_k} \\ &= \varphi_p(\mathbf{x})\prod_{k=1}^{\infty}\left(1+\sum_{i=1}^{\infty} a_{ki}x_i\right)^{p_k} \end{aligned} \tag{4.4}$$

Since p satisfies (4.3), then

$$\prod_{k=1}^{\infty}\left(1+\sum_{i=1}^{\infty} a_{ki}x_i\right)^{p_k} = \prod_{k=1}^{m}\left(1+\sum_{i=1}^{\infty} a_{ki}x_i\right)^{p_k}$$

Due to the facts $\mathbf{x} \in riS$ and $a_{kk} = 0$, we have

$$1+\sum_{i=1}^{\infty} a_{ki}x_i > 0 \text{ for any } k \in \mathbb{N}. \tag{4.5}$$

Therefore Young's Inequality and (4.3) produce

$$\prod_{k=1}^{m}\left(1+\sum_{i=1}^{\infty} a_{ki}x_i\right)^{p_k} \leq \sum_{k=1}^{m} p_k\left(1+\sum_{i=1}^{\infty} a_{ki}x_i\right) = 1+\sum_{k=1}^{\infty} p_k\sum_{i=1}^{\infty} a_{ki}x_i \tag{4.6}$$

Further, let show that $\sum_{k=1}^{\infty} p_k \sum_{i=1}^{\infty} a_{ki}x_i \leq 0$. Indeed, the series $\sum_{k=1}^{\infty} p_k \sum_{i=1}^{\infty} a_{ki}x_i$ converges absolutely. From simple algebra and the fact $a_{ki} = -a_{ik}$, we have

$$\sum_{k=1}^{\infty} p_k \sum_{i=1}^{\infty} a_{ki}x_i = \sum_{i=1}^{\infty} x_i \sum_{k=1}^{\infty} a_{ki}p_k = -\sum_{i=1}^{\infty} x_i \sum_{k=1}^{\infty} a_{ik}p_k$$

Since $p \in \mathbb{P}$, then we infer that $\sum_{k=1}^{\infty} a_{ik}p_k \geq 0$ for any $i \in \mathbb{N}$. Hence, we show that

$$\sum_{k=1}^{\infty} p_k \sum_{i=1}^{\infty} a_{ki}x_i = -\sum_{i=1}^{\infty} x_i \sum_{k=1}^{\infty} a_{ik}p_k \leq 0 \tag{4.7}$$

Equation (4.6) and (4.7) yield

$$\prod_{k=1}^{m}\left(1+\sum_{i=1}^{\infty} a_{ki}x_i\right)^{p_k} \leq 1$$

Hence, from (4.4) one concludes that

$$\varphi_p\left(V(\mathbf{x})\right) \leq \varphi_p(\mathbf{x})$$

for any $\mathbf{x} \in riS$. Therefore, $\lim_{n\to\infty} \varphi_p\left(V^{(n)}(\mathbf{x})\right)$ exist. This completes the proof. □

From the last theorem, it seems hard to choose $p \in \mathbb{P}$ that satisfied condition (4.3) due to infiniteness. This problem is solved by the next theorem.

Theorem 4.3. *Let V be an infinite dimensional Volterra QSO associated with a skew-symmetric matrix $A = (a_{ki})$. Assume there exists m such that*

$$a_{ij} \geq 0 \quad for\ i \geq n+1, j = \overline{1,m} \tag{4.8}$$

Let $\mathbf{x} \in riS$, then for any $p \in \mathbb{P}^{m-1}$, the following functional

$$\varphi_p(\mathbf{x}) = \prod_{k=1}^{m} x_k^{p_k}$$

is a Lyapunov function for V.

Proof. For any $\mathbf{x} \in riS$, $\varphi_p\left(V(\mathbf{x})\right)$ can be written as follows (see (4.4))

$$\varphi_p\left(V(\mathbf{x})\right) = \varphi_p(\mathbf{x}) \prod_{k=1}^{m}\left(1+\sum_{i=1}^{\infty} a_{ki}x_i\right)^{p_k}$$

Keeping in mind (4.5), hence using Young's Inequality and fact $a_{ki} = -a_{ik}$, we have

$$\begin{aligned}\prod_{k=1}^{m}\left(1+\sum_{i=1}^{\infty} a_{ki}x_i\right)^{p_k} &\leq \sum_{k=1}^{m} p_k\left(1+\sum_{i=1}^{\infty} a_{ki}x_i\right)\\ &= 1-\sum_{i=1}^{m} x_i \sum_{k=1}^{m} a_{ik}p_k - \sum_{i=m+1}^{\infty} x_i \sum_{k=1}^{m} a_{ik}p_k \end{aligned} \tag{4.9}$$

Due to $p \in \mathbb{P}^{m-1}$ one infers

$$\sum_{i=1}^{m} x_i \sum_{k=1}^{m} a_{ik}p_k \geq 0 \tag{4.10}$$

and assumption (4.8) yields

$$\sum_{i=m+1}^{\infty} x_i \sum_{k=1}^{m} a_{ik}p_k \geq 0 \tag{4.11}$$

From (4.9), (4.10) and (4.11) we conclude that

$$\varphi_p\left(V(\mathbf{x})\right) \leq \varphi_p(\mathbf{x})$$

Therefore, the limit

$$\lim_{n\to\infty} \varphi_p\left(V^{(n)}(\mathbf{x})\right)$$

exists. This completes the proof.

Indeed, the set of Volterra QSOs that satisfies Theorem 4.3 is non-empty. Consider the following example.

Example 4.4. *Let V be a Volterra QSO associated with the following skew-symmetric matrix*

$$A = \begin{bmatrix} 0 & -a & b & a_{14} & \cdots \\ a & 0 & -c & a_{24} & \cdots \\ -b & c & 0 & a_{34} & \cdots \\ a_{41} & a_{42} & a_{43} & 0 & \cdots \\ \vdots & \vdots & \vdots & \vdots & \ddots \end{bmatrix}$$

where $0 < a, b, c \leq 1$ and $a_{ij} \geq 0$ for all $i \geq 4, j = \overline{1,3}$. One sees that the set

$$\mathbb{P}^2 = \left\{ \frac{c}{a+b+c}, \frac{b}{a+b+c}, \frac{a}{a+b+c} \right\}$$

Hence,

$$\varphi_p(\mathbf{x}) = \left(x_1^c, x_2^b, x_3^a\right)^{\frac{1}{a+b+c}}$$

is a Lyapunov function for V.

5. Acknowledgments

The present work is supported by the UAE "Start-Up" Grant, No. 31S259.

References

[1] Bernstein S 1924 *Uch. Zap. n.-i. kaf. Ukrainy* **1** 83
[2] Lyubich Y I 1992 *Mathematical structures in population genetics* (Springe, Berlin)
[3] Hofbauer J and Sigmund K 1998 *Evolutionary games and population dynamics* (Cambridge University Press)
[4] Plank M and Losert V 1995 *J. Math. Phys.* **36** 3520
[5] Takeuchi Y 1996 *Global dynamical properties of Lotka-Volterra systems* (World Scientific, Singarure)
[6] Mukhamedov F, Akin H and Temir S 2005 *J. Math. Anal. Appl.* **310** 533
[7] Mukhamedov F and Embong A F 2017 arXiv:1704.04804
[8] Ganikhodzhaev R, Mukhamedov F and Rozikov U 2011 *Inf. Dimen. Anal. Quantum Probab. Related Topics* **14** 279
[9] Mukhamedov F and Ganikhodjaev N 2015 *Quantum quadratic operators and processes*, (Springer, Berlin))
[10] Ganikhodzhaev R 1993 *Sbornik Math* **76** 489
[11] Ganikhodzhaev R N 1994 *Mateh. Notes* **56** 40
[12] Ganikhodzhaev R N and Eshmamatova D B 2006 *Vladikavkazs Math. Jour.* **8** 12
[13] Jamilov U 2012 *TWMS J. Pure Appl. Math.* **3** 28

IOP Conf. Series: Journal of Physics: Conf. Series **949** (2017) 012023 doi:10.1088/1742-6596/949/1/012023

Ergodicity of power series-maps on the simplexes of group algebras of finite groups

U. Bekbaev[1], **M. A. M. Johari**[2]

Deparment of Science in Engineering, KOE, IIUM, 50728, Malaysia[1],
Deparment of Mathematics, FS, UPM, 43400, Malaysia[2].

E-mail: `bekbaev@iium.edu.my; mamj@upm.edu.my`

Abstract. For every finite group G the ergodicity of any map $p : S \to S$ on S is shown, where $\mathcal{R}[G]$ is the real group algebra of G ,

$$S = \{x = \sum_{g\in G} x_g g \in \mathcal{R}[G] : \sum_{g\in G} x_g = 1, x_g \geq 0 \text{ for any } \quad g \in G\} - \text{the simplex},$$

$p(x) = a_0 + a_1x + a_2x^2 + ...,\ x \in S,\ a_i \geq 0$ and $\sum_{i=0}^{\infty} a_i = 1$. The regularity of this map at a given point $x \in S$ is investigated as well.

1. Introduction and preliminaries

In applications (see, for example,[1], [2]) the following questions are important. Let

$$S = \{(x_1, x_2, ..., x_m) \in \mathcal{R}^m :\ x_1 \geq 0, x_2 \geq 0, ..., x_m \geq 0,\ \sum_{i=1}^{m} x_i = 1\}$$

be the simplex and $p : S \to S$ be a map, where $\mathcal{R}$ stands for the field of real numbers. If $p^{[1]}(x) = p(x),\ p^{[i+1]}(x) = p(p^{[i]}(x))$ ($q^{(j)}(x) = \frac{1}{j}\sum_{i=1}^{j} p^{[i]}(x)$) for $j \in \mathcal{N}$) does the sequence $\{p^{[i]}(x)\}_{i\in N}$ (respectively, $\{q^{(j)}(x)\}_{j\in N}$) converge, with respect to the Euclidean topology, at a given $x \in S$? If "Yes" the map p is said to be regular (respectively, ergodic) at x. We study these problems when S is the natural simplex in the group algebra of a finite group G and p is a power series. The polynomial case of $p(t)$ has been considered in [3] and [4]. Results of this paper were announced in [5].

First let us prove some common results which will be used further. In this section let $\{p^{[i]}(t) = \sum_{k=0}^{\infty} a_k^{[i]} t^k\}_{i\in\mathcal{N}}$ be any sequence of power series such that $\sum_{k=0}^{\infty} |a_k^{[i]}| < \infty$ and $\lim_{i\to\infty} a^{[i]} = 0$, where $a^{[i]} = \sup\{|a_k^{[i]}| : k \in \mathcal{N}\}$. It should be noted that the sequence of series $\{q^{(j)}(t)\}_{j\in\mathcal{N}} = \{\sum_{k=1}^{\infty} b_k^{(j)} t^k\}_{j\in\mathcal{N}}$ also holds the same property i.e., $\lim_{j\to\infty} b^{(n)} = 0$, where $b^{(j)} = \sup\{|b_k^{(j)}| : k \in \mathcal{N}\}$ and $q^{(j)}(t) = \frac{1}{j}\sum_{i=1}^{j} p^{[i]}(t)$. Let $p_0^{[i]}(t)$ stand for $p^{[i]}(t) - a_0^{[i]}$.

For the next two Lemmas A stands for a finite dimensional real mono associative algebra that is, any sub algebra of A generated by a single element is associative.

Lemma 1. *If* $x \in A$ *and* $\lim_{i\to\infty} x^i = 0$ *then* $\lim_{i\to\infty} p_0^{[i]}(x) = 0$ *as well.*

Proof of this lemma is not difficult due to the finite dimension of A, the assumption that $\lim_{i\to\infty} a^{[i]} = 0$ and the equality $\lim_{i\to\infty} x^i = 0$ or see [3].

Lemma 2. *If $\lim_{i\to\infty} x^i = x_0$, where $x \in A$, then the existence of*

$$\lim_{i\to\infty} p^{[i]}(x)$$

is equivalent to the existence of $\lim_{i\to\infty}[a_0^{[i]}e + (p^{[i}(1) - a_0^{[i]})x_0]$, where e stands for the identity element of A. Moreover, when it is the case then

$$\lim_{i\to\infty} p^{[i]}(x) = \lim_{i\to\infty} [a_0^{[i]}e + (p^{[i]}(1) - a_0^{[i]})x_0].$$

Proof. Due to $x_0 = \lim_{m\to\infty} x^{m+1} = x \lim_{m\to\infty} x^m = (\lim_{m\to\infty} x^m)x$ and $x_0 = \lim_{m\to\infty} x^{2m} = \lim_{m\to\infty} x^m \lim_{m\to\infty} x^m$ one has $x_0 = xx_0 = x_0x = x_0^2$, $x_0(x-x_0) = (x-x_0)x_0 = 0$. Therefore,

$$x^i = (x_0 + (x - x_0))^i = x_0^i + (x - x_0)^i = x_0 + (x - x_0)^i$$

and $\lim_{i\to\infty}(x - x_0)^i = 0$. Moreover,

$$p^{[i]}(x) = a_0^{[i]}e + \sum_{j=1}^{\infty} a_j^{[i]}(x_0 + (x - x_0))^j = a_0^{[i]}e + \sum_{j=1}^{\infty} a_j^{[i]}(x_0 + (x - x_0)^j) =$$

$$a_0^{[i]}e + p_0^{[i]}(1)x_0 + p_0^{[i]}(x - x_0) = a_0^{[i]}e + p_0^{[i]}(1)x_0 + p_0^{[i]}(x - x_0)$$

and according to Lemma 1 $\lim_{i\to\infty} p_0^{[i]}(x - x_0) = 0$. Therefore the existence of $\lim_{i\to\infty} p^{[i]}(x)$ and the existence of $\lim_{i\to\infty}[a_0^{[i]}e + (p^{[i]}(1) - a_0^{[i]})x_0]$ are same.

Further it is assumed that G is a finite group, $A = \mathcal{R}[G]$ is its group algebra [6].

Define $L : \mathcal{R}[G] \to \mathcal{R}$ as $L(x) = \sum_{g\in G} x_g$ for any $x = \sum_{g\in G} x_g g \in \mathcal{R}[G]$ and

$$S = \{x \in \mathcal{R}[G] : L(x) = 1, x_g \geq 0 \text{ for any } \quad g \in G\}.$$

Note that S is closed with respect to the multiplication and $L(xy) = L(x)L(y)$ for any $x, y \in \mathcal{R}[G]$.

Let $\mathrm{Supp}(x) = \{g : x_g \neq 0\}$ and c stand for $\frac{1}{|G|}\sum_{g\in G} g \in \overset{\circ}{S}$, where

$$\overset{\circ}{S} = \{x = \sum_{g\in G} x_g g \in S : x_g > 0, g \in G\},$$

and $|G|$ stands for the cardinality of G. It is clear that $cS = Sc = \{c\}$.

The following result is evident.

Proposition 1. *For any two elements $y, z \in \mathcal{R}[G]$ with nonnegative components the following relations hold true.*

1. $Supp(y) \cup Supp(z) = Supp(y + z)$,

2. $Supp(y)Supp(z) = Supp(yz)$, where $Supp(y)Supp(z)$ means $\{gh : y \in Supp(y), h \in Supp(z)\}$. In particular, $max\{|Supp(y)|, |Supp(z)|\} \leq |Supp(yz)|$, and, if $y_e \neq 0, z_e \neq 0$ then $Supp(y) \cup Supp(z) \subset Supp(yz)$.

Further x stands for a fixed element of S,

$$n_x = \min\{k \in \mathcal{N} : (x^k)_e \neq 0\} = \min\{k \in \mathcal{N} : e \in \mathrm{Supp}(x^k)\},$$

the notation $G_x = \langle \mathrm{Supp}(x^{n_x}) \rangle$ means the subgroup generated by $\mathrm{Supp}(x^{n_x})$, $c_x = \frac{1}{|G_x|}\sum_{g\in G_x} g$ and

$$m_x = \min\{k \in \mathcal{N} : x^k \in \mathcal{R}[G_x]\} = \min\{k \in \mathcal{N} : \mathrm{Supp}(x^k) \subset G_x\}.$$

Note that n_x is divisible by m_x whenever $x \in S$.

Proposition 2. *1. The system $\{e, x, x^2, ..., x^{n_x-1}\}$ is linearly independent.*
2. For any nonnegative integers r_1, r_2 the equality $c_x x^{r_1} = c_x x^{r_2}$ holds true if and only if $r_1 = r_2 mod m_x$.

Proof. If the system $\{e, x, x^2, ..., x^{n_x-1}\}$ is linear dependent then there exist $m < n_x$ and $a_{m-1}, a_{m-2}, ..., a_0 \in \mathcal{R}$ such that

$$x^m = a_{m-1}x^{m-1} + a_{m-2}x^{m-2} + ... + a_1 x + a_0 e.$$

Multiplication it by x^{n_x-m} implies that x^{n_x} is a linear combination of $\{x, x^2, ..., x^{n_x-1}\}$ and it is a contradiction due to $(x^{n_x})_e \neq 0$ and $x_e = (x^2)_e = ... = (x^{n_x-1})_e = 0$. It is the proof of the first part of Proposition 2.

If $n = km_x + r$, where $0 \le r < m_x$, then $c_x x^n = c_x (x^{m_x})^k x^r = c_x x^r$ because of $x^{m_x} \in \mathcal{R}[G_x]$, $x \in S$. If $0 \le r_1 < r_2 < m_x$ and $c_x x^{r_1} = c_x x^{r_2}$ then

$$c_x x^{r_1+m_x-r_2} = c_x x^{r_2+m_x-r_2} = c_x x^{m_x} \in \mathcal{R}[G_x]$$

which evidently implies that $r_1 + m_x - r_2 < m_x$ and $x^{r_1+m_x-r_2} \in \mathcal{R}[G_x]$. This contradiction completes the proof of the second part of Proposition 2.

Proposition 3. *For $y \in S$ with $y_e \neq 0$ the following are equivalent.*

$$1.\ \ G_y = G.\ \ 2.\ \ y^n \in \overset{\circ}{S}\ \text{for some}\ \ n \in \mathcal{N}.$$

Proof. Due to Proposition 1 one has $\mathrm{Supp}(y^n) \subset \mathrm{Supp}(y^m)$ whenever $n \le m$. If $G_y = G$ then for every $g \in G$ one can find such $n \in \mathcal{N}$ and $g_1, g_2, ..., g_n \in \mathrm{Supp}(y)$ for which $g = g_1 g_2 ... g_n$. Therefore, $(y^n)_g \neq 0$. Due to finiteness of G it implies that $y^m \in \overset{\circ}{S}$ for all big enough $m \in \mathcal{N}$. The implication of second condition the first one is evident.

Further $S(x)$ stands for the set of limits of all converging subsequences of $\{x^i\}_{i\in\mathcal{N}}$.

Theorem 1. *The equalities*

$$xc_x = c_x x, \quad S(x) = \{c_x x^r : 0 \le r < m_x\}$$

hold true whenever $x \in S$. In particular, $S(x) = \{c_x\}$ if and only if $x \in \mathcal{R}[G_x]$.

Proof. Consider the following expansion of the algebra

$$\mathcal{R}[G] = \mathcal{R}c \oplus I,$$

where $I = \{x \in \mathcal{R}[G] : L(x) = 0\}$. With respect to this expansion one has the following representations $S = c + S_0$, $\overset{\circ}{S} = c + \overset{\circ}{S}_0$, where

$$S_0 = \{x = \sum_{g\in G} x_g g \in I : -\frac{1}{|G|} \le x_g \le 1 - \frac{1}{|G|} \quad \text{for any} \quad g \in G\} \text{ and}$$

$$\overset{\circ}{S}_0 = \{x = \sum_{g\in G} x_g g \in I : -\frac{1}{|G|} < x_g < 1 - \frac{1}{|G|} \quad \text{for any} \quad g \in G\}.$$

First let us consider the case $x \in \overset{\circ}{S}$. In this case, $G_x = G$ and $x = c + x_0$, where $x_0 \in \overset{\circ}{S}_0$. It is clear that

$$x^m = c + x_0^m$$

for any $m \in \mathcal{N}$. Due to the compactness of S there is a subsequence $\{m_k\}_{k\in\mathcal{N}}$ of $\{m\}_{m\in\mathcal{N}}$ such that $\{x^{m_k}\}_{k\in\mathcal{N}} = \{c + x_0^{m_k}\}_{k\in\mathcal{N}}$ converges. But for the given x_0 one can find number $\lambda > 1$ such that $\lambda x_0 \in \overset{\circ}{S}_0$ so that $\overline{x} = c + \lambda x_0 \in \overset{\circ}{S}$. Once again due to the compactness of S one can find subsequence $\{m_{k_l}\}_{l\in\mathcal{N}}$ of $\{m_k\}_{k\in\mathcal{N}}$ such that the sequence $\{\overline{x}^{m_{k_l}}\}_{l\in\mathcal{N}} = \{c + \lambda^{m_{k_l}} x_0^{m_{k_l}}\}_{l\in\mathcal{N}}$ also converges. But the convergence of sequences $\{x^{m_{k_l}}\}_{l\in\mathcal{N}} = \{c + x_0^{m_{k_l}}\}_{l\in\mathcal{N}}$, $\{\overline{x}^{m_{k_l}}\}_{l\in\mathcal{N}} = \{c + \lambda^{m_{k_l}} x_0^{m_{k_l}}\}_{l\in\mathcal{N}}$ and the inequality $\lambda > 1$ imply that

$$\lim_{k\to\infty} x_0^{m_k} = 0, \qquad \lim_{k\to\infty} x^{m_k} = c.$$

Now assume that the sequence $\{x^m\}_{m\in\mathcal{N}}$ does not converge to c. In this case one can find a subsequence $\{x^{m_s}\}_{s\in\mathcal{N}}$ of it which itself has no subsequence converging to c. But due to the compactness of S one can find a converging subsequence $\{x^{m_{s_k}}\}_{k\in\mathcal{N}}$ of $\{x^{m_s}\}_{s\in\mathcal{N}}$. As it has been shown already in such a case

$$\lim_{k\to\infty} x^{m_{s_k}} = c.$$

This contradiction shows that $\lim_{m\to\infty} x^m = c$ and completes the proof of the theorem in the case $x \in \overset{\circ}{S}$.

Now let us consider case $x \in S \setminus \overset{\circ}{S}_0$ such that $x^n \in \overset{\circ}{S}_0$ for some fixed $n \in \mathcal{N}$. Due to the proof of the previous case $\lim_{m\to\infty} x^{nm} = c$. Let $\{m_k\}_{k\in\mathcal{N}}$ be any subsequence of $\{m\}_{m\in\mathcal{N}}$, for which $\{x^{m_k}\}_{k\in\mathcal{N}}$ converges, and $m_k = nq_k + r_k$, where q_k is an integer and $0 \le r_k < n$ for any $k \in \mathcal{N}$. In this case $\lim_{k\to\infty} x^{m_k} = \lim_{k\to\infty} x^{nq_k + r_k} = \lim_{k\to\infty}[(x^n)^{q_k} x^{r_k}] = \lim_{k\to\infty}[x^{r_k}(x^n)^{q_k}]$. Without loss of generality one can assume that $r_k = r$ for all $k \in \mathcal{N}$. Therefore $\lim_{k\to\infty} x^{m_k} = (\lim_{k\to\infty}(x^n)^{q_k})x^r = cx^r = x^r c = c$ as far as $x = c + x_0$ and $cx_0 = x_0 c = 0$.

Now let x be any element of S. One has $x^{m_x} \in \mathcal{R}[G_x]$ and due to Proposition 3 there is $n \in \mathcal{N}$ for which

$$x^{nn_x} \in \overset{\circ}{S}_{G_x} = \{y = \sum_{g\in G_x} y_g g : L(y) = 1 \quad \text{and} \quad y_g > 0 \text{ for any } \quad g \in G_x\}.$$

It implies that $x^{mm_x} \in \overset{\circ}{S}_{G_x}$ for some $m \in \mathcal{N}$ a far as m_x divides n_x. Therefore due to the first part of the proof

$$\lim_{k\to\infty} (x^{m_x})^k = c_x.$$

Assume that a sequence $\{x^{m_k}\}_{k\to\infty}$ converges and $m_k = m_x q_k + r_k$, where $0 \le r_k < m_x$ for any $k \in \mathcal{N}$. One can assume that $r_k = r$ for all $k \in \mathcal{N}$. Therefore

$$\lim_{k\to\infty} x^{m_k} = \lim_{k\to\infty} [(x^{m_x})^{q_k} x^r] = \lim_{k\to\infty} [(x^{m_x})^{q_k}]x^r = x^r \lim_{k\to\infty} [(x^{m_x})^{q_k}] = c_x x^r = x^r c_x$$

So

$$S(x) = \{c_x x^r : 0 \le r < m_x\} = \{x^r c_x : 0 \le r < m_x\}$$

and due to Proposition 2 one has $|S(x)| = 1$ if and only if $m_x = 1$ i.e. $x \in \mathcal{R}[G_x]$. This is the end of proof of Theorem 1.

Proposition 4. *1. The inequality $n_{xc_x} \le m_x$ is true for any $x \in S$.*
2. The equalities $n_x = m_x = m_{xc_x}$ hold true whenever $n_{xc_x} = n_x$.

Proof. To show the inequality $n_{xc_x} \leq m_x$ assume that $\mathrm{Supp}(x^m) \subset G_x$ for some $m \in \mathcal{N}$. In this case $e \in \mathrm{Supp}((xc_x)^m)$ as far as $xc_x = c_x x$ due to Theorem 1 and $e \in \mathrm{Supp}(x)^m G_x = \mathrm{Supp}((xc_x)^m)$ which implies that $n_{xc_x} \leq m$ i.e. $n_{xc_x} \leq m_x$.

Now let us show the equalities $n_x = m_x = m_{xc_x}$. According to the first part one has $n_{xc_x} \leq m_x \leq n_x$ and therefore $n_{xc_x} = m_x = n_x$. Due to $n_{xc_x} = n_x$ one has $G_{xc_x} = \langle \mathrm{Supp}((xc_x)^{n_{xc_x}})\rangle = \langle \mathrm{Supp}(x^{n_x})\mathrm{Supp}(c_x)\rangle = \langle \mathrm{Supp}(x^{n_x})G_x\rangle = G_x$ as far as $G_x = \langle \mathrm{Supp}(x^{n_x})\rangle$. The equality

$$\mathrm{Supp}((xc_x)^n) = \mathrm{Supp}(x^n)G_x$$

for any $n \in \mathcal{N}$ implies that inclusion $\mathrm{Supp}((xc_x)^n) \subset G_{xc_x} = G_x$ holds if and only if $\mathrm{Supp}(x^n) \subset G_x$ and therefore $m_x = m_{xc_x}$.

Lemma 3. *For any $x \in S$ the existence of $\lim_{n\to\infty} p^{[n]}(x)$ and $\lim_{n\to\infty} p^{[n]}(xc_x)$ are same and*

$$\lim_{n\to\infty} p^{[n]}(x) = \lim_{n\to\infty} p^{[n]}(xc_x),$$

provided that the limits exist.

Proof. One can represent $p^{[n]}(t)$ in the following form

$$p^{[n]}(t) = \sum_{k=0}^{\infty} a_k^{[n]} t^k = \sum_{r=0}^{m_x-1} a_r^{[n]} t^r + \sum_{r=0}^{m_x-1} t^r \sum_{k=1}^{\infty} a_{km_x+r}^{[n]} t^{km_x}$$

Therefore if $x^{m_x} = c_x + x_1$ then

$$p^{[n]}(x) = \sum_{r=0}^{m_x-1} a_r^{[n]} x^r + \sum_{r=0}^{m_x-1} x^r \sum_{k=1}^{\infty} a_{km_x+r}^{[n]} (c_x + x_1)^k =$$

$$a_0^{[n]} e + \sum_{r=1}^{m_x-1} a_r^{[n]} x^r + c_x \sum_{r=0}^{m_x-1} x^r \sum_{k=1}^{\infty} a_{km_x+r}^{[n]} + \sum_{r=0}^{m_x-1} x^r \sum_{k=1}^{\infty} a_{km_x+r}^{[n]} x_1^k =$$

$$c_x \sum_{r=0}^{m_x-1} x^r \sum_{k=0}^{\infty} a_{km_x+r}^{[n]} + a_0^{[n]}(e - c_x) + (e - c_c) \sum_{r=1}^{m_x-1} a_r^{[n]} x^r + \sum_{r=0}^{m_x-1} x^r \sum_{k=1}^{\infty} a_{km_x+r}^{[n]} x_1^k$$

and

$$p^{[n]}(xc_x) = \sum_{r=0}^{m_x-1} a_r^{[n]} (xc_x)^r + \sum_{r=0}^{m_x-1} (xc_x)^r \sum_{k=1}^{\infty} a_{km_x+r}^{[n]} ((c_x + x_1)c_x)^k =$$

$$a_0^{[n]} e + c_x \sum_{r=1}^{m_x-1} a_r^{[n]} x^r + c_x \sum_{r=0}^{m_x-1} x^r \sum_{k=1}^{\infty} a_{km_x+r}^{[n]}$$

$$= c_x \sum_{r=0}^{m_x-1} x^r \sum_{k=0}^{\infty} a_{km_x+r}^{[n]} + a_0^{[n]}(e - c_x)$$

Due to Lemma 1 one has

$$\lim_{n\to\infty} \sum_{r=1}^{m_x-1} a_r^{[n]} x^r = \lim_{n\to\infty} \sum_{r=0}^{m_x-1} x^r \sum_{k=1}^{\infty} a_{km_x+r}^{[n]} x_1^k = 0$$

as far as $\lim_{k\to\infty} x_1^k = 0$ and $\lim_{n\to\infty} a^{[n]} = 0$.

Therefore either

$$\lim_{n\to\infty} p^{[n]}(x) = \lim_{n\to\infty} p^{[n]}(xc_x) = \lim_{n\to\infty} (c_x \sum_{r=0}^{m_x-1} x^r \sum_{k=0}^{\infty} a_{km_x+r}^{[n]} + a_0^{[n]}(e - c_x))$$

IOP Conf. Series: Journal of Physics: Conf. Series **949** (2017) 012023 doi:10.1088/1742-6596/949/1/012023

or none of these limits exist.

This Lemma, due to $n_{xc_x} \leq m_x$, shows that after finite number of steps one is able to reduce the convergence of $\{p^{[n]}(t)\}_{n\in\mathcal{N}}$ at a given point of S to its convergence at a point $x \in S$ for which $n_{xc_x} = n_x$.

Theorem 2. *The existence of limit* $\lim_{n\to\infty} p^{[n]}(x)$, *where* $x \in S$, *such that* $c_x \neq e$ *and* $n_{xc_x} = n_x$, *is equivalent to the existence of limits* $\lim_{n\to\infty} a_0^{[n]}$, $\lim_{n\to\infty} \sum_{k=0}^{\infty} a_{km_x+r}^{[n]}$ *for all* $0 \leq r < m_x$. *In the case of* $x \in S$, *where* $c_x = e$, $n_{xc_x} = n_x$, *the existence of* $\lim_{n\to\infty} p^{[n]}(x)$ *is equivalent to the existence of* $\lim_{n\to\infty} \sum_{k=0}^{\infty} a_{km_x+r}^{[n]}$ *for all* $0 \leq r < m_x$ *and in this case*

$$\lim_{n\to\infty} p^{[n]}(x) = c_x \sum_{r=0}^{m_x-1} x^r \lim_{n\to\infty} \sum_{k=0}^{\infty} a_{km_x+r}^{[n]} + (e - c_x) \lim_{n\to\infty} a_0^{[n]}.$$

Proof. One has $m_x = n_x = n_{xc_x} = m_{xc_x}$ due to Proposition 4. In $c_x \neq e$ case the system

$$e - c_x, c_x, xc_x, x^2c_x, ..., x^{n_x-1}c_x$$

is linear independent. Indeed, if

$$x^m c_x = a_{m-1}x^{m-1}c_x + a_{m-2}x^{m-2}c_x + ... + a_1 xc_x + a_0 c_x + d(e - c_x)$$

, where $m < n_x$ and $a_{m-1}, a_{m-2}, ..., a_0, d \in \mathcal{R}$, then multiplication it by $(e - c_x)$ implies that $d = 0$ and

$$x^m c_x = a_{m-1}x^{m-1}c_x + a_{m-2}x^{m-2}c_x + ... + a_1 xc_x + a_0 c_x.$$

Multiply it by x^{n_x-m} to see that $x^{n_x}c_x$ is a linear combination of $\{xc_x, x^2c_x, ..., x^{n_x-1}c_x\}$, a contradiction due to $(xc_x)_e^{n_x} = (x^{n_x}c_x)_e \neq 0$,

$$(xc_x)_e = (x^2c_x)_e = ... = (x^{n_x-1}c_x)_e = 0.$$

Note that $c_x = e$ case one still has the independence of

$$c_x, xc_x, x^2c_x, ..., x^{n_x-1}c_x$$

Due to Lemma 3 the existence of $\lim_{n\to\infty} p^{[n]}(x)$ and the existence of

$$\lim_{n\to\infty} (c_x \sum_{r=0}^{m_x-1} x^r \sum_{k=0}^{\infty} a_{km_x+r}^{[n]} + (e - c_x)a_0^{[n]})$$

are same. Therefore, due to the linear independence of $c_x, xc_x, x^2c_x, ..., x^{n_x-1}c_x$, the last limit exists if and only if $\lim_{n\to\infty} a_0^{[n]}$, $\lim_{n\to\infty} \sum_{k=1}^{\infty} a_{km_x+r}^{[n]}$ for all $0 \leq r < m_x$ exist.

2. Ergodicity of a power series-map

Lemma 4. *For nonnegative* $a_1, a_2, a_3, ... \in \mathcal{R}$, *for which* $a_1 + a_2 + a_3 + ... = 1$, *and any* $k \geq 2$, $2 \leq i \leq k$ *the following inequality*

$$\sum_{j_1+j_2+...+j_i=k} a_{j_1}a_{j_2}...a_{j_i} \leq (1 - a_k)a$$

is valid, where $a = sup\{a_i\}_{i\in\mathcal{N}}$, *the summation is taken over all* $(j_1, j_2, ..., j_i) \in \mathcal{N}^i$ *for which* $j_1 + j_2 + ... + j_i = k$. *In particular*

$$\sum_{j_1+j_2+...+j_i=k} a_{j_1}a_{j_2}...a_{j_i} \leq a$$

whenever $k \geq 1$ *and* $1 \leq i \leq k$.

Proof. $\sum_{j_1+j_2+...+j_i=k} a_{j_1}a_{j_2}...a_{j_i} =$

$$\sum_{j_1=1}^{k-(i-1)} a_{j_1} \sum_{j_2=1}^{k-(i-2)-j_1} a_{j_2} \sum_{j_3=1}^{k-(i-3)-(j_1+j_2)} a_{j_3}... \sum_{j_{i-1}=1}^{k-1-(j_1+j_2+...+j_{i-2})} a_{j_{i-1}} a_{k-(j_1+j_2+...+j_{i-1})}$$

$$\leq a \sum_{j_1=1}^{k-(i-1)} a_{j_1} \sum_{j_2=1}^{k-(i-2)-j_1} a_{j_2} \sum_{j_3=1}^{k-(i-3)-(j_1+j_2)} a_{j_3}... \sum_{j_{i-1}=1}^{k-1-(j_1+j_2+...+j_{i-2}} a_{j_{i-1}}$$

Due to $a_1 + a_2 + a_3 + ... = 1$ one has

$$\sum_{j_1=1}^{k-(i-1)} a_{j_1} \sum_{j_2=1}^{k-(i-2)-j_1} a_{j_2} \sum_{j_3=1}^{k-(i-3)-(j_1+j_2)} a_{j_3}... \sum_{j_{i-1}=1}^{k-1-(j_1+j_2+...+j_{i-2})} a_{j_{i-1}}$$

$$\leq \sum_{j_1=1}^{k-(i-1)} a_{j_1} = 1 - \sum_{j=k-i+2}^{\infty} a_j \leq 1 - a_k$$

, that is

$$\sum_{j_1+j_2+...+j_i=k} a_{j_1}a_{j_2}...a_{j_i} \leq (1-a_k)a$$

Now the inequality

$$\sum_{j_1+j_2+...+j_i=k} a_{j_1}a_{j_2}...a_{j_i} \leq a$$

is clear.

For any power series $p(t) = \sum_{k=0}^{\infty} a_k t^k = a_0 + a_1 t + a_2 t^2 + ...$, where $0 \leq a_i < 1$, $p(1) = a_0 + a_1 + a_2 + ... = 1$, one can consider its iterations $p^{[n+1]}(t) = p(p^{[n]}(t))$, where $n \in \mathcal{N}$, $p^{[1]}(t) = p(t)$. It is evident that coefficients of power series

$$p^{[n]}(t) = \sum_{k=0}^{\infty} a_k^{[n]} t^k$$

are nonnegative and $p^{[n]}(1) = \sum_{k=0}^{\infty} a_k^{[n]} = 1$. The following result tells us about the behavior of $a_0^{[n]}$ and $a^{[n]} = Sup\{a_k^{[n]} : k \in \mathcal{N}\}$.

Theorem 3. *The sequence* $\{a_0^{[n]}\}_{n\in\mathcal{N}}$ *is monotone increasing,* $\{a^{[n]}\}$ *is a monotone decreasing sequence and* $\lim_{n\to\infty} a^{[n]} = 0$

Proof. Let $p^{[n]}(t) = a_0^{[n]} + p_0^{[n]}(t)$, where $a_0^{[n]}$ stands for the constant term of $p^{[n]}(t)$. Due to

$$p^{[n+1]}(t) = p(p^{[n]}(t)) = \sum_{i=0}^{\infty} a_i (a_0^{[n]} + p_0^{[n]}(t))^i$$

and the binomial formula, one has

$$p^{[n+1]}(t) = \sum_{i=0}^{\infty} a_i \sum_{j=0}^{i} a_0^{[n]i-j} p_0^{[n]}(t)^j \binom{i}{j} = \sum_{i=0}^{\infty} a_i [a_0^{[n]i} + \sum_{j=1}^{i} a_0^{[n]i-j} p_0^{[n]}(t)^j \binom{i}{j}]$$

$$= \sum_{i=0}^{\infty} a_i a_0^{[n]i} + \sum_{i=1}^{\infty} a_i \sum_{j=1}^{i} a_0^{[n]i-j} p_0^{[n]}(t)^j \binom{i}{j} = \sum_{i=0}^{\infty} a_i a_0^{[n]i} + \sum_{j=1}^{\infty} (\sum_{i=j}^{\infty} a_i a_0^{[n]i-j} \frac{i!}{(i-j)!}) \frac{p_0^{[n]}(t)^j}{j!}$$

$$= p(a_0^{[n]}) + \sum_{j=1}^{\infty} \frac{p^{(j)}(a_0^{[n]})}{j!} p_0^{[n]}(t)^j.$$

So

$$a_0^{[n+1]} = p(a_0^{[n]})$$

and

$$a_k^{[n+1]} = \sum_{i=1}^{\infty} \frac{p^{(i)}(a_0^{[n]})}{i!} \sum_{j_1+j_2+...+j_i=k} a_{j_1}^{[n]} a_{j_2}^{[n]} ... a_{j_i}^{[n]}, \qquad (1)$$

where $p^{(i)}$ stands for i-th derivative of $p(t)$. The sequence $\{a_0^{[n]}\}_{n\in\mathcal{N}}$ is monotone increasing due to $a_0^{[n+1]} = p(a_0^{[n]})$ and the monotonicity of $p(t)$ on [0,1). The equality $p(a) = a$ for $a = \lim_{n\to\infty} a_0^{[n]}$ is evident. In $a = 1$ case $\lim_{n\to\infty} a^{[n]} = 0$ as far as $a_k^{[n]} \leq 1 - a_0^{[n]}$ for any $k \in \mathcal{N}$, so the theorem is true.

For the function $f(t) = p(t) - t$ one has $f'(t)$ is a monotone increasing on [0,1), $f(0) = a_0 \geq 0$, $f(1) = 0$, $f(a) = 0$ and $f'(0) = a_1 - 1 < 0$. In the case of $f'(t) \leq 0$ for all $t < 1$ the function $y = f(t)$ is a positive, strictly decreasing on $[0, 1)$ and $\lim_{n\to\infty} a_0^{[n]} = a = 1$, therefore in this case the theorem is true. So it is enough to consider $\lim_{t\to 1} f'(t) > 0$ case. In this case there exists an unique $0 < b < 1$ such that $f'(b) = 0$ and $0 \leq a < b$. It implies that in this case $f'(a) < 0$ i.e. $p'(a) < 1$.

For any $k \geq 2$ due to (1) one has

$$a_k^{[n+1]} = p'(a_0^{[n]})a_k^{[n]} + \sum_{i=2}^{\infty} \frac{p^{(i)}(a_0^{[n]})}{i!} \sum_{j_1+j_2+...+j_i=k} a_{j_1}^{[n]} a_{j_2}^{[n]} ... a_{j_i}^{[n]} =$$

$$p'(a_0^{[n]})a_k^{[n]} + \sum_{i=2}^{\infty} \frac{p^{(i)}(a_0^{[n]})}{i!} (1 - a_0^{[n]})^i \sum_{j_1+j_2+...+j_i=k} \frac{a_{j_1}^{[n]}}{1 - a_0^{[n]}} \frac{a_{j_2}^{[n]}}{1 - a_0^{[n]}} \cdots \frac{a_{j_i}^{[n]}}{1 - a_0^{[n]}}$$

and therefore due to Lemma 4, one has

$$a_k^{[n+1]} \leq p'(a_0^{[n]})a_k^{[n]} + \sum_{i=2}^{\infty} \frac{p^{(i)}(a_0^{[n]})}{i!} (1 - a_0^{[n]})^i (1 - \frac{a_k^{[n]}}{1 - a_0^{[n]}}) \sup_{i\geq 1} \frac{a_i^{[n]}}{1 - a_0^{[n]}} =$$

$$p'(a_0^{[n]})a_k^{[n]} + (1 - \frac{a_k^{[n]}}{1 - a_0^{[n]}}) \frac{a^{[n]}}{1 - a_0^{[n]}} \sum_{i=2}^{\infty} \frac{p^{(i)}(a_0^{[n]})}{i!} (1 - a_0^{[n]})^i$$

But

$$\sum_{i=2}^{\infty} \frac{p^{(i)}(a_0^{[n]})}{i!} (1 - a_0^{[n]})^i = 1 - p(a_0^{[n]}) - (1 - a_0^{[n]})p'(a_0^{[n]}),$$

in particular $1 - p(a_0^{[n]}) - (1 - a_0^{[n]})p'(a_0^{[n]}) \geq 0$ according to
$1 = p(1) = p(a_0^{[n]} + 1 - a_0^{[n]}) = \sum_{i=0}^{\infty} a_i (a_0^{[n]} + (1 - a_0^{[n]}))^i = \sum_{i=0}^{\infty} a_i \sum_{j=0}^{i} \binom{i}{j} (a_0^{[n]})^{i-j} (1 - a_0^{[n]})^j$
$= \sum_{i=0}^{\infty} (\sum_{i=j}^{\infty} \binom{i}{j} a_i (a_0^{[n]})^{i-j})(1 - a_0^{[n]})^j = \sum_{j=0}^{\infty} \frac{p^{(j)}(a_0^{[n]})}{j!} (1 - a_0^{[n]})^j$ and therefore

$$a_k^{[n+1]} \leq p'(a_0^{[n]})a_k^{[n]} + (1 - \frac{a_k^{[n]}}{1 - a_0^{[n]}}) \frac{a^{[n]}}{1 - a_0^{[n]}} (1 - p(a_0^{[n]}) - (1 - a_0^{[n]})p'(a_0^{[n]})) \qquad (2).$$

Due to (1) one has

$$a_1^{[n+1]} = p'(a_0^{[n]})a_1^{[n]} \leq p'(a_0^{[n]})a_1^{[n]} + (1 - \frac{a_1^{[n]}}{1-a_0^{[n]}})\frac{a^{[n]}}{1-a_0^{[n]}}(1 - p(a_0^{[n]}) - (1-a_0^{[n]})p'(a_0^{[n]}))$$

, so inequality (2) is valid at $k = 1$ as well.

Let us show that the sequence $\{a^{[n]}\}_{n\in\mathcal{N}}$ is monotone decreasing. Due to (2) one has

$$a_k^{[n+1]} \leq p'(a_0^{[n]})a^{[n]} + \frac{a^{[n]}}{1-a_0^{[n]}}(1 - p(a_0^{[n]}) - (1-a_0^{[n]})p'(a_0^{[n]})) = a^{[n]}\frac{1-p(a_0^{[n]})}{1-a_0^{[n]}} \leq a^{[n]},$$

where $k \in \mathcal{N}$, and therefore $a^{[n+1]} \leq a^{[n]}$. So there exists

$$\lim_{n\to\infty} a^{[n]} = \lambda \geq 0.$$

Similarly, $a_k^{[n+1]} \leq a_k^{[n]}(p'(a_0^{[n]})-$

$$\frac{\lambda}{(1-a_0^{[n]})^2}(1 - p(a_0^{[n]}) - (1-a_0^{[n]})p'(a_0^{[n]}))) + \frac{a^{[n]}}{1-a_0^{[n]}}(1 - p(a_0^{[n]}) - (1-a_0^{[n]})p'(a_0^{[n]})). \quad (3)$$

It is known that $a_0^{[n]}$ tends to a from the left side. Therefore for all n big enough either

$$\text{I.} \quad g(a_0^{[n]}) \geq 0 \quad \text{or} \quad \text{II.} \quad g(a_0^{[n]}) < 0,$$

where $g(t) = p'(t) - \lambda\frac{(1-p(t)-(1-t)p'(t))}{(1-t)^2}$.

I. In this case for n big enough inequality (3) says that

$$a^{[n+1]} \leq a^{[n]}g(a_0^{[n]}) + \frac{a^{[n]}}{1-a_0^{[n]}}(1 - p(a_0^{[n]}) - (1-a_0^{[n]})p'(a_0^{[n]}))$$

and therefore

$$\lambda \leq \lambda g(a) + \frac{\lambda}{1-a}(1-p'(a))(1-a) = \lambda(1 - \frac{\lambda}{1-a}(1-p'(a)))$$

If $\lambda \neq 0$ one gets contradiction $0 \leq -\frac{\lambda}{1-a}(1-p'(a))$ as far as $1 - p'(a) > 0$.

II. Similarly, for n big enough inequality (3) implies that

$$a^{[n+1]} \leq \frac{a^{[n]}}{1-a_0^{[n]}}(1 - p(a_0^{[n]}) - (1-a_0^{[n]})p'(a_0^{[n]}))$$

and therefore

$$\lambda \leq \lambda(1 - p'(a)).$$

Once again, if $\lambda \neq 0$ one gets contradiction $0 \leq -p'(a)$.

Corollary 1. *If $p(t)$ is as in Theorem 3 and $q^{(n)}(t) = \frac{1}{n}\sum_{i=1}^{n} p^{[i]}(t) = \sum_{i=0}^{\infty} q_i^{(n)}t^i$ then the sequence $\{q_0^{(n)}\}_{n=1}^{\infty}$ is monotone increasing and $\lim_{n\to\infty} q^{(n)} = 0$, where*

$$q^{(n)} = Sup\{q_i^{(n)} : i \in \mathcal{N}\}$$

Corollary 2. *If $-1 < a_i < 1$ for all $i \in \mathcal{N} \cup \{0\}$, $\sum_{i=0}^{\infty} |a_i| \leq 1$, $p(t) = \sum_{k=0}^{\infty} a_k t^k$, $p^{[1]}(t) = p(t)$, $p^{[n+1]}(t) = p(p^{[n]}(t))$, $p^{[n]}(t) = \sum_{k=0}^{\infty} a_k^{[n]} t^k$ and $a^{[n]} = Sup\{|a_k^{[n]}| : k \in \mathcal{N}\}$ then $\lim_{n\to\infty} a^{[n]} = 0$. If $\lim_{n\to\infty} p^{[n]}(0) = \lim_{n\to\infty} a_0^{[n]} = a$ exists then $\lim_{n\to\infty} p^{[n]}(t) = a$ for any $t \in (-1,1)$*

Theorem 2 justifies investigation of the following question. Let $p(t) = \sum_{i=0}^{\infty} a_i t^i$, where $0 \leq a_i < 1$ whenever $i \in \mathcal{N} \cup \{0\}$ and $p(1) = 1$, be a power series, m be a fixed natural number. When does

$$\lim_{n\to\infty} \sum_{k=0}^{\infty} a_{km+r}^{[n]}$$

exist for all $0 \leq r < m$? In other words, when does $\lim_{n\to\infty} p^{[n]}(\bar{t})$ exist, where $\bar{t} = t\mathrm{mod}(t^m - 1)$?

Let $\mathrm{Supp}(p(t)) = \mathrm{Supp}(\sum_{i=0}^{\infty} a_i t^i) = \{t^i : a_i \neq 0 \ , i \in \mathcal{N} \cup \{0\}\}$. In future it is assumed that $p(t) = t^r p_0(t)$, where $r \geq 0$, $\mathrm{Supp}(p_0(t)) = \{t^{q_i} : i < l\}$, where $2 \leq l \leq \infty$ and $0 = q_0 < q_1 < q_2 < \ldots$. It is not difficult to see that

$$p^{[n]}(t) = t^{r^n} p_0^{(n)}(t) \quad \text{and} \quad p_0^{(n+1)}(t) = p_0^{(n)}(t)^r p_0(t^{r^n} p_0^{(n)}(t))$$

, which, due to Proposition 1, implies that

1. $\mathrm{Supp}(p_0^{(n)}(t)) \subset \langle\{t^{q_i} : i < l\}\rangle$, where $\langle\{t^{q_i} : i < l\}\rangle$ stands for the semigroup generated by $\{t^{q_i} : i < l\}$.
2. $t^{q_i}\mathrm{Supp}(p_0^{(n)}(t)) \subset \mathrm{Supp}(p_0^{(n+1)}(t))$ for any $i < l$

The second property implies, as far as $\mathrm{Supp}(p_0^{(n)}(t)) \subset \mathrm{Supp}(p_0^{(n+1)}(t))$, that the sets $\mathrm{Supp}(p_0^{(n)}(\bar{t}))$ are the same sets for all big enough $n \in \mathcal{N}$. Let us denote it by $G_{p_0(t)}$. Moreover $\bar{t}^{q_i} G_{p_0(t)} \subset G_{p_0(t)}$ and $\{\bar{t}^{q_i} : i < l\} \subset G_{p_0(t)}$. Therefore taking into account the first property one can conclude that $G_{p_0(t)}$ is the subgroup of $\{\bar{t}^0, \bar{t}^1, \ldots, \bar{t}^{m-1}\}$ generated by $\{\bar{t}^{q_i} : i < l\}$. Assume that it is as a cyclic group generated by $\bar{t}^q$, where $0 \leq q < m$.

It is clear that the sequence $\{r^k \mathrm{mod} m\}_{k\in\mathcal{N}}$ is a repeating sequence i.e. there are $d \in \mathcal{N}$ and different numbers $m_0, m_1, \ldots, m_{d-1}$ between 0 and $m-1$ for which

$$r^{kd+i} = m_i \mathrm{mod} m$$

whenever $i \in \{0, 1, \ldots, d-1\}$ and k is big enough.

Theorem 4. *1. If $\cap_{i=0}^{d-1} \bar{t}^{m_i} G_{p_0(t)} = \emptyset$ then $\lim_{n\to\infty} p^{(n)}(\bar{t})$ does not exist.*
2. If $\cap_{i=0}^{d-1} \bar{t}^{m_i} G_{p_0(t)} \neq \emptyset$ then $\bar{t}^{m_i} G_{p_0(t)} = \bar{t}^{m_j} G_{p_0(t)}$ whenever $i, j \in \{0, 1, \ldots, d-1\}$, the limit $\lim_{n\to\infty} p^{[n]}(\bar{t})$ exists and

$$\lim_{n\to\infty} p^{[n]}(\bar{t}) = \bar{t}^{m_0} c_{p_0(t)} + (e - c_{p_0(t)})a$$

, where $c_{p_0(t)} = \frac{1}{|G_{p_0(t)}|} \sum_{g \in G_{p_0(t)}} g$ and $a = \lim_{n\to\infty} a_0^{[n]}$

Proof. Assume that $\cap_{i=0}^{d-1} \bar{t}^{m_i} G_{p_0(t)} = \emptyset$, $\lim_{n\to\infty} p^{[n]}(\bar{t})$ exists and the coefficient at $\bar{t}^{m_0+h}$ in $\lim_{k\to\infty} p^{[kd]}(\bar{t}) = \lim_{n\to\infty} p^{[n]}(\bar{t})$ is not zero. Due to $\cap_{i=0}^{d-1} \bar{t}^{m_i} G_{p_0(t)} = \emptyset$ one can find $0 \leq j < m$ for which $\bar{t}^{m_0+h} \notin \bar{t}^{m_j} G_{p_0(t)}$. It means that if even $\lim_{k\to\infty} p^{[kd+j]}(\bar{t})$ exists the coefficient at $\bar{t}^{m_0+h}$ in $\lim_{n\to\infty} p^{[n]}(\bar{t})$ is zero. This contradiction indicates that $\lim_{n\to\infty} p^{[n]}(\bar{t})$ can not exist.

Now assume that $\cap_{i=0}^{d-1} \bar{t}^{m_i} G_{p_0(t)} \neq \emptyset$. In this case one can find integer numbers $s_0, s_1, \ldots, s_{d-1}$ such that $\bar{t}^{m_0} \bar{t}^{q s_0} = \bar{t}^{m_1} \bar{t}^{q s_1} = \ldots = \bar{t}^{m_{d-1}} \bar{t}^{q s_{d-1}}$ and therefore $\bar{t}^{m_i} G_{p_0(t)} = \bar{t}^{m_j} G_{p_0(t)}$ whenever $i, j \in \{0, 1, \ldots, d-1\}$. In particular $\bar{t}^{r^n} c_{p_0(t)} = \bar{t}^{m_0} c_{p_0(t)}$ for all $n \in \mathcal{N}$ big enough.

We know that for $n_0 \in \mathcal{N}$ big enough $p_0^{(n_0)}(\bar{t}) \in \overset{\circ}{S}_{G_{p_0(t)}}$ and due to Theorem 1 one has $\lim_{k\to\infty} p_0^{(n_0)}(\bar{t})^k = c_{p_0(t)}$. Let $p_0^{(n_0)}(\bar{t}) = c_{p_0(t)} + x_1$. Due to $\bar{t}^{r^{n_0}} p_0^{(n_0)}(\bar{t}) = \bar{t}^{r^{n_0}} c_{p_0(t)} + \bar{t}^{r^{n_0}} x_1$ and $c_{p_0(t)} x_1 = 0$ one has

$$p^{[n]}(\bar{t}^{r^{n_0}} p_0^{(n_0)}(\bar{t})) = p^{[n]}(\bar{t}^{r^{n_0}} c_{p_0(t)} + \bar{t}^{r^{n_0}} x_1) = p^{[n]}(\bar{t}^{r^{n_0}})c_{p_0(t)} - a_0^{[n]} c_{p_0(t)} + [p^{[n]}(\bar{t}^{r^{n_0}} x_1) - a_0^{[n]} e] + a_0^{[n]} e$$

But $\lim_{k\to\infty} x_1^k = \lim_{k\to\infty} (\bar{t}^{r^{n_0}} x_1)^k = 0$ and therefore due to Lemma 2, Theorem 3 one has $\lim_{n\to\infty} p^{[n]}(\bar{t}^{r^{n_0}} x_1) - a_0^{[n]} e = 0$. Therefore $\lim_{n\to\infty} p^{[n]}(\bar{t}) = \lim_{n\to\infty} p^{[n]}[\bar{t}^{r^{n_0}} p_0^{(n_0)}(\bar{t})]$ exists if and only if $\lim_{n\to\infty}(p^{[n]}(\bar{t}^{r^{n_0}})c_{p_0(t)} - a_0^{[n]} c_{p_0(t)} + a_0^{[n]} e)$ exists. Now to finish the proof notice that for any $n \in \mathcal{N}$ one has $p_0^{(n)}(\bar{t}^{r^{n_0}})c_{p_0(t)} = c_{p_0(t)}$ and $p^{[n]}(\bar{t}^{r^{n_0}}) = \bar{t}^{r^{n_0+n}} p_0^{(n)}(\bar{t}^{r^{n_0}})$ which implies that $\lim_{n\to\infty} p^{[n]}(\bar{t}) = \bar{t}^{m_0} c_{p_0(t)} + (e - c_{p_0(t)})a$, where $a = \lim_{n\to\infty} a_0^{[n]}$.

Examples. 1. Let $p(t) = \frac{1}{2}t^3(1+t^4)$, $m = 12$. In this case $r = 3$, $p_0(t) = \frac{1}{2}(1+t^4)$, $\bar{t}^{12} = 1$, $G_{p_0(t)} = \{1, \bar{t}^4, \bar{t}^8\}$, $d = 2$ and $m_0 = 3$, $m_1 = 9$. The intersection $\cap_{i=0}^{d-1} \bar{t}^{m_i} G_{p_0(t)}$ is empty as far as $m_0 G_{p_0(t)} = \{\bar{t}^3, \bar{t}^7, \bar{t}^{11}\}$ and $m_1 G_{p_0(t)} = \{\bar{t}, \bar{t}^5, \bar{t}^9\}$.

2. Now let $p(t) = \frac{1}{2}t^3(1+t^4)$ be same, $m = 10$. In this case $r = 3$, $p_0(t) = \frac{1}{2}(1+t^4)$, $\bar{t}^{10} = 1$, $G_{p_0(t)} = \{1, \bar{t}^2, \bar{t}^4, \bar{t}^6, \bar{t}^8, \}$, $d = 4$ and $m_0 = 3$, $m_1 = 9$, $m_2 = 7$, $m_3 = 1$. The intersection $\cap_{i=0}^{d-1} \bar{t}^{m_i} G_{p_0(t)}$ is equal to $\{\bar{t}, \bar{t}^3, \bar{t}^5, \bar{t}^7, \bar{t}^9\}$.

Theorem 5. *The map $p : S \to S$ is ergodic on S.*

Proof. Due to Theorem 2, Theorem 3 and Lemma 3, to show the ergodicity of p at $x \in S$, one can assume that $n_x = n_{xc_x}$ and prove the existence of

$$\lim_{n\to\infty} \sum_{k=0}^{\infty} q_{km_x+r}^{(n)}$$

for all $0 \le r < m_x$, where $q^{(n)}(t) = \frac{1}{n}\sum_{i=1}^{n} p^{[i]}(t)$, $n \in \mathcal{N}$. We consider any fixed $m \in \mathcal{N}$ and show that $\lim_{n\to\infty} q^{(n)}(\bar{t})$ exists. It is easy to see that $q^{(n)}[p^{[n_0]}(t)] = q^{(n+n_0)}(t) + \frac{n_0}{n}(q^{(n+n_0)}(t) - q^{(n_0)}(t))$ and
$\lim_{n\to\infty} \frac{n_0}{n}(q^{(n+n_0)}(\bar{t}) - q^{(n_0)}(\bar{t})) = 0$. So if $\lim_{n\to\infty} q^{(n)}(p^{[n_0]}(\bar{t}))$ exists then

$$\lim_{n\to\infty} q^{(n)}[(p^{[n_0]}(\bar{t})) = \lim_{n\to\infty} q^{(n)}(\bar{t})$$

For $n_0 \in \mathcal{N}$ big enough $p_0^{(n_0)}(\bar{t}) \in \overset{\circ}{S}_{G_{p_0(t)}}$ and due to Theorem 1 one has $\lim_{k\to\infty} p_0^{(n_0)}(\bar{t})^k = c_{p_0(t)}$. If $p_0^{(n_0)}(\bar{t}) = c_{p_0(t)} + x_1$ then one has

$$q^{(n)}(p^{([n_0]}(\bar{t})) = q^{(n)}(\bar{t}^{r^{n_0}} p_0^{(n_0)}(\bar{t})) = q^{(n)}(\bar{t}^{r^{n_0}} c_{p_0(t)} + \bar{t}^{r^{n_0}} x_1) =$$

$$q^{(n)}(\bar{t}^{r^{n_0}})c_{p_0(t)} - q_0^{(n)} c_{p_0(t)} + [q^{(n)}(\bar{t}^{r^{n_0}} x_1) - q_0^{(n)} e] + q_0^{(n)} e$$

and $\lim_{n\to\infty}[q^{(n)}(\bar{t}^{r^{n_0}} x_1) - q_0^{(n)}] = 0$ due to $\lim_{k\to\infty} x_1^k = \lim_{k\to\infty}(\bar{t}^{r^{n_0}} x_1)^k = 0$. Moreover $q^{(n)}(\bar{t}^{r^{n_0}})c_{p_0(t)} = \frac{1}{n}\sum_{i=1}^{n} \bar{t}^{r^{n_0+i}} p_0^{(i)}[(\bar{t}^{r^{n_0}})c_{p_0(t)}$ and $p_0^{(i)}(\bar{t}^{r^{n_0}})c_{p_0(t)} = c_{p_0(t)}$ whenever $i \in \mathcal{N}$. Therefore

$$\lim_{n\to\infty} q^{(n)}(\bar{t}) = \lim_{n\to\infty} q^{(n)}[p^{(n_0)}(\bar{t})] = \frac{1}{d}\sum_{i=0}^{d-1} \bar{t}^{m_i} c_{p_0(t)} + a(e - c_{p_0(t)})$$

To complete the picture let us consider the extremal case $p(t) = t^r$ as well, where $r > 1$. The sequence $\{r^n \mathrm{mod} m_x\}_{n\in\mathcal{N}}$ is a repeating sequence. Let $(m_o, m_1, ..., m_{d-1})$ be its

IOP Conf. Series: Journal of Physics: Conf. Series **949** (2017) 012023 doi:10.1088/1742-6596/949/1/012023

repeating part, where $m_0, m_1, ..., m_{d-1}$ are different numbers between 0 and $m_x - 1$. By the use of Theorem 1 the following result can be proved easily.

Theorem 1'. *1. The equalities*

$$S_r(x) = \{c_x x^{m_i} : 0 \leq i < d\} = \{x^{m_i} c_x : 0 \leq i < d\}$$

are valid for any $x \in S$, where $S_r(x)$ stands for the set of all limits of converging subsequences of $\{p^{(n)}(x)\}_{n \in \mathcal{N}} = \{x^{r^n}\}_{n \in \mathcal{N}}$. In particular, $|S_r(x)| = 1$ if and only if m_x divides $r^k(r-1)$ for some $k \in \mathcal{N}$.

2. $\lim_{n \to \infty} q^{(n)}(x) = \lim_{n \to \infty} \frac{1}{n} \sum_{i=0}^{n} x^{r^n} = \frac{1}{d} \sum_{i=0}^{d-1} x^{m_i} c_x$.

References

[1] H. Kesten 1970, Quadratic transformations: a model for population growth. I,II. *Adv. Appl. Probab.* **2**, 1-82; 179-228.

[2] Yu.I. Lyubich 1992, *Mathematical structures in population genetics*, (Springer-Verlag, Berlin).

[3] U. Bekbaev 2005 *Proceedings of The second International Conference on Research and Education in Mathematics (ICREM 2)*, pp.134–144.

[4] U. Bekbaev, M. A. M. Johari 2005 *Proceedings of The second International Conference on Research and Education in Mathematics (ICREM 2)*, pp.144–153.

[5] U. Bekbaev, M. A. M. Johari 2006 *arXiv: math/0608207v1 [math.DS]*

[6] S. Lang 1972, *Algebra*, (Columbia University, Addison-Wesley).

IOP Conf. Series: Journal of Physics: Conf. Series **949** (2017) 012024 doi:10.1088/1742-6596/949/1/012024

Uniform Convergence of the Eigenfunction Expansions of Distributions Associated with the Polyharmonic Operator on Closed Domain

Abdumalik Rakhimov[1] **and Siti Nor Aini Mohd Aslam**[2]

[1]Department of Science, Kulliyyah of Engineering, International Islamic University Malaysia, 53100 Jalan Gombak, Kuala Lumpur, Malaysia.hang, Malaysia.
[2]Institute for Mathematical Research (INSPEM), Universiti Putra Malaysia, 43400 UPM Serdang, Selangor, Malaysia.

E-mail : [1] abdumalik@iium.edu.my, [2] aini_lily@yahoo.com

Abstract. In the solving the boundary value problems of equations of mathematical physics in bounded domains it is important to adjust the solution on the closed domain. This leads to the investigation of convergence and sumability problems near the boundary of the domain. In the present paper we study this problem for the eigenfunction expansions associated with the polyharmonic operator.

1. Introduction

In [1] Il'in studied uniform convergence of the spectral expansions $\boldsymbol{E_\lambda}$ associated with the Laplace operator in the Sobolev classes $\boldsymbol{W_p^a(\Omega)}$ and found sufficient conditions of uniform convergence for the Riesz means $\boldsymbol{E_\lambda^s f}$ as $\boldsymbol{a+s \geq \frac{N-1}{2}}$, $\boldsymbol{pa > N}$. Later, Alimov in [2] established the uniform convergence of the Riesz means for the functions from the Nikol'skii classes $\boldsymbol{H_p^a(\Omega)}$. The convergence of the spectral decompositions of the Laplace operator on closed domain firstly investigated by Il'in (see in [1]) and then Moiseev [3] proved uniformly convergence of the eigenfunction expansions of the functions from $\boldsymbol{W_p^{\left(\frac{N}{2}+1\right)}(\Omega)}$ on closed domain $\overline{\Omega}$. Furthermore, in [4] Eskin considered the $2m$ order elliptic differential operator with the Lopatinsky boundary condition and proved uniformly convergent of the spectral expansions of the functions from $\boldsymbol{W_p^{\left(\frac{N}{2}+1\right)}(\Omega)}$, $\boldsymbol{\varepsilon > 0}$ on closed domain $\overline{\Omega}$. The uniform convergence of the eigenfunction expansions of the Laplace operator in closed domain was considered by Rakhimov [5], where he showed that the conditions $\boldsymbol{a+s > \frac{N-1}{2}}$, $\boldsymbol{pa = N}$ secure the uniform convergence of the expansions in closed domain for the continuous functions from the Nikolskii classes $\boldsymbol{H_p^a(\Omega)}$. Uniform convergence of the Riesz means of eigenfunction expansions of countinuous function in case of Laplace operator studied in [6] and for the general elliptic operator in [7]. The problems of the uniformly the Riesz summability on a closed domain for the Schredinger operator with the singular potential studied in [8] and [9].

2. The spaces of test functions

Let $\Omega \subset R^2$ be a domain with smooth boundary $\partial\Omega$. Let define the classes of the functions (see in [10]). We say that a function $f(x,y) \in L_p(\Omega)$ belongs to the $H_p^a(\Omega)$, if for any $h=(h,k) \in R^2$ and for all integers α, β satisfying $\alpha+\beta=l$:

$$\left|\partial_x^\alpha \partial_y^\beta f(x+h,y+k) - 2\partial_x^\alpha \partial_y^\beta f(x,y) + \partial_x^\alpha \partial_y^\beta f(x-h,y-k)\right|_{L_p(\Omega_{\sqrt{h^2+k^2}})} \le C\left(h^2+k^2\right)^{\frac{\sigma}{2}}.$$

where a is written as $a=l+\sigma$, $l-$positive integer and $0<\sigma\le 1$. Using the notation $\Delta_{h,k}^2 f(x,y) = f(x+h,y+k) - 2f(x,y) + f(x-h,y-k)$, a norm in $H_p^\alpha(\Omega)$ is defined by

$$\|f\|_{p,\alpha} = \|f\|_{L_p(\Omega)} + \sum_{\alpha+\beta=l} \sup_{h,k} \left(h^2+k^2\right)^{-\frac{\sigma}{2}} \left\|\Delta_{h,k}^2 \partial_x^\alpha \partial_y^\beta f(x,y)\right\|_{L_p(\Omega_{\sqrt{h^2+k^2}})}.$$

The closure of the space $C_0^\infty(\Omega)$ in the norm of $H_p^a(\Omega)$ denoted by $\dot{H}_p^a(\Omega)$.

Let $a \ge 0$ an integer and a function $\gamma(t)$ defined and continuous for $1 \le t \le \infty$, positive and vanishing in $t \to \infty$. Consider an operator

$$I_{r,\gamma} f = F^{-1}\left[(1+|\omega|^2)^{\frac{-a}{2}} \cdot \gamma\left(\sqrt{1+|\omega|^2}\right) \cdot Ff\right]$$

Then the generalized Besov spaces $B_{p,\theta}^{r,\gamma}(R^N)$ defined as

$$B_{p,\theta}^{r,\gamma}(R^N) = I_{r,\gamma} B_{p,\theta}^0(R^N)$$

Here F and F^{-1} direct and inverse Fourier transformations in $J'(R^N)$.

Now we define the generalized Sobolev spaces in the finite domain Ω. Unless whole space R^N in the finite domain the definition not involving the Fourier transformation.

For any square integrable in the domain Ω function u and for any positive number r by $\omega(u,r)$ we denote the second order modulus of continuity in the metric of the space $L_2(\Omega)$

$$\omega(u,r) = \sup_{|h|<r} \|u(x+h) - 2u(x) + u(x+h)\|_{L_2(\Omega_h)}$$

where $\Omega_h = \{y \in \Omega : \mathrm{dist}(y, \partial\Omega) < h\}$.

Let $a > 0$, $a = \mu + \varepsilon$, where $\mu \ge 0$ integer number and $0 < \varepsilon \le 1$. Let $\beta(\delta) = ln\delta$, $1 \le \delta < \infty$. Then by $W_2^{\alpha,\beta}(\Omega)$ we denote the space of functions from $L_2(\Omega)$ with the finite norm

$$\|u\|_{a,\beta} = \|u\|_0 + \sum_{|\alpha|=\mu} \sup_{0<r<1} \frac{\omega(D^\alpha u, r)}{\delta^\varepsilon \beta\left(\frac{1}{r}\right)}$$

here α is multiindex, i.e. N - dimensional vector with non negative integer components $\alpha = (\alpha_1, \alpha_2, \ldots\ldots, \alpha_N)$ and $|\alpha| = \alpha_1 + \alpha_2 + \cdots + \alpha_N$ called length of multiindex. Then $D^\alpha = \frac{\partial^{|\alpha|}}{\partial x_1^{\alpha_1} \ldots \partial x_N^{\alpha_N}}$ denote the generalized partial derivative.

3. Eigenfunction expansion associated with the polyharmonic operator

Let consider the polyharmonic operator $(-\Delta)^p$ with the domain $D_B = \{ u \in W_2^{2p}(\Omega):\ u|_{\partial\Omega} = \Delta u|_{\partial\Omega} = ...\Delta^{p-1}u|_{\partial\Omega} = 0 \}$. The polyharmonic operator $(-\Delta)^p$ with the domain of definition D_B is symmetric and nonnegative. By Fredrichs Theorem it can be extended to $L_2(\Omega)$ as self-adjoint operator, which again denote by $(-\Delta)^p$. It is well known that self-adjoint polyharmonic operator $(-\Delta)^p$ has a set of eigenfunctions $\{u_{nm}(x, y)\}$, that is complete in $L_2(\Omega)$. We denote by $\{\lambda_{nm}\}$ the set of eigenvalues of polyharmonic operator in Ω then we have

$$(-\Delta)^p u_{nm}(x, y) - \lambda_{nm} u_{nm}(x, y) = 0, \qquad (x, y) \in \Omega,$$

with the boundary conditions

$$u_{nm}|_{\partial\Omega} = \Delta u_{nm}|_{\partial\Omega} = ... = \Delta^{p-1} u_{nm}|_{\partial\Omega} = 0.$$

Let $E_\lambda^s f$ be the Riesz means of the eigenfunction expansions associated with the operator above

$$E_\lambda^s f(x, y) = \sum_{\lambda_{nm} \le \lambda}\sum (1 - \frac{\lambda_{nm}}{\lambda})^s f_{nm} u_{nm}(x, y), \qquad s \ge 0,$$

where f_{nm} is the Fourier coefficients of the function f:

$$f_{nm} = \iint_\Omega f(x, y)\, u_{nm}(x, y)\, dx\, dy, \qquad n, m = 1,2,... .$$

The following estimation is established evaluation for the package of eigenfunction on closed domain with respect to the $\lambda \to \infty$ (Anvarjon, Siti Nor Aini and Gafurjan Ibragimov).

For the eigenfunctions $\{u_{nm}(x, y)\}$ and eigenvalues λ_{nm} of the polyharmonic operator $(-\Delta)^p$, corresponding to the boundary conditions $u_{nm}|_{\partial\Omega} = \Delta u_{nm}|_{\partial\Omega} = ... = \Delta^{p-1} u_{nm}|_{\partial\Omega} = 0$ *we have*

$$\sum_{|2m\sqrt{\lambda_{nm}} - \lambda| \le 1}\sum u_{nm}^2(x, y) \le O(1)\, \lambda^{N-1} \ln^2 \lambda, \qquad \lambda > 1, \tag{1}$$

uniformly for all $(x, y) \in \overline{\Omega} = \Omega \cup \partial\Omega$.

From this it follows that if $a + s > \frac{1}{2}, pa > 2, p \ge 1$, then the Riesz means $E_\lambda^s f(x, y)$ uniformly convergence for the functions from Nikolskii classes H_p^a in closed domain.

4. Spectral expansions associated with the biharmonic operator

Denote by $W_2^{-(a,\beta)}(\Omega)$ the space of the linear continuous functionals on $W_2^{\alpha,\beta}(\Omega)$. The main results of this paper are the proof of the following theorems.

IOP Conf. Series: Journal of Physics: Conf. Series **949** (2017) 012024 doi:10.1088/1742-6596/949/1/012024

Theorem 1. Let a distribution $f \in W_2^{-(a,\beta)}(\Omega)$, $a > 0$, has a compact support in the two dimensional bounded domain . If $s \geq \frac{N-1}{2} + a$. Then the relation

$$\lim_{\lambda\to\infty} E_\lambda^s f(x,y) = 0$$

holds uniformly in any compact set $K \subset \bar{\Omega}\backslash \mathrm{supp} f$.

Note that from the embedding $W_2^a(\Omega) \subset W_2^{a,\beta}(\Omega) \subset W_2^{a-\varepsilon}(\Omega)$, $\varepsilon > 0$, it follows that the condition $s \geq \frac{N-1}{2} + a$ is precise (see [1]). More over from this embedding and the Theorem 1 we get the following theorem for the Sobolev spaces

Theorem 2. Let $f \in \mathcal{E}'(\Omega) \cap W_2^{-a}(\Omega)$, $a > 0$, and $s > \frac{N-1}{2} + a$. Then the relation

$$\lim_{\lambda\to\infty} E_\lambda^s f(x,y) = 0$$

holds uniformly in any compact set $K \subset \bar{\Omega}\backslash \mathrm{supp} f$.

Let $f \in C_0^\infty(\Omega)$. Then introduce a function

$$\beta_{\Delta^2}(f)(x) = \sum_{n=1}^{\infty} \lambda_n^{\frac{-a}{2}} \ln\lambda_n f_n u_n(x)$$

which is infinitely differentiable on the closed domain $\bar{\Omega}$.

Lemma 1. For any function $f \in C_0^\infty(\Omega)$ the following inequality is valid

$$\sum_{n=1}^{\infty} \lambda_n^{-a} \ln^2\lambda_n f_n^2 \quad \leq \quad c \; \|f\|_{-(a,\beta)}$$

with the constant in depended from f.

The Lemma 1 can be proven using formula (1). From this we get the following:

Lemma 2. Let a function $\varphi(t)$ defined for $t \in [1,\infty)$ and $\sum_{n=1}^{\infty} \varphi^2(n) \leq \mathrm{const}$. Moreover, let $\varphi(t+s) \leq c\,\varphi(t)$, $0 \leq s \leq 1$, $t \geq 1$. Then the estimate

$$\sum_{n=1}^{\infty} \varphi\left(\sqrt{\lambda_n}\right) \lambda_n^{\frac{-2a-\frac{N-1}{2}}{2}} f_n u_n(x) \;\leq\; c\,\|f\|_{-(a,\beta)}, \quad a > 0. \tag{2}$$

is valid uniformly in any compact set $K \subset \bar{\Omega}$ and for any function $f \in C_0^\infty(\Omega)$.

Finally from the Lemma 1 and 2, we obtain that if $f \in C_0^\infty(\Omega)$ and $s = \frac{N-1}{2} + a$, then

$$E_\lambda^s f(x) = O(1)\|f\|_{-(a,\beta)}$$

is valid on an arbitrary compact set $K \subset \bar{\Omega}\backslash \mathrm{supp} f$.

IOP Conf. Series: Journal of Physics: Conf. Series **949** (2017) 012024 doi:10.1088/1742-6596/949/1/012024

Then the statement of the Theorem 1 follows from the fact that a distribution has a compact support and can be approximated by the functions from the space $C_0^\infty(\Omega)$ in the norm of corresponding generalized Sobolev spaces.

Note that, in the paper [11] it is proved necessary conditions for the summability of the eigenfunction expansions of the distributions from the generalized Sobolev spaces. These questions also discussed in [12]. In the papers [13] and [16] spectral expansions associated with elliptic differential operators in the negative Sobolev spaces are studied. In case of the Hilbert spaces in [13] sharp conditions for the localization are established. In [14], [15] and [17] special methods of summations associated with the elliptic differential operators are studied. In [18] the uniform convergence of the Fourier series studied on a closed domain.

5. Summary

In this paper the uniform convergence of eigenfunction expansions associated with the polyharmonic operator for the distributions from the generalized Sobolev spaces is studied and sharp theorems on the localization on the closed domain proved. In this we studied expansions in eigenfunction with the Navier boundary conditions. Note, that study of the problem for the first and second boundary problems remains open even for the bi-harmonic operator. Moreover, the problem is still open even for the Laplace operator in the case of second and third boundary conditions.

Acknowledgments

This research has been supported by the IIUM Fundamental Research Grant Scheme, FRGS14-142-0383.

6. References

[1] Il'in V A 1957 On the uniform convergence of expansions in eigenfunctions for summation in the order of the eigenvalues *Dokl. Akad. Nauk SSSR* **114:4** p 698– 701.

[2] Alimov Sh A and Ilin V A 1970 Spectral decompositions corresponding to an arbitrary nonnegative selfadjoint extension of Laplace's operator *Dokl. AN SSSR* **193** p 9–12.

[3] Moiseev E 1977 The uniform convergence of certain expansions within a closed domain, *Dokl. Akad. Nauk SSSR* **233:6** p 1042-45.

[4] Eskin G 1975 Asymptotics near the boundary of spectral functions of elliptic self-adjoint boundary problems *Isr. J.Math.* **22:3-4** p 214-46.

[5] Rakhimov A A 1987 About uniform convergence of the spectral decomposition for the functions from Nikolskii classes in closed domain *VINITI* 1564-B87 pp 17.

[6] Rakhimov A A 1987 On uniform convergence of spectral resolutions of a continuous function in a closed domain *Izvestiya Akademya Nauk UZSSR* **6** p 17-22.

[7] Alimov Sh A and Rakhimov A A 1986 On the uniformly convergence of spectral expansions in a closed domain *Dokl. Acad Nauk UzSSR* **10** p 5-7.

[8] Rakhimov A A and Kamran Zakariya 2010 On an Estimation of Eigenfunctions of Schrödinger's operator on a closed domain *MathDigest. Bulletin of Universiti Putra Malaysia* **June Edition** p 1-4.

[9] Rakhimov A A, Kamran Zakariya and Nazir Alikhan 2010 Uniform convergence spectral expansions connected with Schrodinger's operator of continuous function on a closed domain *J. The Nucleus* **47:4** p 261-265.

[10] Triebel H 1999 The uniform convergence of spectral expansions of the Laplace operator on closed $\overline{\Omega} \subset R^2$ *Wiley - Science* **Jan 6** p 1-532.

[11] Alimov Sh A and Rakhimov A A 1997 On the localization spectral expansions in a closed domain *J. Differential equations* **33:1** p 80–82.

[12] Rakhimov A A 2014 On the Fourier series of continuous functions *Proceedings of USA-Uzb Conference on Mathematical Physics, Californian State University, Fullerton* **May** p 1–3.

[13] Alimov Sh A and Rakhimov A A 1996 Localization of Spectral Expansions of Distributions.

Difference Equations. **32**, 798-802.

[14] Rakhimov A A 2000 On the Localization of Multiple Trigonometric Series of Distributions. Dokl. Math. 62,163-165. (translation from Dokl. Akad. Nauk, Ross. Akad. Nauk, **374,** 20-22).

[15] Rakhimov A A 1996 Localization Conditions for Spectral Decompositions Related to Elliptic Operators from Class Ar. Mathematical Notes. **59**, 298-302. (translation from Mat. Zametki, **59**, 421-427).

[16] Rakhimov A A 1996 Spectral Decompositions of Distributions from Negative Sobolev Classes. Difference Equations. **32**, 1011-1013.

[17] Rakhimov A.A. 2011 On the localization of the riesz means of multiple fourier series of distributions. Abstract and Applied Analysis, DOI: 10.1155/2011/731249.

[18] Rakhimov A.A. 2015 On the uniform convergence of the Fourier series on a closed domain, Eurasian Mathematical Journal, **8 (3),** 60-69.

IOP Conf. Series: Journal of Physics: Conf. Series **949** (2017) 012025 doi:10.1088/1742-6596/949/1/012025

Simulation of disturbance rejection control of half-car active suspension system using active disturbance rejection control with decoupling transformation

Faried Hasbullah[1,*] **and Waleed F. Faris**[2]

[1] Department of Science in Engineering, Kulliyyah of Engineering, International Islamic University Malaysia, 53100 Jalan Gombak, Kuala Lumpur, Malaysia.
[2] Department of Mechanical Engineering, Kulliyyah of Engineering, International Islamic University Malaysia, 53100 Jalan Gombak, Kuala Lumpur, Malaysia.

*Email: faried@iium.edu.my

Abstract. In recent years, Active Disturbance Rejection Control (ADRC) has become a popular control alternative due to its easy applicability and robustness to varying processes. In this article, ADRC with input decoupling transformation (ADRC-IDT) is proposed to improve ride comfort of a vehicle with an active suspension system using half-car model. The ride performance of the ADRC-IDT is evaluated and compared with decentralized ADRC control as well as the passive system. Simulation results show that both ADRC and ADRC-IDT manage to appreciably reduce body accelerations and able to cope well with varying conditions typically encountered in an active suspension system. Also, it is sufficient to control only the body motions with both active controllers to improve ride comfort while maintaining good road holding and small suspension working space.

1. Introduction

Ride comfort is one of the main criteria in assessing the performance of a vehicle. When a vehicle travels across uneven surfaces, the vibration impacts ride comfort in an undesirable way. Suspension system is the main part in a vehicle that functions to facilitate good ride comfort by isolating that vibration from the passengers. Suspension system also functions to support vehicle weight, keep good contact between the tire and the road, and reduce excessive pitching and rolling motion of the vehicle body. A passive suspension system has always been designed to achieve a good compromise between these goals. For instance, a "soft" suspension setting could be used to facilitate a comfortable ride at the expense of increased tire motions and suspension working space as the tire must travel further before it stops. On the other hand, good road handling characteristics and smaller tire motion is an attribute of "hard" suspension setting. Therefore, it is not possible to achieve the best performance for all goals by using a passive suspension alone.

Active suspension system has a better design trade-off than the passive suspension as it makes use of actuator to continuously adjust itself. The desired force is usually employed by pneumatic, hydraulic or electromagnetic actuators which are secured in parallel with a spring and a damper. The force generated by the actuator is controlled based on the states of the vehicle which are acquired from various sensors located at different points of the vehicle.

Active disturbance rejection control (ADRC) has become an attractive control alternative in recent years for its easy applicability and good robustness against varying processes [1]. It is based on the idea that if the internal and external disturbances of a dynamic system can be approximated in real time, then they can be negated without having to know their exact mathematical models. ADRC for improving vehicle ride comfort was proposed by [2] using a quarter car model. The main objective this article is to propose ways to implement ADRC to an active suspension system for improving ride comfort using half-car model as an extension to the work done by [2].

2. Half-car vehicle ride model with active suspension system

The 4-DOF half-car model used in this analysis is shown in Figure 1. This model can capture basic performances of vehicle suspension such as vehicle body heave and pitch, body acceleration, suspension working space, and the displacement of the wheel. The four DOFs are the sprung mass vertical displacement (x_s), sprung mass pitch (θ), front unsprung mass displacement (x_{u1}), and rear unsprung mass displacement (x_{u2}). Both active and passive suspension systems can be obtained in this model, where the passive system is obtained by letting both front actuator force, f_1 and rear actuator force, f_2 equal to zero. The state variable descriptions are provided in Table 1. The model parameters definition and their typical values are shown in Table 2 which are based from [3].

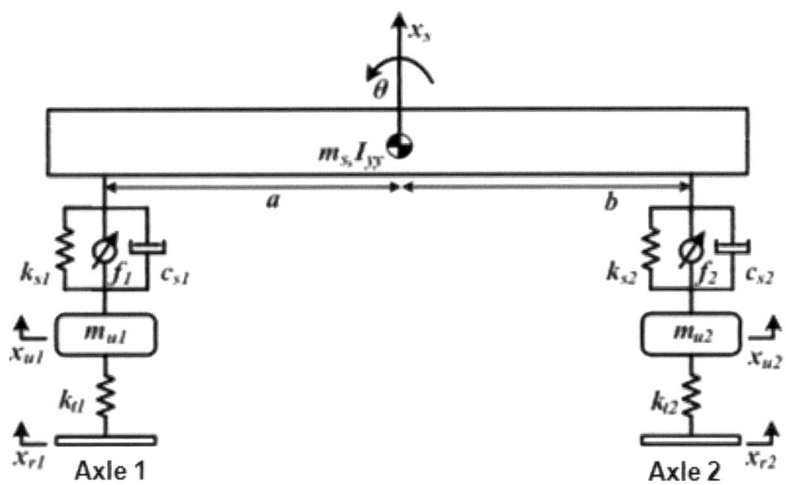

Figure 1. 4-DOF half-car model

Table 1. State variables and inputs description for the half-car vehicle

Variable	Description
x_s	m_s displacement (m)
θ	m_s pitch (rad)
x_{u1}	m_{u1} displacement (m)
x_{u2}	m_{u2} displacement (m)
x_{r1}	Front road input (m)
x_{r2}	Rear road input (m)
f_1	Front actuator input (N)
f_2	Rear actuator input (N)

IOP Conf. Series: Journal of Physics: Conf. Series **949** (2017) 012025 doi:10.1088/1742-6596/949/1/012025

Table 2. Model parameters for the half-car vehicle.

Variable	Description	Value	Unit
m_s	Sprung mass	730	kg
I_{yy}	Pitch moment of inertia	2460	kgm^2
m_{u1}	Front unsprung mass	40	kg
m_{u2}	Rear unsprung mass	35.5	kg
a	Distance from m_s C.G. to front axle	1.011	m
b	Distance from m_s C.G. to rear axle	1.803	m
k_{s1}	Front suspension stiffness	19960	N/m
k_{s2}	Rear suspension stiffness	17500	N/m
k_{t1}	Front tire stiffness	175500	N/m
k_{t2}	Rear tire stiffness	175500	N/m
c_{s1}	Front suspension damping coefficient	1290	N s/m
c_{s2}	Rear suspension damping coefficient	1620	N s/m

Referring to the model in Figure 1, assuming small pitch angle, θ, throughout the vehicle forward travel through the road profile so that sin $\theta = \theta$ and cos $\theta = 1$, the force balance on the sprung mass in the vertical direction is given by

$$m_s\ddot{x}_s = -F_{s,1} - F_{s,2} - F_{d,1} - F_{d,2} + f_1 + f_2 \quad (1)$$

where

$F_{s,n}$: Spring force at the nth side of the vehicle $= k_{s,n}\left(x_{sn} - x_{un}\right)$,

$F_{d,n}$: Damper force at the nth side of the vehicle $= c_{s,n}\left(\dot{x}_{sn} - \dot{x}_{un}\right)$,

f_n : Actuator force at the nth side of the vehicle,

$n = 1, 2$: Front and rear location of the vehicle, respectively.

Similarly, the moment balance equation is given as:

$$I_{yy}\ddot{\theta} = F_{s,1}a - F_{s,2}b + F_{d,1}a - F_{d,2}b - f_1a + f_2b \quad (2)$$

and the force balance on the front and rear unsprung mass is given as:

$$m_{u1}\ddot{x}_{u1} = F_{s,1} + F_{d,1} - F_{t,1} - f_1 \quad (3)$$

$$m_{u2}\ddot{x}_{u2} = F_{s,2} + F_{d,2} - F_{t,2} - f_2 \quad (4)$$

where

$F_{t,n}$: Tire force at the nth side of the vehicle = $k_{tn}\left(x_{un} - x_{rn}\right)$

The displacement of the sprung mass at the front wheel is given by

$$x_{s,2} = x_s - a\theta \quad (5)$$

and the displacement of the sprung mass at the rear wheel is given by

$$x_{s,2} = x_s + b\theta \quad (6)$$

The system state variables are defined as follows:

$$\mathbf{x} = \begin{bmatrix} \mathbf{X}_1 \\ \dot{\mathbf{X}}_1 \end{bmatrix}, \quad (7)$$

where

$$\mathbf{X}_1 = [x_s, \theta, x_{u1}, x_{u2}]^T = [x_1, x_2, x_3, x_4]^T . \tag{8}$$

These equations can be written in standard state-space form as:

$$\dot{\mathbf{x}} = \mathbf{Ax} + \mathbf{Bu} + \mathbf{Gw}, \tag{9}$$

where $\mathbf{x}$ is the state vector, $\mathbf{u}$ is the controlled input that represents the two actuator forces,

$$\mathbf{u} = [f_1, f_2]^T , \tag{10}$$

and $\mathbf{w}$ is the uncontrolled input that represents the road disturbance,

$$\mathbf{w}_{hc} = [x_{r1}, x_{r2}]^T . \tag{11}$$

The passive system is described as:

$$\dot{\mathbf{x}} = \mathbf{Ax} + \mathbf{Gw} . \tag{12}$$

The responses of interest are (i) sprung mass vertical acceleration, ($\ddot{x}_1$), which is the main parameter for evaluating ride comfort, (ii) sprung mass pitch angular acceleration, ($\ddot{x}_2$), (iii) front suspension deflection, $(x_1 - ax_2 - x_3)$ and rear suspension deflection which represents the suspension working space, $(x_1 + bx_2 - x_3)$, as well as (iv) front tire deflection, $(x_3 - x_{r1})$ and rear tire deflection, $(x_4 - x_{r2})$ in which the road holding performance of the vehicle are measured. Active controllers based on ADRC are designed to minimize the sprung mass vertical acceleration as the main objective while keeping other responses small.

3. Active disturbance rejection controller (ADRC)

ADRC consists of three main parts [1]: extended state observer (ESO), feedback controller, and disturbance rejection law.

3.1 Extended state observer (ESO)

The ESO is designed to estimate the dynamics of the system including the unknown dynamics such as the dynamics of the actuator (internal disturbance), and uneven road surface (external disturbance). In this article, a linear extended state observer (LESO) is used.

In general, the half-car active suspension system in Section 2 can be written as:

$$\ddot{y} = g(t, \dot{y}, y) + bu + r. \tag{13}$$

where $g(\cdot)$ is the dynamics of the system (including internal disturbances), y is the response signal, u is the input signal generated by the actuator, b is the system parameter, and r is the external disturbances. $g(\cdot)$, b, and r are usually not precisely known. By merging the internal and external disturbances into the system dynamics in one function, equation (13) can be rewritten as:

$$\ddot{y} = f(t, \dot{y}, y, r) + bu. \tag{14}$$

where the function $f(\cdot)$ is now the total disturbance. Equation (14) can be written in state space form as:

$$\begin{cases} \dot{x}_1 = x_2, \\ \dot{x}_2 = f(t, x_1, x_2, r) + bu \\ y = x_1 \end{cases} \tag{15}$$

Before the LESO is constructed, the total disturbance is first augmented in the system's state,

$$\begin{cases} \dot{x}_1 = x_2, \\ \dot{x}_2 = x_3 + bu, \\ \dot{x}_3 = \dot{f}(t, x_1, x_2, r), \\ y = x_1 \end{cases} \tag{16}$$

IOP Conf. Series: Journal of Physics: Conf. Series **949** (2017) 012025 doi:10.1088/1742-6596/949/1/012025

Then, the LESO can be designed to estimate the total disturbance so that its impact can be compensated through disturbance rejection. The LESO is designed as:

$$\begin{cases} \dot{x}_1 = x_2 - \beta_{o1}\hat{e}, \\ \dot{x}_2 = x_3 - \beta_{o2}\hat{e} + b_0 u, \\ \dot{x}_3 = -\beta_{o3}\hat{e}, \end{cases} \tag{17}$$

where x_1, x_2, and x_3 are the estimated values of states x_1, x_2, and x_3 respectively, β_{o1}, β_{o2}, and β_{o3} are the observer gains, and $\hat{e} = y_{ref} - x_1$ is the approximated error of x_1. b_0 is the approximated value of b from equation (14).

The estimated variables, x_1, x_2, and x_3 as well as b_0 are then used to implement the actual feedback controller and the disturbance rejection law as shown in Figure 2.

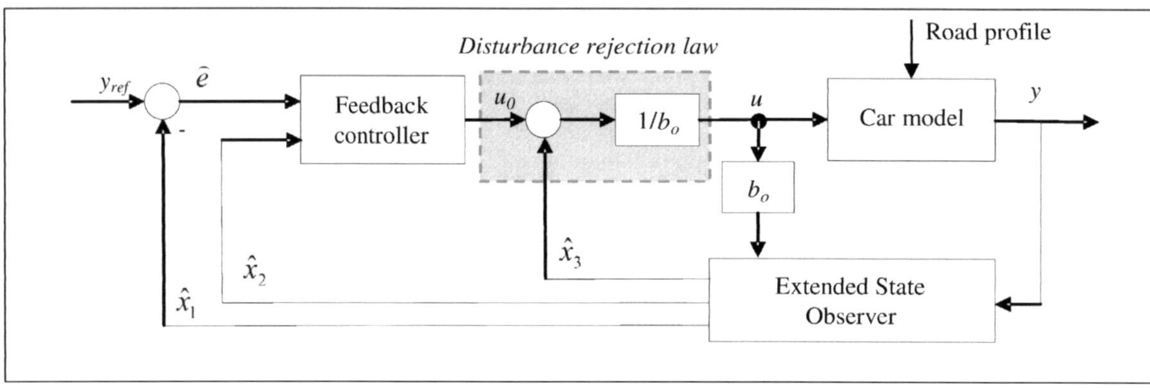

Figure 2. ADRC Control Structure

3.2 Feedback Controller

A typical proportional-derivative (PD) controller is implemented as the feedback controller. The feedback controller output signal, u_0 is defined as:

$$u_0(t) = K_p\left(y_{ref}(t) - z_1(t)\right) + K_d z_2(t) \tag{18}$$

where K_p and K_d are the proportional gain and the derivative gain respectively, and are chosen as [1]:

$$K_P = \omega_{CL}{}^2 \text{ and } K_d = -2\omega_{CL} \tag{19}$$

where $\omega_{CL1,2} = \omega_{CL}$ is the desired close-loop pole.

To guarantee that the observer dynamics are fast enough, the observer poles, ω_{ESO} are placed to the left of the close loop pole. The observer poles are chosen as [4]:

$$\omega_{ESO,1} = \omega_{ESO,2} = \omega_{ESO,3} = \omega_{ESO} = 3 \cdot \omega_{CL} \tag{20}$$

ω_{CL} is usually chosen empirically as a trade-off between the influence of sampling time and states estimation convergence speed [5]. The observer poles are placed at the same location (ω_{ESO}) for simplicity.

The observer gains can be obtained from the characteristic equation of the LESO described in equation (17). The solutions for β_{o1}, β_{o2}, and β_{o3} are:

$$\beta_{o1} = -3\omega_{ESO},\ \beta_{o2} = -3\omega_{ESO}^2, \text{ and } \beta_{o3} = -\omega_{ESO}^3 \tag{21}$$

3.3 Disturbance rejection law

A The disturbance rejection law is defined as:

$$u = \frac{u_0 - z_3}{b_0} \tag{22}$$

Equation (14) can now be defined with equation (22) as the control signal:

$$\ddot{y} = f(\cdot) + b\left(\frac{u_0 - z_3}{b_0}\right) \tag{23}$$

As long as the observer delivers good approximates, $b_o \approx b$ and $z_3 \approx f(\cdot)$, equation (23) is simplified into:

$$\ddot{y} \approx u_0 \tag{24}$$

The system defined by equation (15) is then reduced to a double integrator model:

$$\begin{cases} \dot{x}_1 = x_{2,} \\ \dot{x}_2 = u_0 \\ y = x_1 \end{cases} \tag{25}$$

4. ADRC for half-car active suspension system.

The ADRC is implemented into a half-car active suspension system by making use of an input decoupling transformation (IDT). The IDT is used to stabilize heave and pitch motions of the vehicle. The inputs to the IDT are the heave and pitch corresponding forces to stabilize these motions, which are the outputs of the two (heave and pitch) ADRCs respectively. These forces can be decoupled into forces of the front and rear actuators by using the following IDT scheme:

$$f_z = f_1 + f_2 \tag{26}$$

for heave control, and

$$f_\theta = -af_1 + bf_2 \tag{27}$$

for pitch control. These relations can be rearranged in matrix form as

$$\begin{bmatrix} f_z \\ f_\theta \end{bmatrix} = \begin{bmatrix} 1 & 1 \\ -a & b \end{bmatrix} \begin{bmatrix} f_1 \\ f_2 \end{bmatrix}. \tag{28}$$

The forces for front and rear actuators can be obtained from the following inverse transformation relationship

$$\begin{bmatrix} f_1 \\ f_2 \end{bmatrix} = \frac{1}{a+b} \begin{bmatrix} b & -1 \\ a & 1 \end{bmatrix} \begin{bmatrix} f_z \\ f_\theta \end{bmatrix}. \tag{29}$$

The ADRC design procedures in section 3 are repeated for both heave and pitch control respectively. Based on these procedures, the controller parameters for heave and pitch ADRCs are obtained as listed in Table 3. The final ADRC-IDT control structure for the half-car active suspension system is illustrated in Figure 3.

Table 3. ADRC-IDT parameters for half car

	b_o
Heave	7.8904×10^{-04}
Pitch	4.3171×10^{-04}

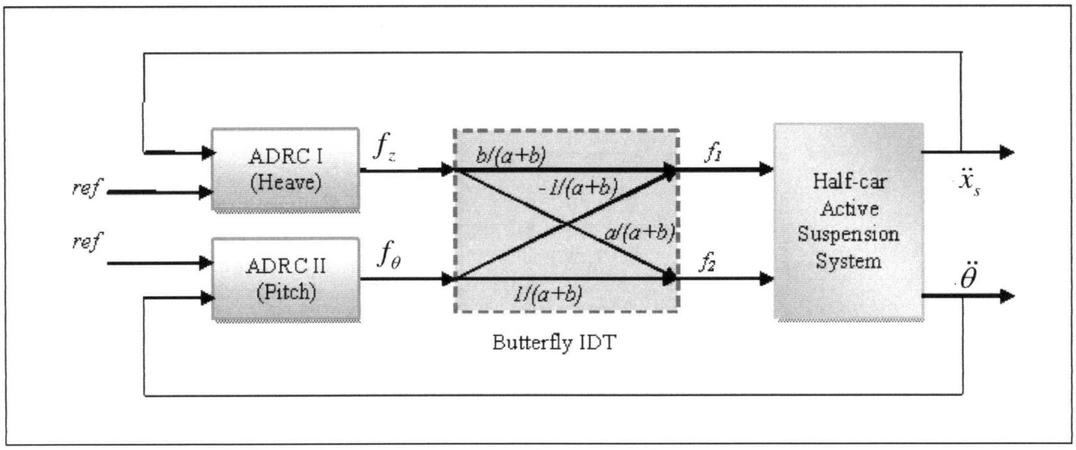

Figure 3. The proposed ADRC-IDT structure

The simulation results for ADRC-IDT structure are compared with decentralized ADRC control structure. Since there are two actuators in a half-car active suspension model (front and rear actuators), two ADRC controllers are design for the front and the rear actuators as shown in Figure 4. x_{s1} and x_{s2} are the sprung mass displacement at the front and rear axles respectively. The design procedures in section 3 are repeated for both front and rear ADRC control. Based on these procedures, the controller parameters for front and rear ADRCs are obtained as listed in Table 4.

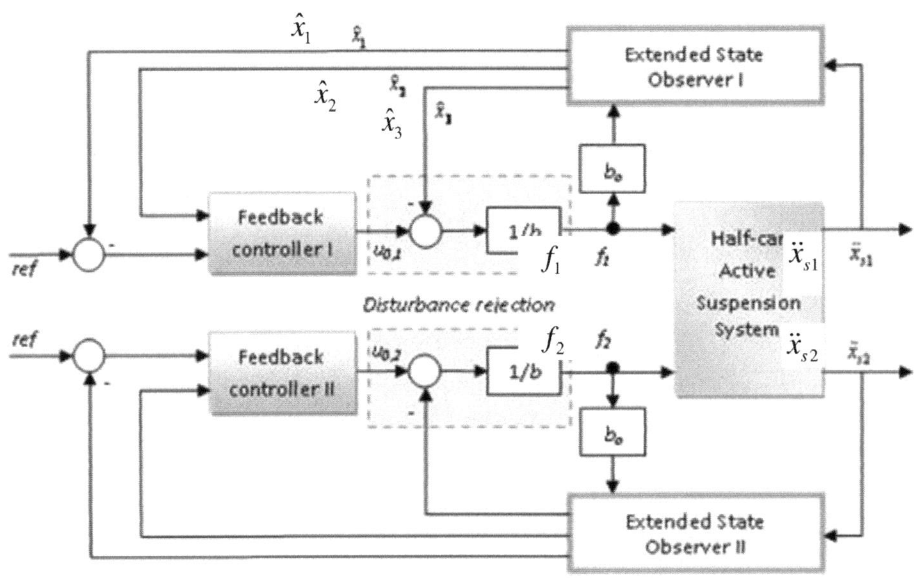

Figure 4. ADRC control structure for half-car active suspension system

Table 4. ADRC parameters for half car

	b_o
Front	0.0012
Rear	0.0022

5. Simulation and results

Simulation of the system was done using Matlab/Simulink. Two deterministic road profiles: speed hump and double bumps were used as the external disturbance (uncontrolled input) to the system. These road profiles were used with the assumptions that the road is perfectly rigid, the velocity of the car, v, is constant throughout the simulation, and the second input (x_{r2}) is a time delay of the first (x_{r1}). The time delay, τ is expressed as:

$$\tau = \frac{L}{v} \tag{30}$$

where L is the length of the car from its front wheel to the rear wheel. The road profiles are described as the following.

5.1 Road profiles

5.1.1 Speed hump

The speed hump is a typical road obstacle that the car travels over. The speed hump used is based on the Institute of Transportation Engineers (ITE) specifications shown in Table 5. The profile is shown in Figure 5 [6], [7].

Table 5. Specifications of speed hump (ITE)

	Minimum	Maximum
Height (m)	0.076	0.09
Travel length (m)	3.7	4.3
Travel velocity (km/h)	32	40

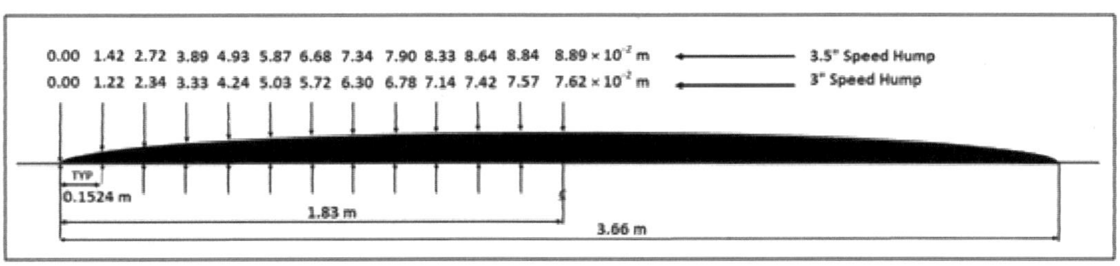

Figure 5. ITE speed hump profile (1″ = 0.0254m).

5.1.2 Double bumps input

A bump input is one of the most common deterministic road profile for evaluating a suspension system. The double bumps used in this simulation are defined by [8] :

$$x_{r,bump}(t) = \begin{cases} 0.5h\left(1-\cos\left(8\pi t\right)\right), & t_1 \le t \le t_2 \\ \dfrac{0.5h\left(1-\cos\left(8\pi t\right)\right)}{2}, & t_3 \le t \le t_4 , \\ 0, & otherwise \end{cases} \tag{31}$$

where h = 0.05 m is the maximum height of the bump, t_1 and t_2 are the lower and upper time limit of the function for the first bump whereas t_3 and t_4 are the lower and upper time limit for the second. This profile is shown in Figure 6.

Peak-to-peak (PTP) value,

$$PTP = \max(y(t)) - \min(y(t)) \quad (32)$$

where $y(t)$ is the response of interest, of each responses of the system are noted and compared for both passive and active systems. The maximums and minimums are defined as the maximums and minimums of the overshoot of the responses. The settling time of the response, t_s, defined as the time when the error of the output, $|y(t) - y_{final}|$ has settled to less than two percent of its peak value is also observed.

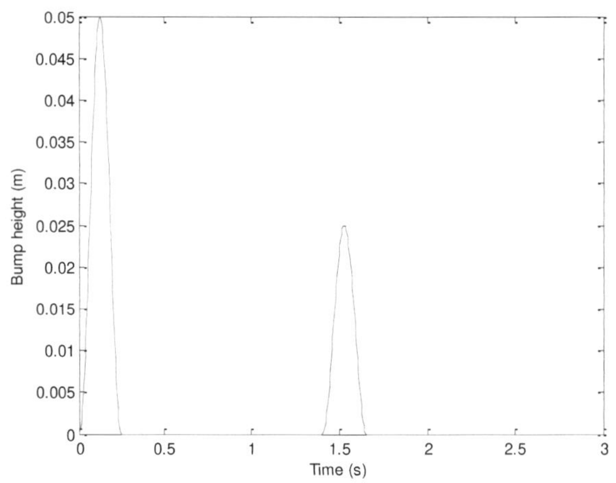

Figure 6. Double bumps disturbance input

5.2 Response to speed hump

Figure 7 to Figure 10 show the PTP values, settling times, and responses of the sprung mass acceleration, suspension deflection and tire deflection when subjected to hump input for the half-car model. For all responses in these figures, it is observed that both ADRC and ADRC-IDT were able to produce identical results. Overall, both ADRC and ADRC-IDT were able to significantly reduce the PTP of sprung mass vertical and pitch acceleration with 30.1% and 25.8% reduction respectively, as well as the settling time of both responses with 38.5% and 64.6% reduction respectively. This is achieved at the same time as maintaining smaller suspension working space and better road holding performance in comparison with the passive system. In producing similar responses, it is apparent that the peak force and average RMS force for each actuator were the same for both ADRC and ADRC-IDT as shown in Figure 11.

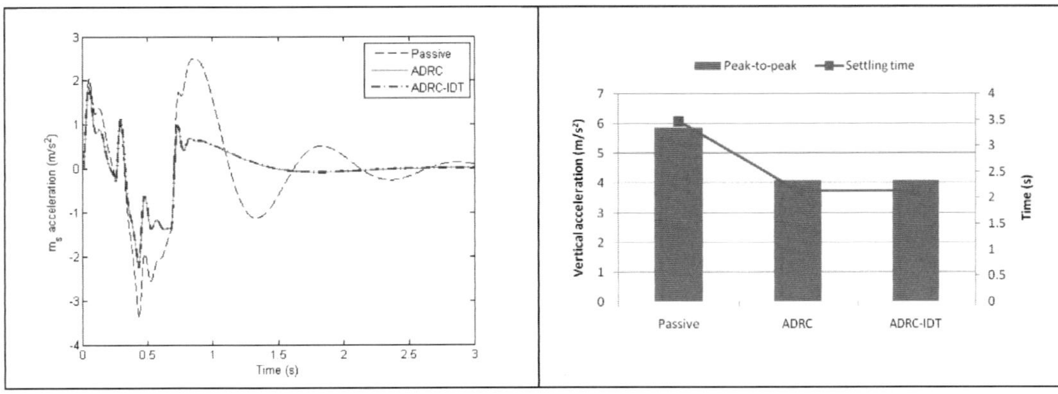

Figure 7. Response to hump input for half-car model - m_s vertical acceleration

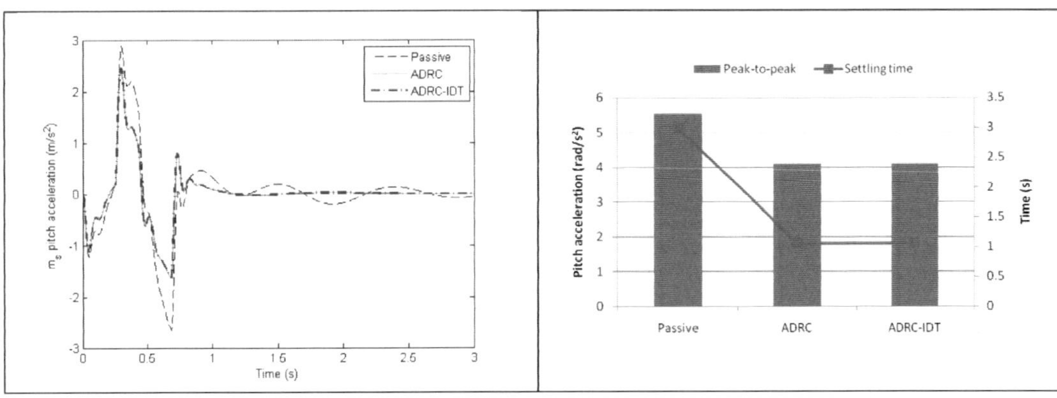

Figure 8. Response to hump input for half-car model - m_s pitch acceleration

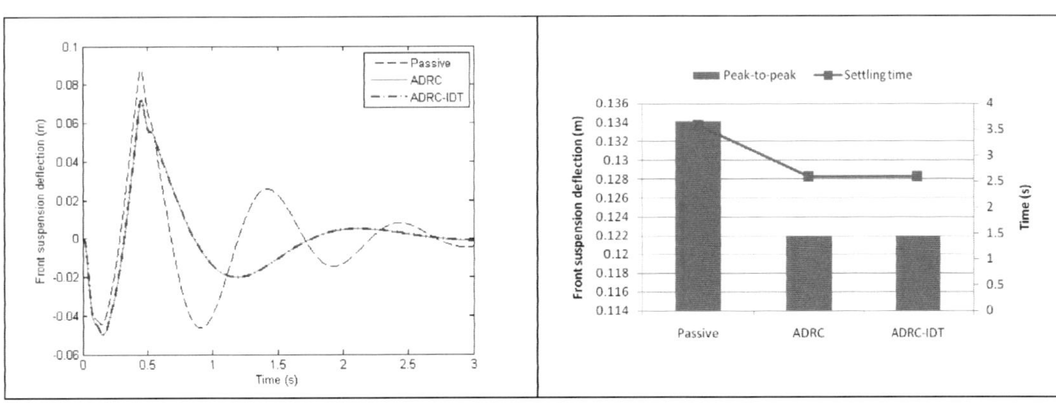

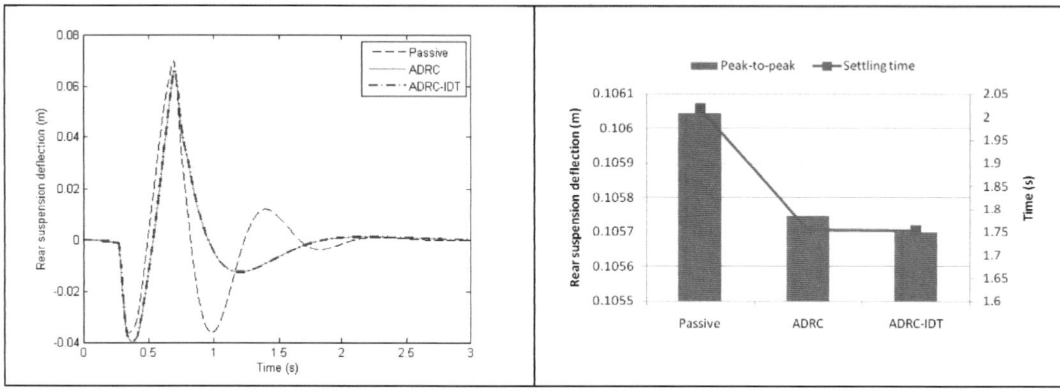

Figure 9. Response to hump input for half-car model - suspension deflection

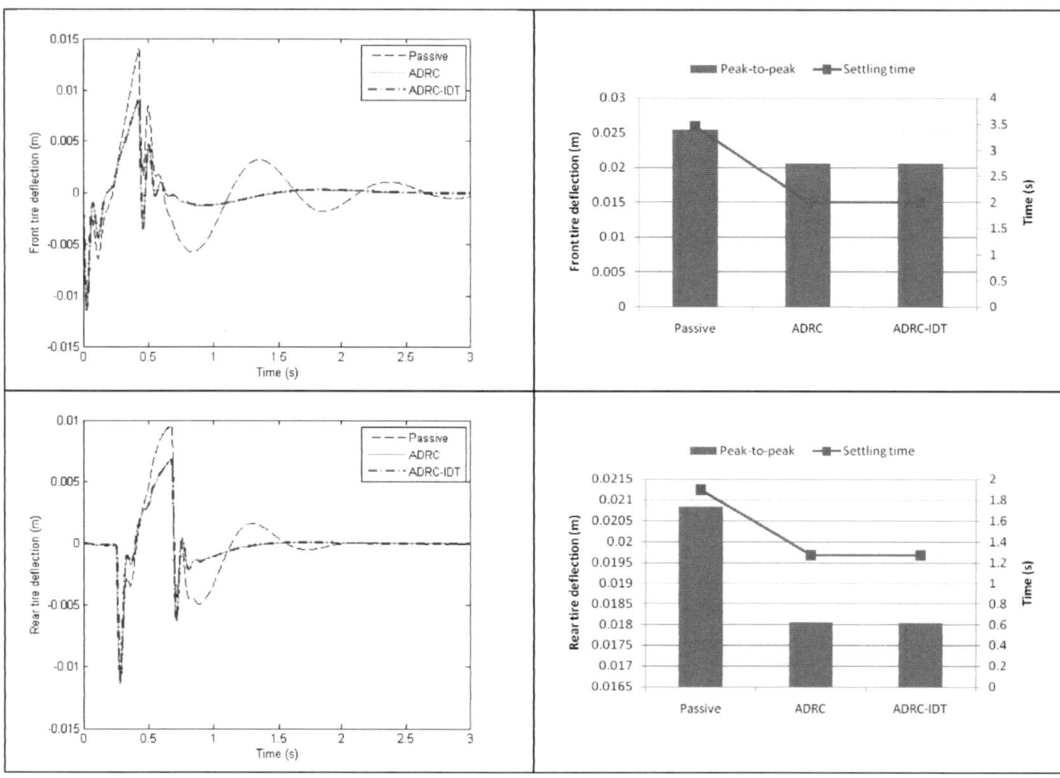

Figure 10. Response to hump input for half-car model - tire deflection

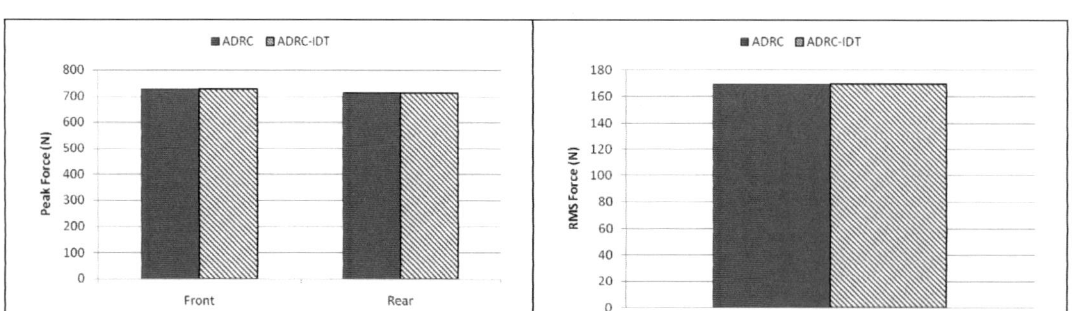

Figure 11. Peak force (left) and RMS force (right) for speed hump input

5.3 Response to double bumps input

Figure 12 to Figure 15 show the PTP values, settling times, and responses of the sprung mass acceleration, suspension deflection and tire deflection when subjected to double bumps input for the half-car model. For all responses in these figures, it is observed that both ADRC and ADRC-IDT were able to produce virtually identical results as well. Overall, both ADRC and ADRC-IDT were able to considerably reduce the PTP of sprung mass vertical and pitch acceleration with 25.2% and 25.5% reduction respectively, in addition to reducing the settling time of both responses by 27.2% and 6.7% respectively. This is accomplished by concurrently reducing the suspension deflection and tire deflection with reference to the passive system. It is also evident from Figure 16 that the peak force and average RMS force for each actuator were the same for both ADRC and ADRC-IDT.

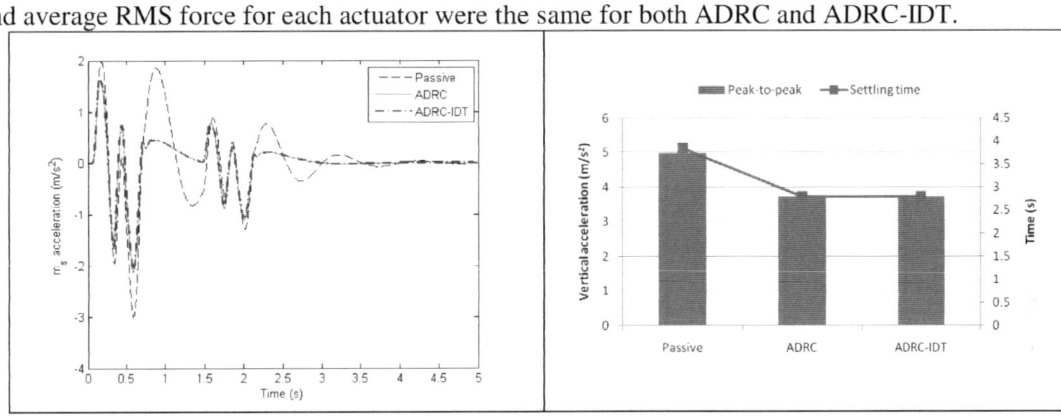

Figure 12. Response to double bumps input for half-car model - m_s vertical acceleration

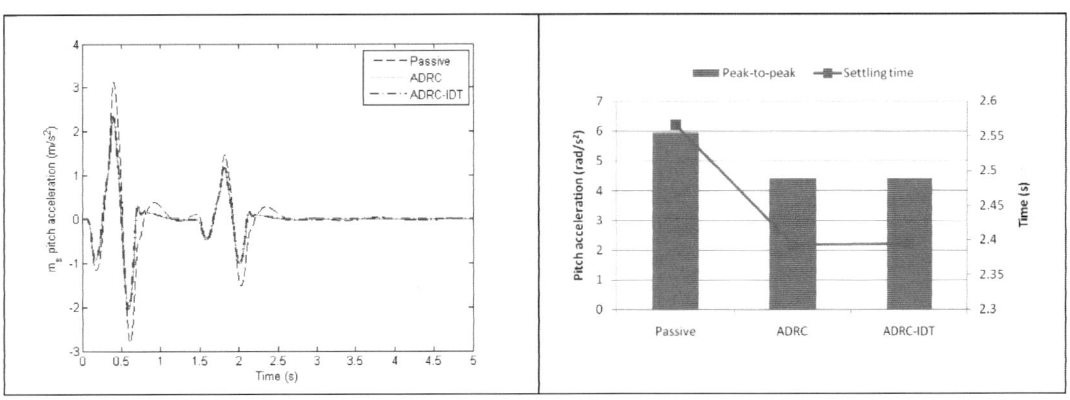

Figure 13. Response to double bumps input for half-car model - m_s pitch acceleration

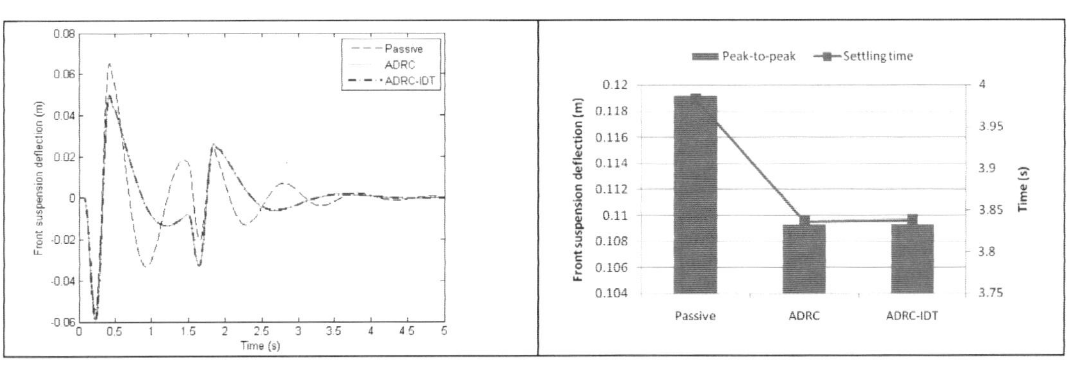

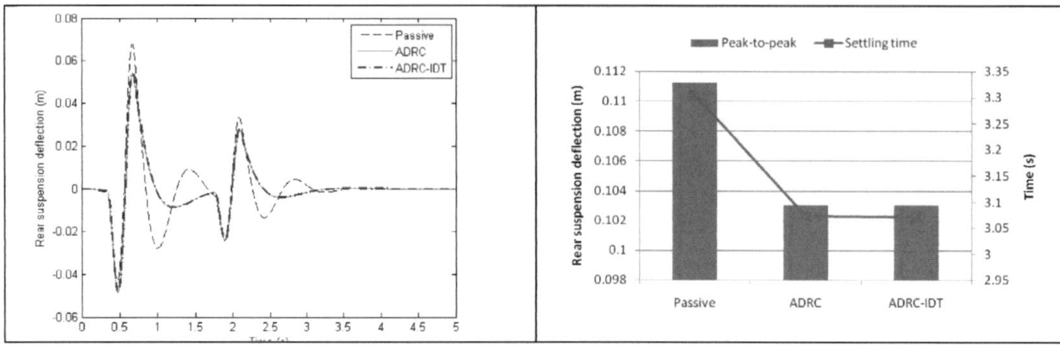

Figure 14. Response to double bumps input for half-car model – suspension deflection

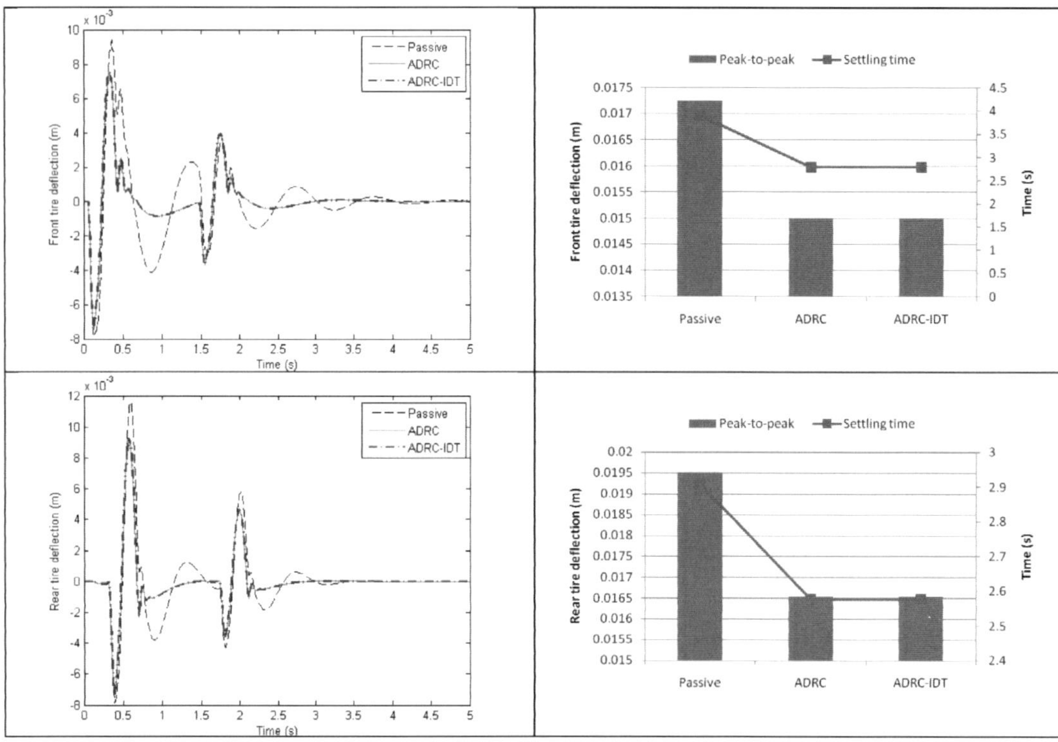

Figure 15. Response to double bumps input for half-car model-tire deflection

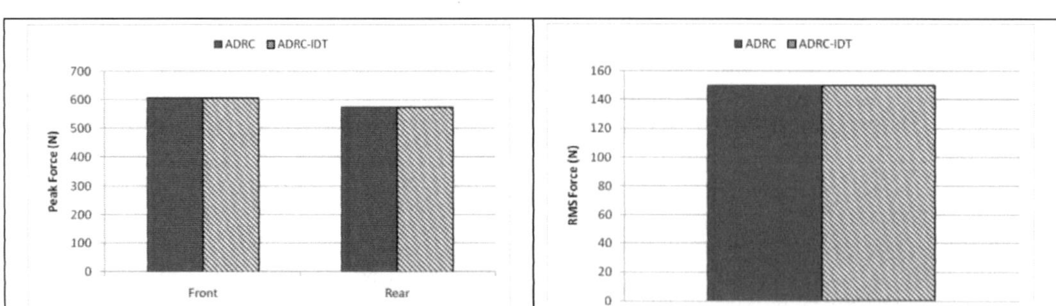

Figure 16. Peak force (left) and RMS force (right) for double bumps input

5.4 Effect of actuator saturation on the half-car model

In this section, the effect of actuator saturation is investigated where the actuator force was limited from 90% to 70% of the peak force required to produce the response in the previous section.

Figure 17 to Figure 19 show the effect of actuator saturation on sprung mass acceleration, suspension deflection, tire deflection, and control effort for both ADRC and ADRC-IDT respectively. Based on these responses, it can be observed that the responses of the ADRC and ADRC are almost identical, as the saturation value is varied. This shows that ADRC and ADRC-IDT recovers very well from the period of actuator saturation.

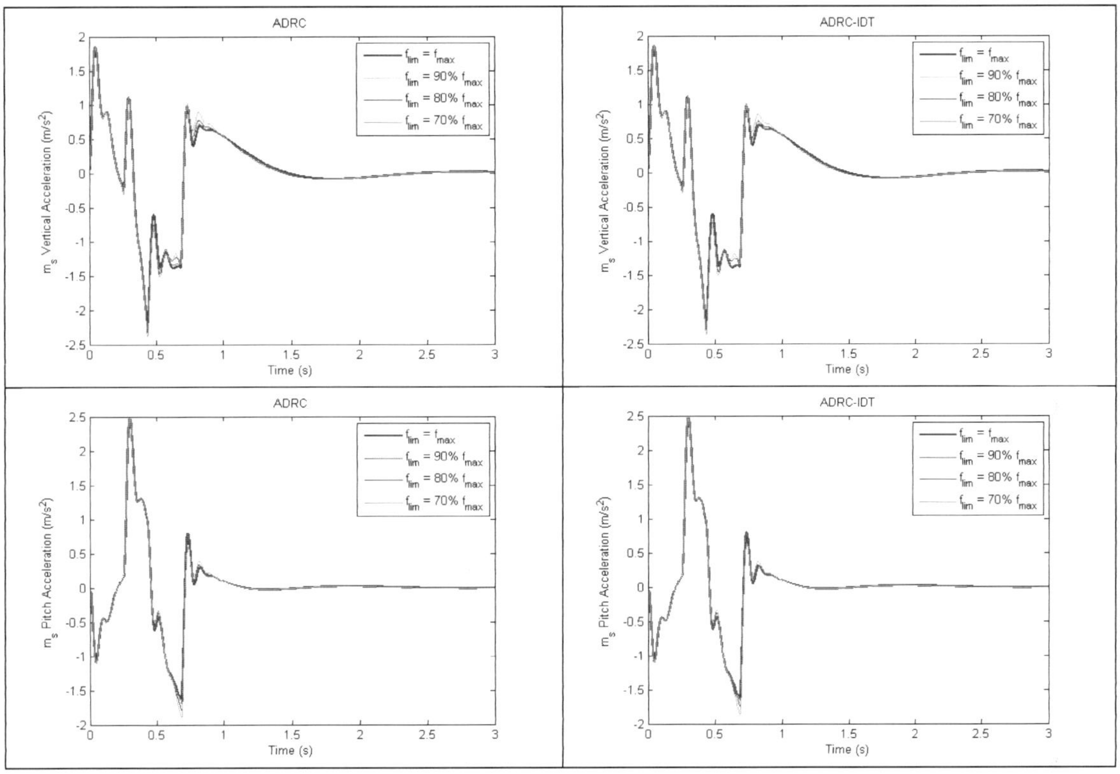

Figure 17. Effect of actuator saturation on vertical (top) and pitch (bottom) body acceleration for half-car model – ADRC (left) and ADRC-IDT (right)

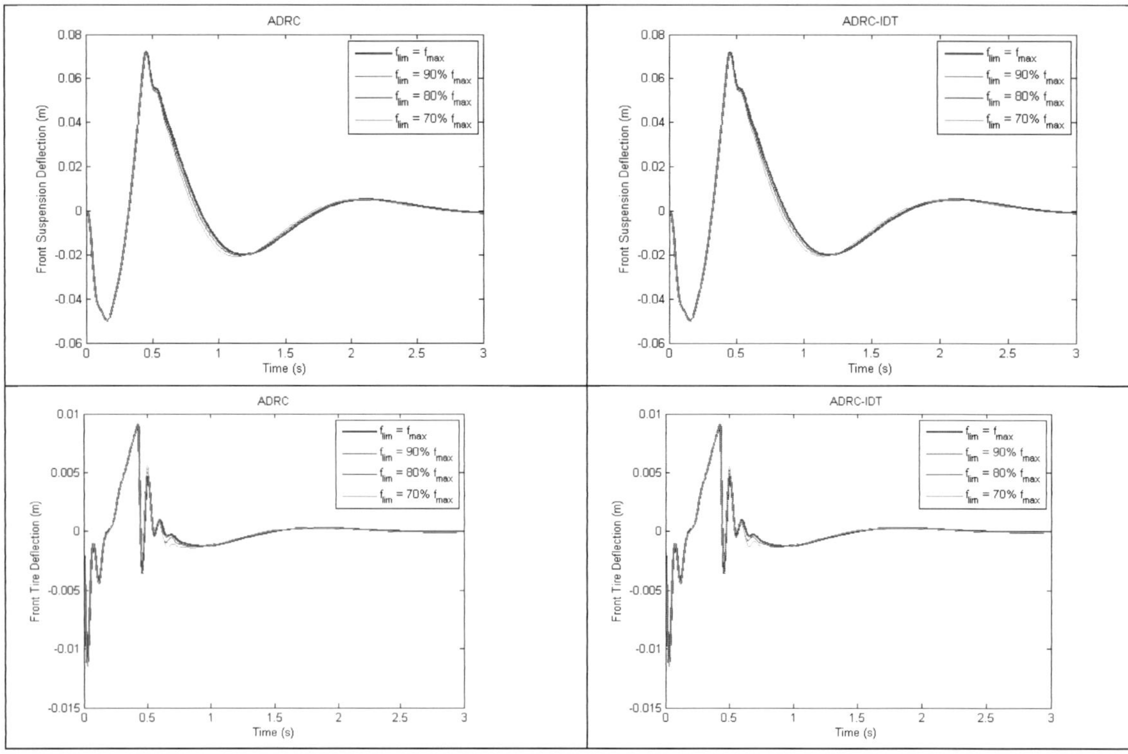

Figure 18. Effect of actuator saturation on front suspension (top) and tire (bottom) deflection for half -car model – ADRC (left) and ADRC-IDT (right)

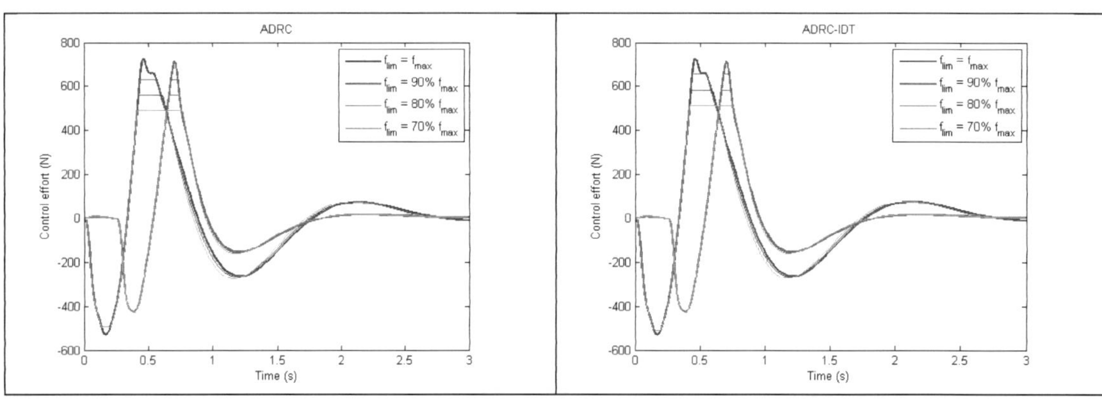

Figure 19. Effect of actuator saturation on control effort for half -car model ADRC (left) and ADRC-IDT (right)

5.5 Sensitivity to sprung mass variations on the half-car model

In this section, the capability of the active suspension system control to compensate increasing payload is studied. Assuming that additional payload is distributed evently on the sprung mass, increment of the sprung mass value was made in a range of 10% to 40% from its original value.

Figure 20 to Figure 22 show the effect of sprung mass variation on sprung mass acceleration, suspension deflection and tire deflection for both ADRC and ADRC-IDT respectively. In Figure 20, it is observed that the peak-to-peak response of the sprung mass vertical acceleration for ADRC and ADRC-IDT are reduced with additional mass. This is considered to be as the vertical acceleration is inversly proportional to the sprung mass value. However, this does not greatly affect the settling time responses. From Figure 21, it is observed that both controllers produced identical sprung mass pitch acceleration response as pitch acceleration is not directly related with the sprung mass value. There is no significant difference for suspension deflection and tire deflection responses in terms of PTP and settling time as shown in Figure 22. Overall, it can be observed that ADRC and ADRC-IDT are able to respond well to increasing payload.

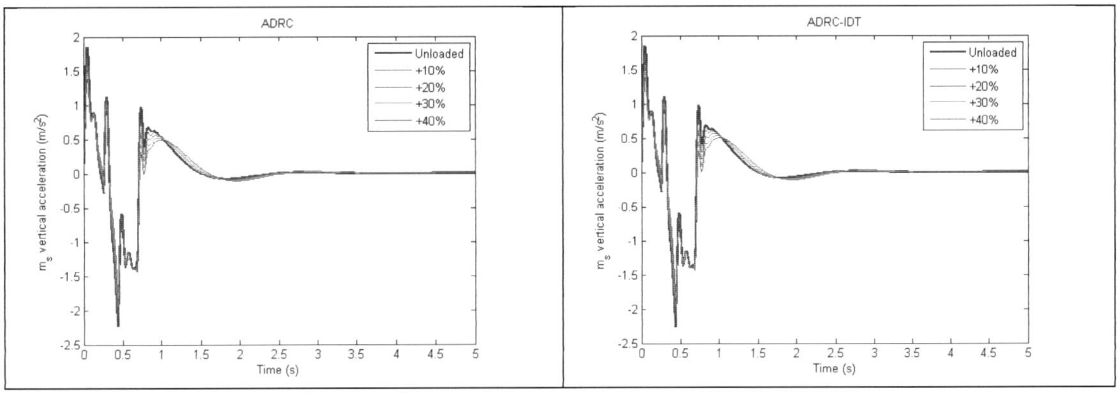

Figure 20. Effect of sprung mass variation on vertical body acceleration for ADRC (left) and ADRC-IDT (right) for half -car model.

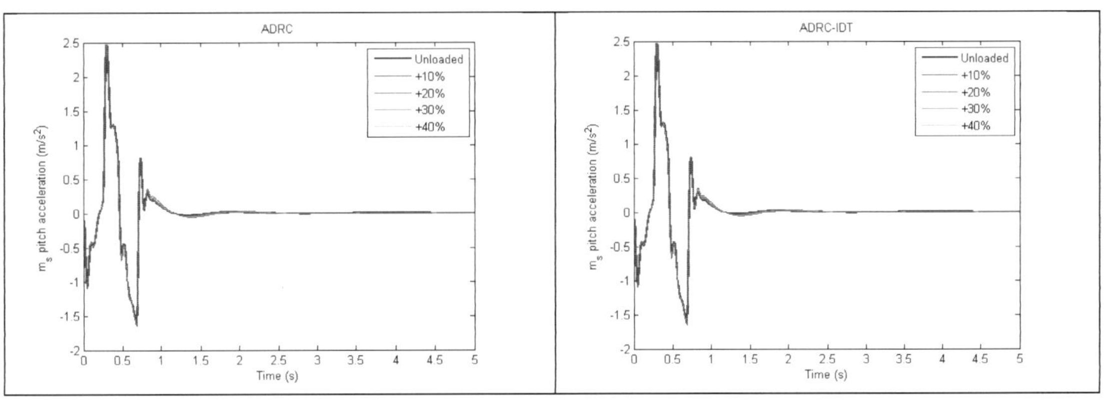

Figure 21. Effect of sprung mass variation on pitch body acceleration for ADRC (left) and ADRC-IDT (right) for half -car model

IOP Conf. Series: Journal of Physics: Conf. Series **949** (2017) 012025 doi:10.1088/1742-6596/949/1/012025

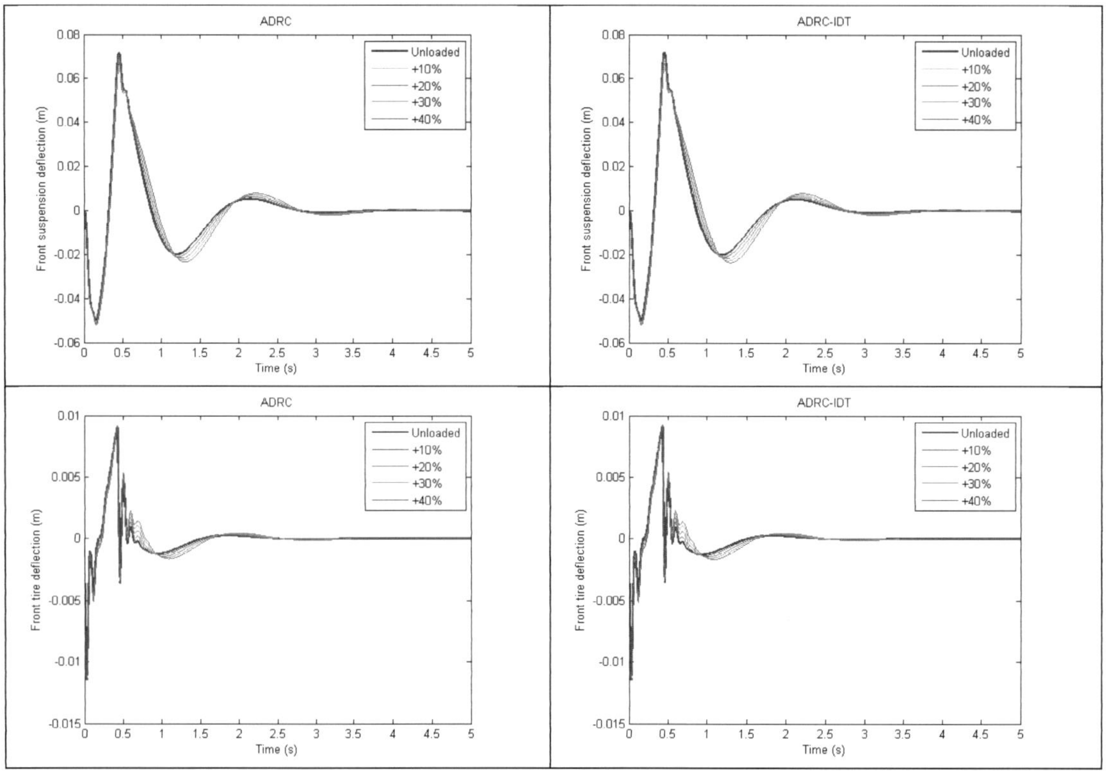

Figure 22. Effect of sprung mass variation on suspension (top) and tire (bottom) deflection for ADRC (left) and ADRC-IDT (right) for half -car model

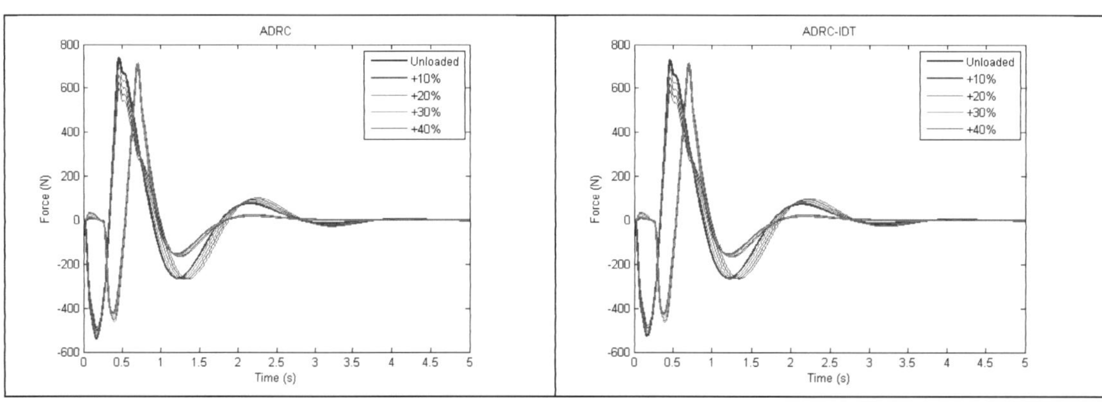

Figure 23. Effect of sprung mass variation on control effort for ADRC (left) and ADRC-IDT (right) for half -car model

5.6 Effect of nonlinearity on the half-car model

Until now, the modelling and simulation of the system were based on the assumption that they can be well described by using linear dampers and springs. These springs and dampers may exhibit nonlinear behaviour in real world. In this section, how both controllers react to changing dynamics of the system, particularly in the presence of nonlinear spring and damper are investigated by replacing the linear spring and damping forces in equation with the nonlinear spring and damper models without changing the parameters of the controllers. The nonlinear spring and damper can are defined as [9]–[11]:

$$F_{s,i} = k_{s,i}\left(x_{s,i} - x_{u,i}\right) + k_{ns}\left(x_{s,i} - x_{u,i}\right)^3 \tag{33}$$

and

$$F_{d,i} = c_{s,i}\left(\dot{x}_{s,i} - \dot{x}_{u,i}\right) + c_{ns}\left(\dot{x}_{s,i} - \dot{x}_{u,i}\right)^2 \operatorname{sgn}\left(\dot{x}_{s,i} - \dot{x}_{u,i}\right) \tag{34}$$

where i = 1 and i = 2 refer to the front and rear location respectively.

Figure 24 to Figure 27 show the sprung mass acceleration, suspension deflection, and tire deflection responses of ADRC, ADRC-IDT, and the passive system with nonlinear spring and damper as well as the comparison of percentage reduction of the PTP responses for both linear and nonlinear system. From these figures, it can be observed that both ADRC and ADRC-IDT maintain good performance similar to the response of the linear model, when compared to the passive system. This demonstrates the ability of both ADRC and ADRC-IDT to perform well with both linear and nonlinear system, as well as able to cope with changes in the process dynamics without having to retune the parameters of the controllers.

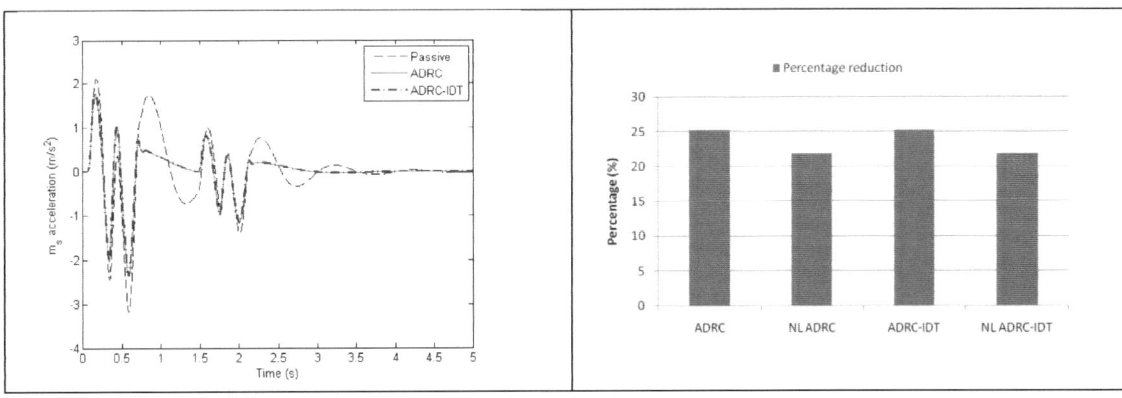

Figure 24. m_s vertical (top) and pitch (bottom) acceleration response for half-car model with nonlinear spring and damper.

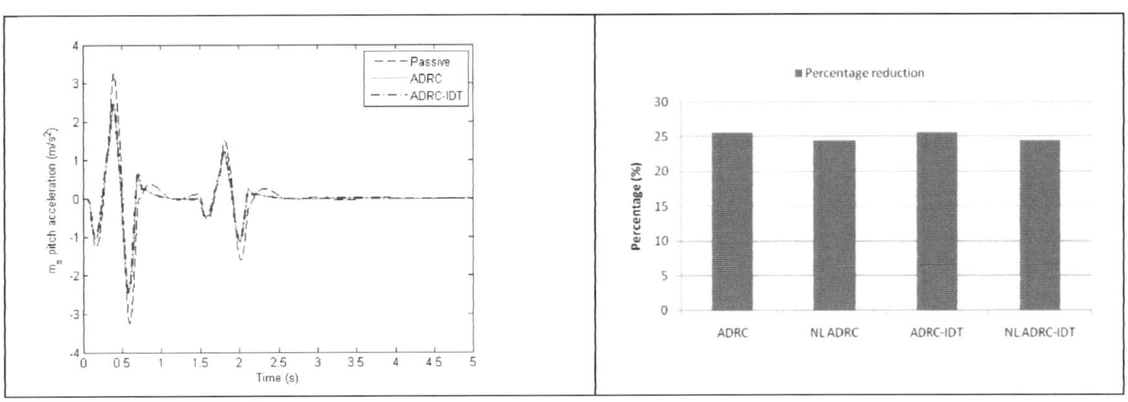

Figure 25. m_s pitch acceleration response for half-car model with nonlinear spring and damper.

IOP Conf. Series: Journal of Physics: Conf. Series **949** (2017) 012025 doi:10.1088/1742-6596/949/1/012025

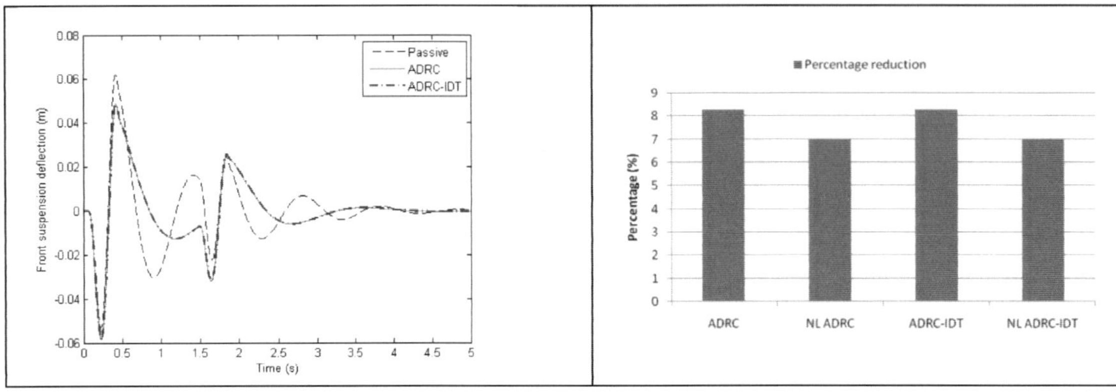

Figure 26. Suspension deflection response for half-car model with nonlinear spring and damper.

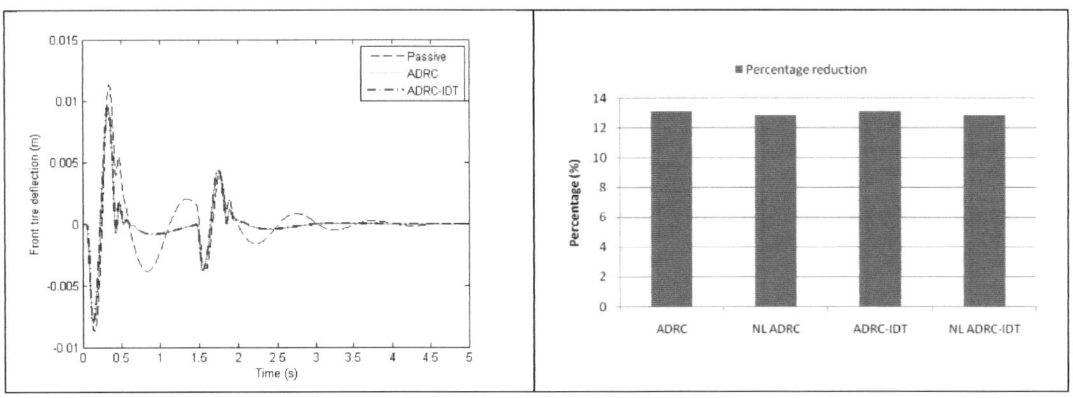

Figure 27. Tire deflection response for half-car model with nonlinear spring and damper

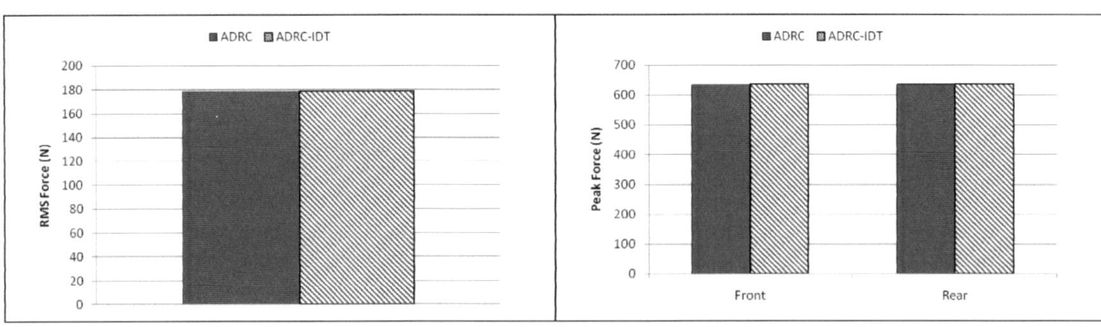

Figure 28. Control effort for half-car model with nonlinear spring and damper - ADRC (left) and ADRC-IDT (right)

The results in this section follow a trend. In general, both ADRC and ADRC-IDT can reduce the PTP and settling time of the responses compared to the passive system. The results also show that both ADRC and ADRC-IDT are robust to changes in disturbance input, able to adapt to increasing mass and changes in dynamics of the system as well as able to recover well from actuator saturation.

Additionally, both ADRC and ADRC-IDT produce similar results and performances. This is because from ADRC point of view, the IDT which is serially connected to the vehicle model can be considered as part of the total disturbance. Furthermore, both controllers are tuned to improve sprung mass accelerations of the half-car system while at least maintaining the same suspension and tire deflection responses as the passive system thus producing comparable results. While it might seem like the additional IDT scheme on the control structure does not contribute to any advantage at this point, it is worth noting that it can simplify the ADRC design when implemented in a full-car model as a decentralized ADRC control structure requires four sets of parameters to be design independently for each of the four actuators (front-left, front-right, rear-left, and rear-right). The ADRC-IDT control structure for full-car model on the other hand will require only 3 sets, that is for heave, pitch, and roll controls.

6. Conclusion

Two active disturbance rejection control approach: with and without decoupling transformation for half-car active suspension system is developed. The nonlinearity of vehicle ride model is investigated from three sources: geometric nonlinearity, nonlinearity of the suspension spring and nonlinearity of the damper. The ride performance of the ADRC-IDT is evaluated and compared with the typical ADRC as well as the passive system. Simulation results show that both ADRC and ADRC-IDT were able to appreciably reduce body accelerations and able to cope well with varying conditions typically encountered in an active suspension system. Also, it is sufficient to control only the body motions with both active controllers to improve ride comfort while maintaining good road holding and small suspension working space.

7. References

[1] G. Herbst, "A simulative study on active disturbance rejection control (ADRC) as a control tool for practitioners," *Electronics*, vol. 2, no. 3, pp. 246–279, 2013.
[2] F. Hasbullah, W. F. Faris, F. J. Darsivan, and M. Abdelrahman, "Ride comfort performance of a vehicle using active suspension system with active disturbance rejection control," *Int. J. Veh. Noise Vib.*, vol. 11, no. 1, pp. 78–101, 2015.
[3] S. I. Ihsan, *Semiactive control policies : ride analysis for passenger vehicles*. Saarbrucken, Germany: LAP LAMBERT Academic Publishing, 2011.
[4] J. Han, "From PID to Active Disturbance Rejection Control," *IEEE Trans. Ind. Electron.*, vol. 56, no. 3, pp. 900–906, Mar. 2009.
[5] R. Madonski and P. Herman, "An experimental verification of adrc robustness on a cross-coupled aerodynamical system," in *Industrial Electronics (ISIE), 2011 IEEE International Symposium on*, 2011, pp. 859–863.
[6] ITE Technical Council Task Force on Speed Humps, "Guidelines for the Design and Application of Speed Humps," *ITE J.*, no. May, pp. 11–15, 1993.
[7] M. Parkhill, R. Sooklall, and G. Bahar, "Updated Guidelines for the Design and Application of Speed Humps," in *ITE 2007 Annual Meeting and Exhibit, Pittsburgh PA, Institute of Transportation Engineers*, 2007.
[8] R. Darus and N. I. Enzai, "Modeling and control active suspension system for a quarter car model," 2010, pp. 1203–1206.
[9] A. A. Aldair and W. J. Wang, "Design an Intelligent Controller for Full Vehicle Nonlinear Active Suspension Systems," *Int. J. Smart Sens. Intell. Syst.*, vol. 4, no. 2, 2011.
[10] Y. Ando and M. Suzuki, "Control of active suspension systems using the singular perturbation method," *Control Eng. Pract.*, vol. 4, no. 3, pp. 287–293, 1996.
[11] F. Beltran-Carbajal, E. Chavez-Conde, G. S. Navarro, B. V. Gonzalez, and A. F. Contreras, "Control of Nonlinear Active Vehicle Suspension Systems Using Disturbance Observers," *Vib. Anal. Control. Trend Dev.*, 2011.

Institute of Physics
Dirac House, Temple Back
Bristol BS1 6BE UK

ISSN: 1742-6588
ISBN 978-1-5108-5521-2

CURRAN ASSOCIATES INC.
proceedings.com